Assembling and Supplying the IS.

The Space Shuttle Fulfills Its Mission

David J. Shayler

Assembling and Supplying the ISS

The Space Shuttle Fulfills Its Mission

Springer

Published in association with
Praxis Publishing
Chichester, UK

David J. Shayler, F.B.I.S.
Astronautical Historian
Astro Info Service Ltd
Halesowen
West Midlands
United Kingdom

SPRINGER PRAXIS BOOKS IN SPACE EXPLORATION

Springer Praxis Books
ISBN 978-3-319-40441-7   ISBN 978-3-319-40443-1   (eBook)
DOI 10.1007/978-3-319-40443-1

Library of Congress Control Number: 2017945669

Front cover: The Space Shuttle Endeavour docked to ISS in August 2007. This image captures the major elements of Shuttle missions to space stations. The robotic arms of both the Shuttle and ISS can be seen, along with the Spacehab logistics module in the payload bay as one of the crewmembers conducts an EVA at bottom left of frame. Back cover right: Endeavour is seen docked to the ISS in November 2002. The station's robotic arm appears in between the Shuttle and Mission Specialist John Harrington making an EVA. (NASA images) Back cover left: The front cover of the companion volume *Linking the Space Shuttle and Space Stations: Early Docking Technologies from Concept to Implementation.*

Cover design: Jim Wilkie
Project Editor: David M. Harland

Printed on acid-free paper

This Springer imprint is published by Springer Nature
The registered company is Springer International Publishing AG
The registered company address is: Gewerbestrasse 11, 6330 Cham, Switzerland

Contents

Preface

For any researcher the history of the American Space Shuttle program is both complex and extensive. It is therefore difficult to condense almost fifty years of development and operations into a single book, or even to a series of volumes. A more suitable aim for a single volume is to provide an in-depth account on a specific topic, a single flight, or a series of related missions. Even then, it is a challenge to condense the voluminous data into a single volume and still present a worthy account. This was the challenge I faced when commencing the research into how the Space Shuttle program was organized and operated to support the delivery, assembly, and resupply of the large structure in Earth orbit that we now know as the International Space Station.

Central to this story were the missions themselves, but the planning which went into each flight, the support provided by numerous ground facilities and teams of engineers, controllers and managers, plus the selection and training of the astronauts that flew the missions were also stories worth investigation. Trying to understand how each Shuttle stack arrived at the launch pad highlighted the extensive and interwoven preparations undertaken at the Kennedy Space Center in Florida, and this prompted further research into the process of integrating the hardware, experiments, and payloads, in addition to the facilities that supported the processing prior to launch. Next I came to the research for the various phases of a mission: the launch, the period of rendezvous and docking, the activities of the Shuttle crews at the station, the transfer of the payloads externally using the Canadian-built Remote Manipulator System, or internally by crewmembers manhandling the supplies into the station and the trash back out. Then there were the spacewalks conducted to support the external assembly and relocation of hardware at the station. Finally, there were the undocking and fly-around phases prior to returning to Earth.

For BIS publications, I had already briefly reviewed the series of Shuttle-Mir flights and the first four years of Shuttle-ISS operations spanning STS-88 to STS-113. For this project, I decided I would bring the story up to date, through to the completion of ISS assembly and the termination of the Space Shuttle program. I included the 1994 flight of the first Russian cosmonaut aboard an American mission (STS-60) that initiated the Shuttle-Mir program. These activities involved forty-eight flights of the Shuttle spread across seventeen years, encompassing one rendezvous and nine dockings with Mir and an impressive thirty-seven missions to the International Space Station. Having briefly covered each of these missions in the two editions of the *Manned Space Flight Log* by Springer-Praxis in 2007 and 2013, I did not wish merely to repeat the flight activities recorded in therein. Equally, the constraints of the original single volume would not allow me to delve as deeply into each mission as I would have liked. I also couldn't merely update the articles in the BIS publications. Hence this project started out as a different approach to recording the extensive program.

From the start, it became apparent that a distinct profile was flown by each Shuttle mission to a space station. Of course a 'typical mission' never really existed, as each flight was unique in its own right. Nevertheless, I have reviewed a number of generic processes and sequences that can be considered to be common ground for each of the missions flown. What was also evident, was the clear division between the story prior to commencing the assembly of the ISS and the period after construction commenced. In researching for this project, I found that the desire to send a Shuttle to a station was almost as old as the program itself, and was one of the original reasons for developing what became the Space Transportation System (STS) concept in the first place.

In fact, the background to blending the Shuttle system with a space station program, and the other way around, was over twenty years in the making. From early conceptual studies for the Space Shuttle through to the dockings of the Orbiter at the Russian Mir space station in the 1990s was a long journey of design, development, cancelation, and eventual success. And this was before a single element of the ISS was launched. After discussing the scope of my research with Springer, they decided that rather than cover the entire story in a single huge volume, I should write two separate books.

In *Linking the Space Shuttle and Space Stations: Early Docking Technologies from Concept to Implementation*, I wound the clock back to the start of the Space Shuttle era and the ambitions to use that system to create and supply a scientific research station in Earth orbit, the background story to the proposed 1981 Shuttle-Salyut docking mission, as well as a simultaneous study by NASA into using the Shuttle to return to the Skylab orbital workshop and restore it to use. The narrative continues with the plans to use the Shuttle to assemble Space Station Freedom, before this morphed into the International Space Station with the inclusion of the Russians enabling NASA to gain much-needed experience of operating the Shuttle in concert with a large space facility by performing the series of Shuttle-Mir docking missions prior to initiating ISS operations. This first volume explains the background to key milestones required to 'link' the Shuttle to the space station, the development of a successful rendezvous and docking system, and an effective method of carrying and transferring tons of hardware to and fro between the two vehicles. The nine dockings by Shuttles at Mir between 1995 and 1998 provided a 'proof of concept' for more extensive operations planned at the ISS.

This second book takes up the story from the end of Shuttle-Mir, at the point where there was some concern that the start of the ISS assembly might be delayed once again due to the unavailability of Russian hardware to initiate assembly. The second chapter briefly reviews the three dozen Shuttle flights between starting to assemble the ISS in December 1998 and the retirement of the Shuttle in July 2011. The third chapter steps away from flight operations to review the fascinating and often complicated world of how crews are assigned. The how and why of a particular astronaut's assignment to a given flight remain a largely mysterious process, a closely guarded secret confined to higher NASA management. Even those in the Astronaut Office who actually flew the missions sometimes have no idea why they were chosen over their colleagues.

Chapter four recalls the myriad procedures, facilities and activities that were needed to get a Shuttle mission off the launch pad, and specifically those to the ISS. Details of the principal facilities at the Kennedy Space Center in Florida which supported Shuttle pre-launch and launch operations were covered in depth in the first volume, and so are only briefly recalled here. Instead, specific to the International Space Station program, is the description of the specialized building that was created in order to handle the vast amount of hardware and cargo that was assigned to the Shuttle missions to the ISS over a period of thirteen years. A 'flight' into space can be viewed as the sequence of events from a spectacular launch through to a safe landing, but a 'mission' often starts months or indeed years before anything reaches the pad, let alone leaves it, and it can continue for some time after landing. Each mission is inseparably linked to the activities on the ground needed to prepare, launch, and return an Orbiter and its payload (including the crew) to Earth…prior to starting the process all over again. It is a fascinating story of management, planning, dedication and sometimes compromise, while also keeping an eye on the weather, not just at the launch site but also at the various potential landing sites around the globe. In fact, Shuttle processing throughout the program, not merely for the missions to Mir and the ISS, deserves a dedicated book!

Chapter five returns to flight operations and the journey from the launch pad to the ISS and, after the loss of Columbia in 2003, the introduction in 2005 of an impressive 'back-flip' maneuver known as the Rendezvous Pitch Maneuver that was executed by each Orbiter to allow ISS crewmembers to conduct a detailed examination, especially of the thermal tiles on its underside. The chapter also recalls the docked phase, during which a visiting Shuttle crew would integrate with the long-duration station crew.

The following two chapters review how the tons of hardware and cargo destined for the space station was delivered and then either transferred inside or bolted on. Chapter six covers the transfer of large elements of hardware internally by the crew, physically moving equipment from the pressurized compartments of the Shuttle. This transfer was often through numerous hatches, tunnels, and crowded compartments into the modules of the space station. By a reverse route unwanted materials, samples from experiments, and trash were loaded aboard the Shuttle for the trip home. This chapter also reviews the various methods and items of crew apparatus that were available to carry logistics between vehicles, and the skills required by the 'loadmaster' to ensure that every item was positioned just where it ought to be. A series of procedures developed during the Shuttle-Mir missions were found to be very effective during the Shuttle-ISS assembly and resupply flights.

The seventh chapter explores, in some depth, the development of using the Shuttle Remote Manipulator System (RMS or Canadarm) in an extensive program of robotics at the expanding station. Operating the RMS was never a large part of Shuttle-Mir but with a wide range of experience from using the arm on several other projects, such as the Hubble servicing missions, NASA was confident it would function satisfactorily during the early stages of ISS assembly. As the station expanded, in 2002 the reach of the RMS on the Orbiter became a limiting factor in continued assembly, requiring the installation of a robotic arm on the station (SSRMS, named Candarm2). Despite these restrictions, the RMS continued to be used on missions to the end of the program for examining the Orbiter, working jointly with the SSRMS, and in supporting teams of astronauts during EVAs. The history of the RMS also warrants a separate story to be written.

The series of EVAs at the ISS are explored over two chapters, and how after many years of study the myriad of tasks were approached, resulting in the highly successful series of spacewalks that supported the assembly and expansion of the facility. Here a more in-depth approach is presented in the build-up to space station EVA activities *by Shuttle flight crew members*, with those conducted by resident crews left to be covered elsewhere. The story begins with plans to use extensive EVA operations to support the assembly of Space Station Freedom and the enormous growth in the projected number of spacewalks that would be required not only to assemble that facility but to maintain it throughout its operational lifetime. It would have been difficult for the astronauts of visiting Shuttles to support station resident teams in a seemingly endless series of back to back EVAs during each year of operations.

The penultimate chapter continues the EVA story at the ISS and the preparations to scale what was known as the 'Wall of EVA' for assembling the ISS, and how, step by step the use of a larger neutral buoyancy training facility and a series of spacewalks to evaluate tools and techniques helped to create the baseline to support an extensive and highly successful series of EVAs at the ISS over the thirteen-year period of assembly. As with other aspects of the Shuttle and ISS programs, it is hoped a far more detailed history of EVA at space stations will appear in due course.

The closing chapter reviews the undocking from the station, fly-around inspection, and return of an Orbiter to Earth. It is remarkable how smoothly missions to Mir and the ISS were executed. Nevertheless there were always plans and procedures available to overcome a variety of potential problems. One of these 'what if' scenarios addressed the difficult situation of an Orbiter that had docked at the ISS and was unable to return safely to Earth. This chapter concludes by looking at how an unmanned Orbiter would be undocked and disposed of. This situation never occurred of course, but it is worth recording here in order to complete the story of the Shuttle activities at space stations.

In researching for my project about the Hubble service missions – which were also instrumental in developing techniques for Shuttle rendezvous and EVA operations at the ISS – I found so much more to study and record. The same is true for this current work, which has prompted more in-depth research into each Shuttle assembly mission and the people who prepared and flew them. There was simply not enough room even within two volumes to tell those stories in great detail, but this work continues…

David J. Shayler, FBIS
Council Member, British Interplanetary Society,
Director, Astro Info Service Ltd.,
www.astroinfoservice.co.uk
Halesowen, West Midlands, UK
March 2017

Acknowledgements

As recorded in the companion title *Linking the Space Shuttle and Space Stations: Early Docking Technologies from Concept to Implementation*, "Each book project undertaken requires not only a significant amount of personal research and input but also a network of contacts, assistance from a number of key individuals, and a great support team to convert the initial ideas and scribbles into the finished product." This volume has been no different.

In researching both books, the network of contacts around the world from these and other activities I have conducted at (and with) the British Interplanetary Society, have been of great help in my ongoing work. I should like to thank the long-term assistance and support of the former and current members the BIS Council and former Executive Secretary Suzann Parry and her successor Gill Norman. And thanks also to Ben Jones and Mary Todd, who were always most helpful and supportive.

I would like to extend my deep appreciation to former NASA astronaut Jerry Ross, veteran of seven Shuttle missions, including the second Shuttle docking flight to Mir and the first and thirteenth Shuttle-ISS assembly missions, for his excellent Foreword and Afterword. Other former astronauts who have assisted with research for this and closely related recent and future books were Tom Akers, Leroy Chiao, Jean-François Clervoy (ESA), Bob 'Crip' Crippen, Robert 'Hoot' Gibson, Steve Hawley, Tom Jones, Janet Kavandi, George 'Pinky' Nelson, Ellen Ochoa, Jerry Ross, Steve Smith, and Joe Tanner. Thank-you one and all for taking the time to tell me of your experiences and also for explaining certain techniques and procedures.

Again, a network of international contacts and colleagues over the years deserve a mention for their continued support and assistance on several projects, including this one: Colin Burgess, Michael Cassutt, Phil Clark, Brian Harvey, Bart Hendrickx, and Bert Vis. Bert also provided several elusive images for this book from his extensive collection and plugged gaps in my records of American astronaut activities in Russia. And mention must also be made of the late Rex Hall and Andy Salmon, both of whom cooperated and supported my early research and publications in this field; their loss is greatly lamented.

The various public affairs and history office staff of NASA and at the University of Houston at Clear Lake and Rice University, also in Houston, who provided assistance and support in my research over a period lasting nearly forty years also deserve thanks. Their help ranged from simply replying to recent E-mail queries (or in the 'old days' to written letters) to assisting and directing my research focus during my visits to NASA JSC and KSC in the period 1988–2002.

My special thanks go to Lynne Vanin, Manager, Public Affairs, MDA Corporation, Ontario, Canada, for detailing the flight assignments and activities of the Shuttle and Station Remote Manipulator Systems on space station missions.

All images are courtesy of NASA via the AIS collection unless otherwise specified. Thanks also to the various contractors and partner agencies for supplying images over the years used in support of research for this and related projects and presentations. I also thank Joachim Becker of *SpaceFacts.de* and Ed Hengeveld for their assistance in providing me some really elusive pictures. The *Space Shuttle Almanac*, painstakingly compiled by Joel W. Powell and Lee Robert Brandon-Cremer between 1992 and 2011 was also a very useful and highly recommended reference source.

On the production side, I am once again indebted to my brother Mike Shayler, who continues to guide me through the quagmire of developing my wordsmithing skills and refined the draft. To David M. Harland for his excellent editorial skills and suggestions that greatly improve the presentation of each book he works on. To Jim Wilkie for his mastery in converting my original notions for a cover to the final product. Thanks also to Clive Horwood at Praxis in England for supporting the proposal, and to his wife, Jo, for letting Clive continue to steer book projects through the refereeing process when by rights he ought to be relaxing and driving their VW camper van around the countryside in a leisurely manner. Maury Solomon, Nora Rawn and Elizabet Cabrera at Springer in New York are also to be thanked for their encouragement and management throughout the production process.

Last, but certainly not least, love and appreciation go to my mother Jean Shayler who once again read the initial draft and offered useful observations and suggestions, and to my wife Bel who used her interview transcribing skills and also assisted with scanning images, collating tables, and checking the final manuscript. Of course it is thanks to Bel that I was given 'parole' from domestic chores to undertake this project. I also welcome our young and rather large German Shepherd 'puppy' Shado, as both the latest addition to the Shayler clan and the new mascot for Astro Info Service. He can finally enjoy the longer walks that he asks for and deserves…at least until the next writing project makes its way to the launch pad.

To each and every one who helped, a very large thank-you.

Also to the memory of our beloved German Shepherd, Jenna (2004–2016), who was the original company mascot and is much missed and will always be remembered.

Foreword

In 1984, four years after I had been selected by NASA to become an astronaut, the agency made the decision to build a large space station in Earth orbit. This was to be assembled and supplied using the capabilities of the Space Shuttle. I was to fly on the Shuttle as a Mission Specialist, trained to operate the robotic arm, deploy and retrieve satellites, perform scientific experiments, and conduct spacewalks. It was exciting to anticipate actually being involved in the assembly of a space station someday.

By the time of my first mission in November-December 1985, NASA was evaluating many options for the configuration of the station and how best to build it. Indeed, on my first mission, STS-61B, Sherwood 'Woody' Spring and I completed two spacewalks to evaluate two different techniques for assembling structures in space. Six years later, on my third mission, STS-37, I again performed a spacewalk associated with space station development. This time Jay Apt and I evaluated a [CETA] rail cart that was intended to facilitate the movement of astronauts and equipment along the main body of the station. The Shuttle was an indispensable tool for developing new techniques and procedures in preparation for assembling a station. But the maturing design of the station, at that time named Freedom, was encountering difficulties concerning its complexity and cost, and was projected to require hundreds of hours of spacewalking activities for its assembly.

After years of debate and redesign, a revised configuration called the International Space Station (ISS) emerged, with Russia as a new partner. The station was still to be assembled and supplied by the Shuttle, but some elements were to be launched by the Russian Space Agency. In addition, a series of Shuttle missions were planned to fly to the Russian Mir space station in order for NASA to gain experience in rendezvous and docking with a large object in space. Meanwhile, during my fourth spaceflight, STS-55 in 1993, I was the Payload Commander for Spacelab-D2, a 10 day US-German science mission with eighty-eight experiments from around the world. The research performed on this mission was a forerunner of the work that was to be conducted on the ISS.

In 1995, I was a member of the second Shuttle crew to dock to Mir. During mission STS-74 we focused on adding a Russian-built docking module to Mir and delivering a significant amount of supplies and equipment to the station. On all subsequent Shuttle missions to the Russian space station, this docking module was used by the Shuttle to mate to Mir.

NASA astronaut Jerry Ross.

Following my only visit to Mir, I served as the Chief of the Astronaut Office's EVA and Robotics Branch and helped to lead the development of the spacewalking hardware, tools, and procedures to be used to build and maintain the International Space Station. Development was proceeding at a hectic pace. It was a challenging effort to make sure that every aspect of every spacewalking task was reviewed, tested, corrected, retested, and verified ready for flight. We also conducted a series of developmental spacewalks that significantly increased the number of astronauts with experience of spacewalking. This was an exciting time, and it was personally very rewarding to be involved in all aspects of the station assembly process. Much to my delight, I was assigned to be the lead space-walker for STS-88, the first ISS assembly mission.

From my previous Shuttle flights and spacewalks, it was clear that the Shuttle was going to be an ideal vehicle to support the assembly of the station. For nearly twenty years, the ability of the Shuttle to carry tons of cargo in its cavernous payload bay, its versatile robotic arm, and its ability to support extensive spacewalking activities had been demonstrated in over ninety missions, including nine docking missions with the Mir station. Though a daunting challenge lay ahead, including a substantial number of spacewalks that we called the 'Wall of EVA,' we were ready and eager to get on with the task in front of us.

As author David J. Shayler explains in this book, assembling the ISS was not just a matter of sending the next Shuttle to space. There had to be an infrastructure in place to build and prepare the ISS hardware, to train the crews, and to conduct over thirty highly successful missions. The story of how the Shuttle program supported the assembly of the International Space Station by applying the rendezvous and docking experience at Mir and by using a blend of robotics and spacewalking, while delivering tons of station elements and supplies, reveals what a complex and involved program it became. It is a real testament to those who designed and built the Shuttle that they had the foresight to provide the vehicle with the awesome capabilities it could draw upon in order to reach 'assembly complete' of the ISS in just thirteen years.

I am honored and proud to have played a part in the creation of the ISS, and pleased that the story of how we used the Space Shuttle to achieve that goal has been expertly presented in this book.

Colonel Jerry L. Ross USAF (retired)
NASA Astronaut (1980–2012)
Mission Specialist, STS-61B, -27, -37, -55, -74, -88 & -110

Author of *Spacewalker: My Journey in Space and Faith as NASA's Record-Setting Frequent Flyer* (2013) and *Becoming a Spacewalker: My Journey to the Stars* (2014).

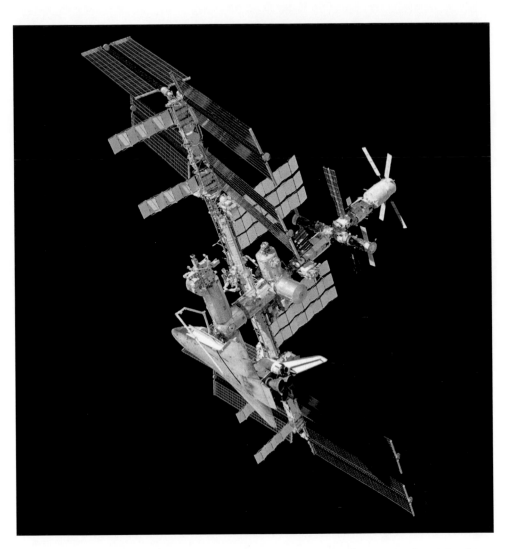

The almost completed ISS, with the docked Space Shuttle Endeavour, as viewed by the Expedition 27 crew on Soyuz TMA-20 shortly after its undocking in May 2011.

Dedication

In memory to my aunt Gwen Waldron (1924–2015) who found it hard to believe that a 'house' could be built in space, even after meeting several astronauts and cosmonauts who had visited one.

Prologue

THE FINAL SHUTTLE MISSION: *STS-135 Atlantis, Flight Day 12, Tuesday, July 19, 2011.* Sixteen years and 31 days since Atlantis' historic first docking with Mir, that same Orbiter was now preparing to leave the completed International Space Station for the final time, flying the very last Shuttle mission. As the complex passed 211 nautical miles (391 km) above and off to the east of Christchurch, New Zealand, Pilot Douglas Hurley flipped the switches to release the docking latches and allow the Orbiter to slip its moorings. In synchrony Commander Chris Ferguson radioed, "Physical Separation. Atlantis departing the International Space Station for the last time." In accordance with tradition on the station, ISS-28 Flight Engineer Ron Garan rang the ship's bell and bid his farewell to his departing colleagues, "Thank-you [Atlantis] for your twelve docked missions to the [station] and for capping off thirty-seven Shuttle missions to construct this incredible orbiting research facility."

Following the undocking, Hurley moved Atlantis into position for the formal fly-around maneuver to take pictures to document the exterior of the station. To provide optimum illumination, the ISS was yawed 90°. During the 27 min photo opportunity, the never-before-seen perspective of the longitudinal axis of the ISS from a departing Shuttle was recorded. Chris Ferguson informed the residents, "We just wanted to give you a final good-bye."

In Mission Control, Houston, Capcom Dan Tani, a veteran of a Shuttle assembly and a residency mission to the station, spoke for all in the flight control room when he told those on board Atlantis and the ISS, "We are proud to be the last in a countless line of Mission Control teams that have had the honor to watch over the ISS while Discovery, and Endeavour and Atlantis have visited during the last thirteen years. From this room, we have watched and supported as the Shuttle has enabled the station to grow from a humble single module that was grappled by the Shuttle's arm to a stunning facility that has grown so large that some astronauts have momentarily got lost in it…you can take it from me. The ISS wouldn't be here without the Space Shuttle, so while we have the communications link up for the last time, we want to say thank-you and farewell to the magnificent machines that delivered, assembled, and staffed our world-class laboratory in space."

With the STS-63 and -71 missions to Mir in 1995, these missions formed the 'book-ends' of a remarkable series of flights that had its origins in the late 1960s and the very beginning of the Shuttle program. If events had turned out differently, the first docking of a Shuttle with a space station could have occurred in the early 1980s but that wasn't to be. Instead of assembling Space Station Freedom, the three Orbiters had visited the Russian Mir space station as a prelude to assembling the ISS.

But there is so much more to the story than the three dozen docking missions which the history books record. The assembly of the station and the flights of the Shuttle may now be over, but the story of that achievement has only just begun to be told.

1

An Alternative Plan

> *I am directing NASA to develop*
> *a permanently manned space station*
> *and do it within a decade.*
>
> President Ronald W. Reagan,
> State of the Union Address,
> January 25, 1984

The 40th President of the United States of America spoke those words at a time when the idea of creating a large research facility in space seemed to be the next natural goal after demonstrating the Space Shuttle could fly, and fly again. The idea of a significant national space station had turned into plans for a largely 'international' station to help with the operating costs and development of associated hardware and logistics supply.

Echoing the May 25, 1961 call by John F. Kennedy to land Americans on the Moon "before the decade is out," some insiders may have foreseen the complexity and costs of such an undertaking, but no one could have guessed that it would take over twenty-seven years and the support and infrastructure of a former Cold War opponent to make that dream a reality.

To the casual observer, it might appear that it was somewhat simpler to develop the capability within eight years to travel a quarter of a million miles out to the Moon than the three decades it took to create a large space station at an altitude of several hundred miles but that surmise would neglect the huge and complicated background story of the years of planning, negotiations, the amendments to those plans, and further negotiating, before junking almost everything in order to pursue a simpler design. Nevertheless, the Shuttle was required to launch the majority of the hardware, support the assembly, and deliver the bulk of the logistics, supplies, and resident crews. It turned out that the new space station program also needed the talents and resources of the former Soviet Union to ensure that the first element of the facility literally got off the ground.

As documented in my *Linking the Space Shuttle and Space Stations: Early Docking Technologies from Concept to Implementation*, the journey from the early suggestions that

© Springer International Publishing Switzerland 2017
D.J. Shayler, *Assembling and Supplying the ISS*, Springer Praxis Books,
DOI 10.1007/978-3-319-40443-1_1

the Shuttle could support the creation of a large research facility in Earth orbit to actually visiting one was a matter of repeated delay, re-design, disappointment, huge expense, and compromise. Despite these struggles, the addition of the Russians in the station program was both fortuitous and timely. The interim Shuttle-Mir program was mostly highly successful for both the Americans and the Russians, with seven NASA astronauts spending many months aboard Mir, and despite challenges that included a fire and a collision, the prospects for continued cooperation with the assembly of the International Space Station looked promising.

There were, naturally, some doubters, mainly in the United States, who questioned what would happen if Russia proved unable to deliver the promised commitment. The question then arose of how should NASA and its other international partners fulfill the commitment to launch, assemble, and operate the space station.

To address this dilemma there had to be a back-up plan, a second 'tier' of assembly designed to ensure that at least *something* made it into orbit, because further delay and expense would be unacceptable not only to the other international partners but also the US Congress.

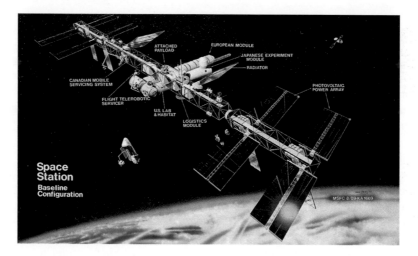

This image shows the baseline configuration of the proposed Space Station Freedom circa 1989. The design featured the phased approach to building the station, which ensured an initial capability at a reduced cost to the original design, followed by an enhanced capability at a later date.

SHUFFLING THE SHUTTLE TO ASSEMBLE THE STATION

Planning any mission into space can take several years. The drawing up of a workable Shuttle manifest became a daunting task to mission planners. In addition to selecting a crew, the payload had to be received and checked and the vehicle assembled, checked, and loaded. A complication for the scheduling of missions to the space station was the uncertainty in the timing of not only the hardware to be launched but also the funds to support the protracted program. This fluidity also affected the preparation of hardware and the sequence in which elements would be launched. The assistance of the Russians would help the program, but there was no short-term guarantee that they would launch on time. That uncertainty threatened to undermine the planning of Shuttle missions for station assembly.

Furthermore, the Shuttle manifest during the 1990s was now to include at least ten missions to the Russian Mir space station as a sort of stepping stone to the ISS. On top of this were Spacelab missions, servicing flights to the Hubble Space Telescope, a plan to launch and routinely service a commercially operated Industrial Space Facility with the prospect of at least fifteen missions, and a number of satellite deployments and a range of 'observation' missions for the Mission to Planet Earth program.

By 1994 NASA had plans for seventy-eight Shuttle flights running through 2003, including the ten to Mir, no fewer than thirty related to the new space station, and at least two servicing missions to the Hubble Space Telescope. The schedule was by no means an official manifest but a regular internal KSC document aimed at scheduling long range plans, and therefore the resources needed to achieve those missions. Later flights had not yet been assigned specific payloads but the schedule (see Appendix 3) gives an indication of the intensity of Shuttle operations planned for a ten-year period. This was a pace that the space agency came nowhere close to matching. There was a shuffling of the Shuttle vehicles in order to match the changing forecasts almost right up to the end.

To accommodate the Shuttle-Mir program, other intended missions were delayed, canceled, and deleted from the manifest. This process continued as the effort switched to the assembly of the ISS. The missions that lost out included at least three additional ATLAS surveys, the series of Industrial Space Facility deployments, over half a dozen intended Shuttle Radar Laboratory missions, and a number of science flights using the Spacelab module. In scrapping other projects in order to commit the Shuttle to the ISS NASA had to ensure that if Russia was unable to launch its promised elements then at least certain of the hardware envisaged for Space Station Freedom could be pursued to get the ISS assembly started on time.

NASA'S TIER-2: AN ALTERNATIVE ASSEMBLY PLAN

In 1995, in the light of American concerns about Russia being in the 'critical path' of ISS assembly, NASA devised an alternative scenario in case the Russian hardware for the nominal scheme did not materialize for whatever reason.

When the agreement on full cooperation with the redesigned Space Station Freedom was signed between the USA and Russia on September 2, 1993, the name of the station was changed to 'Alpha' because that was a more diplomatic label. The new station had become essentially a slimmed down Freedom configuration that was to be assembled in four phases. The final hybrid configuration included elements from both Option A and Option B, and also incorporated some elements of what had been intended for the now canceled Russian Mir-2 proposal. The renewed partnership with the Russians for Alpha naturally generated a lot of media interest and for a time the new station was referred to as 'RAlph.' But the name Alpha never really caught on, and the project became known simply as the International Space Station.

By 1994 the configuration of Space Station Freedom had been abandoned and instead a mutually agreed stripped down configuration had been accepted. Here the image shows the core configuration using Russian elements and the first Shuttle-delivered components up to the attachment of the Joint Airlock (Phase II).

American doubts of sustained Russian participation in the ISS included the amount of finance NASA was injecting into the ailing Russian space industry. The agreement was for Russia to launch the first two elements of the new station. NASA would add a docking node and a science laboratory to form the core of the station. This plan meant that the major elements that were to be supplied by Canada, Europe, and Japan would be seriously delayed because the truss that carried the solar arrays would require to be in place to support their operation.

As plans for the ISS were defined, the challenge of securing sufficient, sustained funding also remained, even to the point of having to assure the Russians that the US had no intention of pulling out of the project. With budget reductions in Europe and Canada and with continuing fears of the Russians failing to produce their hardware, a back-up scheme was a prudent (and costly) move. This came at a time when issues in the struggling Russian space infrastructure imposed further budgetary pressure which led to requests for even more finance just to survive, let alone prepare for and support ISS operations – even in the short term.

In addition there were concerns regarding Shuttle safety, with news that the strict launch safety regulations would be compromised if the Orbiter had to meet the very narrow 5 min windows for ISS missions heading for the 51.6° orbit, in contrast to the 28.8° inclination that had been planned for Freedom. A direct consequence of this was the development of a lighter aluminum-lithium construction External Tank for Shuttle-ISS assembly flights, as this would allow a greater fuel margin. Another concern with the Shuttle was that the three-Orbiter fleet would not be sufficient for the twenty-three assembly and utilization missions planned at that point. The venerable Columbia was too heavy to lift the large elements to the station's orbital inclination and altitude, so it could not participate in the assembly work. Another worrying factor was the very fluid politics of the former Soviet Union, with the newly independent nations of Kazakhstan and Ukraine in open dispute with Russia. Consequently, there was a lot at stake as the time approached to start the actual assembly of the station.

The Interim Control Module

If the Russians failed to launch the Salyut FGB that was to serve as the Control Module that NASA intended to buy rather than lease, then a Lockheed-built military propulsion module known as Bus-1 would be sent up as a substitute. If the crucial Service Module also failed to appear, was lost at launch, or was unable to dock with the FGB or Bus-1 (as appropriate) then NASA would launch an Interim Control Module, which was also based upon the Bus-1 design, as a stand-in. If all else failed, a much bigger American propulsion unit was a possibility. Although all this alternative hardware would require budgetary approval Congress was eager for NASA to take precautions against Russian failures.

Both Russian modules were successfully delivered, albeit much later than intended. The first, the FGB named Zarya ('Dawn') was launched in November 1998. A month later STS-88 added the first American element. Russia launched the Service Module Zvezda ('Star') in July 2000. With these modules docked on-orbit, the nascent station was able to support its first crew. The expansion of the station progressed remarkably smoothly with only the loss of Columbia in February 2003 imposing a pause between December 2002 (STS-113) and September 2006 (STS-115).

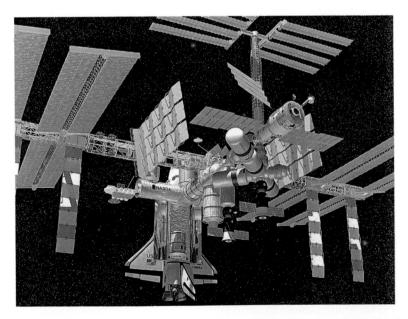

In this 1995 graphic the completed International Space Station is portrayed with the planned Russian segment prominent in the foreground.

But if the worst case scenario had happened, and Russia had failed to deliver, then NASA's plan to assemble the ISS without the Russians was ready to be implemented. The account in the next section is based on the document 'Tier-2 Bus-1 Option of the ISS.'[1]

Tier-2 Assembly

The premise of the Tier-2 study was the assumption: "Russian participation had been eliminated and that the functions that were once supplied by the Russians (propulsion, resupply, initial attitude control, communications, etc.) are now supplied by the United States." The Lockheed-built Bus-1 was to replace the Russian hardware, and the data was adjusted in the report to reflect this. The role of the Data Book was to define some of the issues that this situation would impose and to recommend means by which they could be solved.

Firstly, several of the early assembly flights would deploy station hardware at the designated assembly attitude, but it was found that in doing this the hardware would have an orbital lifetime of less than 90 days without a re-boost capability. The study also identified a significant shortfall in the time available to undertake both assembly and maintenance EVAs on Shuttle missions. During the first half of the construction sequence, the mass properties of the assembly stages would create large flight angles and poor microgravity environments, hence new CMG hardware must be developed to resolve the problem. In addition, there was insufficient re-boost propellant manifested to compensate for expected increases in the density of the upper atmosphere during the part of Solar Cycle 23 (June 1996 through January 2008) when this construction was to take place. With the solar maximum occurring in March 2000, significant delays could be expected in achieving the program's milestones.

One of the most significant issues in the Tier-2 plan was to retain the station at an orbital inclination of 51.6°. According to the report this was to enable the Russians to rejoin the program if they were able to do so at some point in the future. The assembly altitude was restricted to 150 nautical miles (277.8 km) to achieve the minimum of 90 days of orbital life. To reduce the Shuttle flight rate the manifest called for at most six assembly missions per annum, with the option of a seventh flight to supply additional fuel and to replenish the gases used during EVA airlock activities. This plan therefore manifested a Shuttle flight every other month starting in February 1998. This assumed that delivery of the pressurized nodes wouldn't be accelerated, and it took into account any delays caused by the non-participation of the Russians.

Though the Tier-2 assembly sequence included margins, reserves, and overheads similar to the baseline plan and was not meant to deviate far from that sequence, the order of several flights was changed and some of the proposed cargo elements were rearranged to accommodate the alternative planning. The Tier-2 assembly manifest in Table 1.1 details thirty-six Shuttle launches between February 1998 and December 2003 that would have lifted some 760,036 lb (344,752.3 kg) into orbit and involved at least sixty-five EVAs spread across 286 mission days. This daunting program was expected to be accomplished in less than six years.

Ground Rules And Assumptions

It became clear to the authors of the report that a significant amount of EVA would be required by the Tier-2 program. To assess the EVA resources that would be needed to complete assembly in this new scenario, the following assumptions were made:

Table 1.1 Tier-2 Assembly Sequence Manifest Circa 1995

Flt	Launch Date	Flight Name	Delivered Elements	Altitude (Nm)	Mass to Orbit (lb)	Mission Duration (Days)	STS Crew	Scheduled EVA's
1	2/1998	1A	Bus-1, Spacer	210	31,221	7	5	0
2	4/1998	2A	Node 1 (2 storage racks), PMA3, PMA2	205	27,631	7	5	2
3	6/1998	3A	Z1 truss, CMG's, Ku-band, HP Gases, EVAs (Spacelab Pallet)	200	14,004	9	5	3
4	9/1998	4A	P6, PV Array (4 battery sets) / EATCS radiators, S-band	190	32,956	8	5	2
5	11/1998	5A	Lab (4 Lab Sys racks)	205	29,765	9	5	3
6	12/1998	6A	1 Storage, 7 Lab Sys racks (on MPLM), UHF, SSRMS (on Spacelab Pallet)	215	12,852	12	5	3
7	2/1999	UF-1	ISPRs (on MPLM)	215	15,000	12	7	0
8	4/1999	7A	Airlock, HP gas (on Spacelab Pallet)	215	21,609	9	5	2
9	6/1999	8A	SO, MT, GPS, Umbilical's, A/L Spur	215	30,205	7	5	2
10	8/1999	Bus	Bus-1, spacer	215	28,256	9	5	3
11	10/1999	UF-2	ISPRs, 2 Storage Racks (on MPLM), MBS	215	5,615	12	7	1
12	12/1999	9A	S1 (3 rads), TCS, CETA (1), S-band	215	31,026	7	5	2
13	2/2000	10A	Node 2 (4 DDCU racks), Cupola	215	27,359	12	5	4
14	4/2000	11A	P1 (3 rads), TCS, CETA (1), UHF	215	30,720	9	5	3
15	6/2000	UF-3	ISPRs, 1 Storage Rack (on MPLM)	215	12,890	12	7	0
16	8/2000	1 J/A	JEM ELM PS (5 JEM Sys, 2 ISPR, 1 Storage Rack), 2 O2 tanks (on ULC), SPDM	220	22,810	10	7	6

(continued)

Table 1.1 (continued)

Flt	Launch Date	Flight Name	Delivered Elements	Altitude (Nm)	Mass to Orbit (lb)	Mission Duration (Days)	STS Crew	Scheduled EVA's
17	10/2000	12A	P3/4, PV Array (4 battery sets), 2 ULCAS	220	32,781	8	5	2
18	12/2000	12A+	P5, P4/P5 MT/ CETA Rails, P4 PV Battery Sets (2) 16-day EDO Pallet	230	6,083	15	7	7
19	2/2001	UF-4	ISPRs (on MPLM)	230	13,000	12	7	0
20	4/2001	BF-1	Bus-1	230	25,000	7	5	0
21	6/2001	13A	S3/4, PV Array (4 battery sets), 4 PAS	230	31,994	8	5	2
22	8/2001	13A+	S4 PV battery sets (2), S4 & P6 MT/ CETA rails (on ULC), 16-day EDO Pallet	230	2,556	15	7	6
23	10/2001	UF-5	ISPRs on (MPLM), Attached Payloads (on ULC)	230	9,000	12	7	0
24	12/2001	1J	JEM PM (3 JEM Sys racks), JEM RMS	230	30,864	9	5	2
25	2/2002	2E	1 APM Storage, 3 U.S. Storage, 7 JEM racks (on MPLM), S5	230	13,229	9	5	1
26	4/2002	UF-6	ISPRs (on MPLM)	230	13,000	12	7	0
27	6/2002	2J/A	JEM EF, ELM-ES, P6 PV battery sets (2) (on ULC)	230	14,540	7	5	1
28	8/2002	15A	S6, PV Array (4 battery sets)	230	26,886	9	5	3
29	10/2002	BF-2	Bus-1	230	25,000	7	5	0
30	12/2002	UF-7	ISPRs, 1 Storage Rack (on MPLM)	230	14,390	12	7	0
31	2/2003	14A	Centrifuge	230	24,255	9	7	1
32	4/2003	1E	APM (5 Sys, 1 Storage, 5 ISPR racks)	230	26,467	9	5	1
33	6/2003	16A	Hab (6 Hab racks)	230	27,502	8	5	2

(continued)

Table 1.1 (continued)

Flt	Launch Date	Flight Name	Delivered Elements	Altitude (Nm)	Mass to Orbit (lb)	Mission Duration (Days)	STS Crew	Scheduled EVA's
34	8/2003	17A	1 Lab Sys, 8 Hab Sys racks (on MPLM), S6 PV battery sets (2) (on ULC)	230	10,913	8	5	0
35	10/2003	18A	CTV#1	230	24,255	8	5	0
36	12/2003	19A	3 Hab Sys, 11 U.S. Storage racks (on MPLM)	230	14,402	7	7	0

- Checkout of the SSRMS would be undertaken by the flight that delivered it to orbit, ensuring its operational readiness for the next flight. [In 2001 the actual checkout of Canadarm2 continued after STS-100 had delivered it to the ISS.]
- Every EVA during the assembly phase would be undertaken from the Shuttle airlock rather than the station airlock. [This was not the case during the actual assembly of the ISS. Quest was the primary airlock from 2002. The Orbiter airlock was used by only one subsequent assembly flight (STS-114 in 2005) through to 2011.]
- A baseline of two spacewalks per flight would be established (plus a nominal contingency EVA). Additional spacewalks would be feasible by placing extra tanks of gas in the payload bay. [Consumables were an issue on each mission and reserves had to be husbanded in case a late contingency EVA was called for, such as manually closing the payload bay doors after departing the station. On average, three EVAs were accomplished per ISS assembly mission during the actual program.]

For the purpose of this study the ground rules for the EVA baseline included:

- Maintenance EVAs were not addressed during the reference version of the study. [Presumably time was the critical issue here. It was easier and more important to plan the assembly of the station before setting out to review its maintenance requirements, because until the station was actually built these could only be notional at best. Detailed timelines of known procedures had been studied over the years and there had been extremely detailed time and motion exercises (see Chapter 8).]
- There would be no planned EVAs during utilization flights. [Again the focus was on the assembly missions; the role of the utilization flights during Tier-2 was to stock the station with internal supplies and apparatus. In reality EVAs were scheduled for every assembly and utilization flight. There were several reasons for

this, chiefly that they reduced the pressure on the assembly flights to achieve their primary tasks and they also addressed maintenance issues and any delayed or get-ahead tasks.]

For the Tier-2 studies, electronic models were used to determine the mass and aerodynamic properties for the various flights, while several models of the Earth's atmosphere were used to assess flight characteristics. In addition, CMG and RCS attitude control simulations were performed, using a peak solar cycle worst-case scenario. This predicted a denser atmosphere and a probable slip in the schedule. Attitude control models were studied, both with and without a docked Shuttle. The CMG control of the station was simulated with the units of the Bus serving prior to activating the Z1 truss containing the station's own CMGs. Re-boost analysis used a Freedom RCS control algorithm which once again used the Bus to initiate the burns. Under the Tier-2 planning, propellant resupply for ISSA (International Space Station Alpha) would be by "an as yet undetermined delivery mechanism aboard the Shuttle." This was not an entirely new proposal because the STS-41G mission in October 1984 had demonstrated automated transfer of hydrazine for an experiment. Nothing further had been attempted since, but several NASA field centers had made studies of orbital refueling during space servicing missions, most notably by the Goddard Space Flight Center for Nimbus and Landsat satellites and some of the Great Observatories which NASA was developing.

The minimum operating altitude of the station during its assembly was required to offer "90 days of gradual orbital decay to a low point of 278 km [150 n. miles]." The minimum operational attitude thereafter was given at 180 days of orbital decay down to an altitude of 278 km. During both phases the station would hold a reserve propellant that included a 'skip cycle' (which essentially meant missing a refueling mission slot) in order to re-boost to an altitude that could provide at least 360 days of orbital decay to 278 km under nominal operations, thereby gaining sufficient time to launch a refueling mission.

Systems And Logistical Impacts

In the Tier-2 report, the authors highlighted the changes to the various ISSA system capabilities. They compared the baseline capability from Russian-provided services versus the Tier-2 capabilities without the Russians. This included studies of thermal control; command and data handling; communications and tracking; environmental control and life support; electrical power systems; guidance, navigation and control; propulsion; and EVA and robotics.

Most of these areas were worked around in the Tier-2 model but some couldn't be satisfactorily addressed. The use of the Lockheed Bus-1 prior to activating the US Lab, for example, would have prevented the launching of a resident crew in the early stages. The Tier-2 scenario also ruled out station-based crews being available for maintenance and assembly EVA support activities. It was also noted that when a Shuttle was docked with Node 1, the safe entry of astronauts into that module would require the Orbiter to provide ventilation and atmospheric control. Extra high pressure oxygen and nitrogen would need to be manifested early in the sequence to allow for gas seepage out of the pressurized modules. And during the second phase of the assembly, the Orbiter would have to provide waste management. But atmospheric control and supply, fire detection and suppression, atmosphere revitalization, temperature, humidity, and water recovery and management would all be managed through the laboratory in which the crew were to live. To address the requirement for consumables, any additional gases would have been delivered using modified Gas Conditional Assembly (GCA) tanks on EDO-type pallets in the payload bay.

A Useful But Redundant Study

Whilst useful and thought provoking, the Tier-2 document was happily discarded after the launch of the Russian Zarya and Zvezda modules. Despite being a dead-end study, it did raise a few issues and instigated a useful second look at the assembly plan in the event of Russian involvement being either delayed or withdrawn. Similar studies were performed when the assembly sequence was revised after the loss of Columbia led to a reduction of NASA's overall budget and the decision to retire the Shuttle as soon as the assembly of the ISS was completed.

Studies of this type address something that is often under-reported, namely the huge effort involved in taking into account potential 'what if' situations, and the planning of alternative missions. Alternative, back-up, and contingency plans are part and parcel of space flight, no matter how large or small the project or mission. It is always hoped that such plans won't be enacted, but when they are required (such as switching to the two-person caretaker crews for the ISS between 2003 and 2006) they do provide reassuring breathing space when flight operations go off-nominal and time is needed to work out the next step.

In this image of the completed ISS the American segment, with the European and Japanese laboratories, is shown the foreground, this time with the planned Russian segment in the background.

FROM IMAGINATION TO REALITY

Despite difficulties and delays, the ISS slowly expanded between 1998 and 2002 due to the efforts of sixteen Shuttle crews, six resident crews, four Soyuz visiting crews, and a number of unmanned resupply vehicles.

On February 1, 2003 everything came to a grinding halt again with the tragic loss of Columbia. No Shuttles flew for the next twenty-nine months. When the first Return-To-Flight mission failed to resolve all of the concerns, requalifying the Shuttle system took a further twelve months.

While the Shuttle was grounded, the ISS was kept operating with reduced crews by Russian Soyuz and Progress vehicles. But what remained to be determined was whether the remaining three Orbiters would be able to finish the assembly process in advance of the planned retirement of the Shuttle in 2010. After a second Return-To-Flight mission requalified the Shuttle for operations, twenty further Shuttle flights completed the ISS between September 2006 and the postponed retirement of the Shuttle fleet in July 2011. The ISS was stocked up with supplies and logistics, not only by the Shuttle but also by a variety of resupply vessels from Russia, ESA, Japan, and new American commercial space operators.

When the last Shuttle rolled to a stop on Earth in the summer of 2011 nearly thirty resident crews had manned the station and produced a wealth of scientific data. Five years after the end of the Shuttle program, the work conducted on those missions has enabled the ISS to continue flying, supporting fifty crews so far, including a one-year endurance mission, and plans are in place to extend operations well into the 2020s.[*]

SUMMARY

None of this would have been possible without the vision, half a century ago, to design a vehicle capable of building a space station. The disappointment was that it took nearly thirty years before a Shuttle visited a station, a Russian one at that, and the assembly of a new international station took almost another twenty years.

That task involved many components on each mission, significantly extending (and testing) the system like no other series of missions, except perhaps the Hubble service missions. The assembly and supply of the ISS required skill in rendezvous, proximity operations and docking, the efficient transfer of logistics and supplies, extensive use of robotics, and multiple EVA operations. As the following chapters will show, there was considerably more to a Shuttle mission to the ISS than simply launching hardware and supplies into space.

---

[*] At the time of writing in 2017, members of the 50th ISS Expedition Shane Kimbrough (NASA, Commander), Andrei Borisenko (Russian Flight Engineer) and Sergei Ryzhikov (Russian Flight Engineer) were nearing the end of their tour.

Notes

1. Alternate Assembly Sequence Data Book for the Tier-2 Bus-1 Option of the International Space Station, L.M. Brewer et al, Langley Research Center, Hampton, Virginia and M.A. Garn et al, Analytical Mechanics Associates Inc., Hampton, Virginia. NASA Langley Research Center, Hampton, Virginia, NASA Technical Memorandum TM-110198, August 1995

2

The Shuttle-ISS Assembly Missions

*Once rocket craft capable of ferrying loads to and
from the terminal orbit have been successfully
demonstrated, the building [of a space station] is
estimated to last ten years*

Kenneth W. Gatland and Anthony M. Kunesch,
from *Space Travel*, 1953

A review of human space activities of 2011 reveals that it was a milestone year in many ways. April 12 celebrated the 50th anniversary of the launch of the world's first manned space flight. Since then, more than 500 individuals from around the world had ventured into space on over 300 missions lasting from a few brief hours, to several weeks and up to fourteen months. Five decades after Yuri Gagarin's 108 min, single orbit of Earth, a permanent human presence was being maintained by multi-national crews on board the International Space Station. In September, China, the new player in the field of human space exploration, had launched its first unmanned space laboratory, named Tiangong, and they would later demonstrate their automated rendezvous and docking capabilities during the unmanned Shenzhou 8 mission prior to sending their first crew to the station in 2012. It could be said, therefore, that 2011 was a year that celebrated past successes while looking towards future prospects.

It was also the year in which the majority of the assembly and outfitting of the ISS was finished and the American Space Shuttle made its final flights. After thirty years and 135 missions, Shuttle orbital operations came to a close. Since 1995 a significant number of these missions (over one third of the total) had been associated with space station operations. The Shuttle-Mir program involved one rendezvous mission to the Russian space station, nine docking missions, and a solo flight. Later, a further thirty-seven dockings were executed at the ISS. Across both programs, all of the attempted dockings were successful. It was a remarkable achievement over sixteen years, thanks to the efforts of dedicated teams of workers on the ground and in space.

© Springer International Publishing Switzerland 2017
D.J. Shayler, *Assembling and Supplying the ISS*, Springer Praxis Books,
DOI 10.1007/978-3-319-40443-1_2

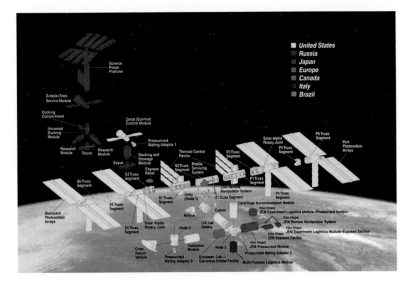

A 1999 image showing the original components intended for the International Space Station prior to budget cuts.

PAGES FROM HISTORY

The ISS has existed for almost two decades. Remarkably, it took almost as long from President Reagan's authorization in January 1984 for NASA to build a space station to the launch of the first hardware in November 1998. Beginning in the 1960s, even as it was developing Apollo to take astronauts to the Moon, and through the 1970s when it was developing the Space Shuttle, the space agency had devoted considerable effort to debating, designing, proposing, costing, and repeatedly redesigning a large variety of stations it imagined might host potentially multi-national crews. At the time of writing this book, the ISS is hosting its 50th long-duration crew, and with agreement to operate it until at least 2024 there is the potential for the 100th crew to take up residence there; a remarkable achievement.

As daring as the pioneering Vostok, Voskhod, Mercury, and Gemini programs were, and as inspiring as Apollo became, it is the experiences of the crews of Salyut, Skylab, Mir and now the ISS that will provide the foundation for even greater achievements in space. There is tremendous potential for sustained human presence in space, for a return to the Moon, and for ventures deeper into the solar system. One vital key for unlocking this potential was the ability of the Shuttle to expand operations involving Mir and then to assemble and maintain the ISS.

Although NASA did not possess extensive space station experience, for several years Shuttle missions had been developing the techniques and procedures required to support station assembly. A three-phase program was devised to address this shortfall. Phase-I was to include Shuttle missions docking at the Russian Mir space station. Phase-II was to see sufficient flights of the Shuttle and other spacecraft to develop the core of the ISS to the point at which a resident crew could remain aboard without requiring a Shuttle to be present with consumables. Finally, Phase-III would see a number of Shuttle missions expand the station, particularly by the addition of power systems, science and habitation modules, crew facilities, and the supplies to make the ISS an operational laboratory for research into the microgravity environment.

SHUTTLE MISSIONS TO CREATE THE ISS (PHASE- II)

In December 1998, Shuttle flight operations under the International Space Station program finally began. This was some four years after the plan agreed following the initial authorization by President Ronald Reagan had scheduled the construction of Space Station Freedom to be *finished*. During the first eighteen months, progress was slow due in part to waiting for Russia to launch Zvezda as the second of the two core modules but also due to ongoing problems with the electrical wiring and main engines of the Shuttle fleet.

The objective of the early missions was to create the core of the station as a facility capable of hosting a three-person resident crew without a Shuttle attached. This core station would include research facilities on board, independent access for EVA at the station, and sufficient power resources and consumables for such activities. The next step would be to expand the 'backbone' truss to support both additional solar arrays to increase the power supply and radiator panels to regulate heat. The third phase would add further habitation and laboratory modules, as well as enhanced robotics, to enable the station to support six-person resident crews. The early crews would be comprised only of Russian cosmonauts and American astronauts. Command of the station would alternate between these two nations. Once the station was expanded with laboratories from Europe and Japan and could support six there would finally be opportunities for representatives from the European, Canadian and Japanese space agencies to serve on the station, making it a truly international resource.

The Shuttle was essential to this plan, primarily owing to its capacity to deliver the required hardware to the station. It could also serve as a major logistics and resupply craft, a resident crew taxi, and a garbage truck for returning unwanted items to Earth. These roles would be supported later in the program by other vessels. From 2000 the Russian Soyuz supplemented crew transport duties and served as an always-available Crew Rescue Vehicle, while the unmanned Progress version of that ship assisted with logistics supply and waste disposal. The latter tasks were further supplemented by the ESA Automated Transfer Vehicle (ATV) beginning in 2008 and by the Japanese H-II Transfer Vehicle (HTV) in 2009. But in the early stages the sheer volume of material that was required to stock up the station, install the scientific hardware, and replenish the supplies could only be transported by the Shuttle.

The other essential aspect of the Shuttle was its reusability. In addition to its cargo capacity of several tons (vastly superior to that of the Soyuz of about 110 lb or 50 kg), the large volume of the payload bay and pressurized modules could transport surplus material and, most importantly, experiment samples and data back to Earth. While the automated Progress, ATV, and HTV resupply ships were extremely useful, they were designed to burn up in the atmosphere during re-entry, destroying anything they were carrying.

Following the retirement of the Shuttle fleet, the US cargo delivery capacity was taken up by commercial enterprises, initially with the Dragon spacecraft in 2012 and the Cygnus in 2013. The Cygnus carrier burns up in the atmosphere. The Dragon can return cargo to Earth, although not as much as the Shuttle could accommodate.

The Shuttle launch manifest and assembly sequence changed several times over the course of the program. There were a variety of reasons, with the foremost being delays in production of hardware for the station, difficulties preparing and launching Shuttles, the loss of Columbia and the long recovery from that tragedy, and budgetary changes across several of the partner nations (as well as the US). There was also a lack of clear direction in America concerning long-term planning for programs beyond the Shuttle.

What follows is a brief summary of Shuttle-ISS missions that made it through this background and actually flew.

1998: Towards the end of the year, the physical assembly of the ISS began with the launch of the first core module by Russia and the first docking node by the USA. From November 1998 through to October 2000, between the launch of the Control Module and the first resident crew (Expedition One, or EO-1), the pre-residency missions were flown under the 'Expedition Zero' banner. For NASA, this included the flights of STS-88, -96, -101, -106 and -92.*

STS-88, Endeavour: 1st ISS Assembly Flight 2A (December 4–15, 1998): This first flight in the Shuttle assembly series was actually the second launch of the ISS program because the Russian Zarya ('Dawn') Control Module had been launched unmanned a fortnight earlier from the Baikonur Cosmodrome in Kazakhstan using a Proton rocket. The primary mission of STS-88 was to rendezvous with Zarya in order to mate it with the 'Unity' docking node. On December 5, Unity, which featured six ports to facilitate expansion of the American segment by subsequent flights, was lifted from its stowage using the RMS and then securely attached to the Shuttle docking system located in the front of the payload bay. The following day the crew guided Endeavour to rendezvous with Zarya, to permanently mount Unity on the forward docking port of that module. Three spacewalks completed the attachment and activation of the combined structure. Inside, the crew removed launch bolts and restraints and inspected the interiors of both modules to confirm their readiness to support further activities. It was a good start to a long and complicated program.

1999: Delays finishing Russian hardware and various problems with the Shuttle fleet meant that only one Shuttle mission visited the ISS during this year, frustrating attempts to complete the core configuration and begin an early residency. The good start seen at the close of 1998 was not followed up throughout 1999 and most of 2000.

STS-96, Discovery: 2nd ISS Assembly Flight 2A (May 27– June 6, 1999): Six months after the initial Shuttle-ISS mission, this second launch was the first dedicated logistics mission to start the delivery of additional hardware and supplies. It was a direct follow-on to STS-88, which could not have carried both the Unity Node and all the necessary early logistics. STS-96 became the first vehicle to physically dock with the station. It was also the only Shuttle-ISS mission in 1999, primarily due to delays in launching the Russian

---

*Note that Shuttle missions were not flown in the strict order of their numerical designations.

1998: STS-88, the first ISS assembly mission berths the US Unity Node to the Russian Zarya Control Module to start the assembly of ISS.

Zvezda ('Star') Service Module that would host resident crews. The result was a frustrating wait between this mission and the next of almost a year. The STS-96 crew had a busy program of transferring hardware and supplies around both the interior and exterior of the station. A single EVA supported the transfer of the US Orbital Transfer Device and the first parts of the Russian Strela crane that would be assembled by later crews. There were some concerns reported over air quality when astronauts worked in Zarya. This was attributed to open panel doors interrupting the flow of air around the module, enabling carbon dioxide to accumulate. This type of 'contamination' was not uncommon in commissioning Russian modules, and in later interviews the astronauts played down the 'bad air' reports.

2000: The primary goal of this year was to get the station to the point where the first resident crew could be launched in a Soyuz spacecraft. For this to occur, Zvezda had to be launched and docked automatically to Zarya. The expanded core station would also require the delivery of a large amount of supplies and logistics, as well as internal and external outfitting before the first resident crew arrived. This would enable Expedition One to focus on creating the foundations for prolonged habitability aboard the station and the start of scientific operations.

STS-101, Atlantis: 3rd ISS Assembly Flight 2A.2a (May 19–29, 2000): With the ISS assembly placed on hold, NASA cleared some of the backlog of Shuttle missions in its manifest by flying three non-ISS missions: STS-93 which deployed the Chandra X-Ray Observatory, the STS-103 Hubble servicing (designated SM-3A), and STS-99 with the Shuttle Radar Topographic Mission. Meanwhile, further slippage in preparing Zvezda for launch caused maintenance concerns for the Unity-Zarya combination already on-orbit. The batteries on Zarya had not been designed to function for so long without the more capable power systems offered by the Service Module. The setback also affected planning for the next Shuttle flight. The crew for STS-101 had trained to work with the Service Module after its docking at the ISS. With that module still slipping, a decision was made to split the original STS-101 (2A.2) mission over two flights and divide up the crew between the two missions. The revised -101 mission (now designated 2A.2a) would fly first, taking fresh logistics to the station and making a single EVA to finish work on the external cranes that was started by STS-96. The rest of the preparations to make the station ready to receive the first resident crew would be completed by a new mission, STS-106 (2A.2b), that would launch after Zvezda had docked at the rear port of Zarya. The new Shuttle crew was supplemented by the three ISS-2 crewmembers to give them experience aboard the station a year *before* their official residency.

STS-106, Atlantis: 4th ISS Assembly Flight 2A.2b (September 8–20, 2000): The successful launch of Zvezda and subsequent docking with Zarya on July 26, 2000, and then the docking of the first unmanned Progress mission (Progress M1-3) with Zvezda, came as a huge relief to the ISS planners. At last, the station had its own flight control and orbital maintenance functions. Before a crew could live on board the ISS without a Shuttle being present it needed to be outfitted both internally and externally. The STS-106 mission became the second logistics flight of the year, finishing the work that had originally been assigned to STS-101. The single EVA of the new mission ran cabling between the two Russian modules. The way was now clear for a fast expansion of the ISS to host a resident crew of three that could undertake significant scientific research, robotics, and station-based EVAs without requiring a Shuttle in attendance.

STS-92, Discovery: 5th ISS Assembly Flight 3A (October 11–24, 2000): This flight included the largest piece of hardware delivered to the station by the Shuttle since the Unity Node nearly two years earlier. The Zenith (Z1) truss became the focal point for the expansion of the truss structure which would support a number of solar arrays and radiators to power and regulate the temperature of the station. The full structure was so large that it needed to be split up and assigned to several flights and, as it turned out, a number of years. Installing the Z1 truss segment at the uppermost (zenith) port of Unity was accomplished using the RMS and the support of four EVAs. The crew also finished work within the station, continuing the delivery of supplies and logistics to support the first residents who were scheduled to arrive at the end of the month, barely a week after the departure of the STS-92 crew. The third Pressurized Mating Adapter (PMA-3) was installed to provide an alternative location at the nadir (lower) port of Unity to receive a Shuttle. Although STS-92 was docked for a week, the majority of the time was devoted to installing the Z1 truss and the crew only entered the station for about 24 hr at the end of their stay.

STS-97, Endeavour: 6th ISS Assembly Flight 4A (November 30–December 11, 2000): This was the first Shuttle mission to dock at the ISS with a resident crew already on board, namely the EO-1 trio of William M. ('Bill') Shepherd, Yuri P. Gidzenko and Sergei K. Krikalev. They had arrived on November 2 on Soyuz TM-31 after their two day rendezvous. STS-97 docked at the newly installed PMA-3 and carried the initial set of solar arrays (Port 6 or P6). The installation of this unit on the Z1 truss was supported by three spacewalks. This was the first of eleven Shuttle missions over the next two years that would enable the ISS to reach a point where it could be sustained (albeit only just) when the Shuttle fleet was grounded between 2003 and 2005 in the wake of the loss of Columbia.

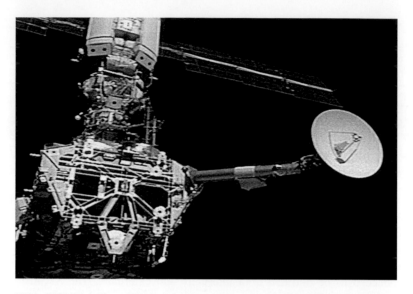

2000: STS-92 delivers Z1, the initial segment of the Integrated Truss Structure.

2001: In the year in which Stanley Kubrick's 1968 movie 'A Space Odyssey' was set, partly inspired by the story 'The Sentinel' published by Arthur C. Clarke in 1951, there was a real space odyssey underway in Earth orbit. It was a great year for the ISS. The station grew in size and capability and the rotation of resident crews began to take on a routine pattern. During a very busy twelve months, six Shuttle missions delivered more hardware and supplies. After the essentially quiet period of Expedition Zero (November 1998–October 2000) in advance of the arrival of the first resident crew, the pace began to step up.

STS-98, Atlantis: 7th ISS Assembly Flight 5A (February 7–20, 2001): The objective of this first mission of the year was to add NASA's science laboratory, named Destiny. Once it was in place on the Unity Node the Shuttle crew, assisted by the resident crew, began the process of configuring and outfitting of the large pressurized module to open the station's first real scientific facility for business. The installation was supported by the Orbiter crew undertaking three spacewalks. Although it was still a very early stage in the assembly process, there was a sense of growing maturity in orbital operations at the station.

In March 2001, Mir was finally de-orbited amid much regret from most of those in the Russian program. That same month the first resident crew of the ISS landed after a highly successful 141 day mission. The baton had now clearly been handed over to the new facility.

STS-102, Discovery: 8th ISS Assembly Flight 5A.1 (March 8–21, 2001): As the first mission fully to draw upon the experience gained from Shuttle-Mir, Discovery arrived with the ISS-2 crew and a significant amount of supplies in the payload bay aboard the first Multi-Purpose Logistics Module (MPLM) known as Leonardo. The visit included two EVAs, one of which was carried out by two members of the new ISS crew and its purpose was to continue the reconfiguration of the fittings, power and utility cables on the exterior

2001: The US Destiny laboratory is delivered by STS-98.

of the station. The MPLM was temporarily berthed on the side of Unity to transfer 5 tons of materials into the ISS and then over a ton of unwanted hardware and miscellaneous trash, much of it packaging, into the module for return to Earth. After a single expedition these operations confirmed the logistical challenge that the program would face in maintaining the station in the state needed to support a crew comfortably and also support their research work. Mindful of ensuring a smooth transition from one expedition crew to the next, this first handover saw the individual members of the two crews exchange roles in stages over an interval of several days. In particular, each pair had to transfer their Sokol pressure suits and Soyuz seat liners. This was the first of a series of exchanges of complete expeditions on the station via Shuttle missions, a feat achieved only once on Mir during STS-71.

STS-100, Endeavour: 9th ISS Assembly Flight 6A (April 19–May 21, 2001): With astronauts from the US, Canada and Italy working with the US-Russian residents this mission gave the ISS a truly international feel. The second logistics mission delivered 3.4 tons of cargo within the MPLM Raffaello and, outside, the Space Station Remote Manipulator Systems (SSRMS or 'Canadarm2'). This improved version of the Shuttle RMS was also the first part of what would become the Mobile Base System (MBS) of the station. The increased pace of activity at the ISS was highlighted a few hours after Endeavour undocked, when the first short-term visiting crew arrived in a replacement Soyuz. This capacity to utilize the Soyuz as a resident crew ferry and stand-by rescue ship would prove vital in the three years that followed the grounding of the Shuttle in 2003 and then again following the retirement of the fleet in 2011.

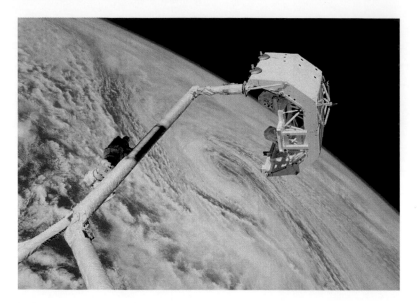

2001: STS-100 transfers a pallet containing the station's robotic arm, known as Canadarm2.

STS-104, Atlantis: 10th ISS Assembly Flight 7A (July 12–24, 2001): The task of this mission was to add the Joint Airlock named Quest to the Unity Node of the ISS. There was now a permanent three-person crew aboard the station, the Destiny laboratory for research, the Canadarm2 robotic arm for external manipulation, and an adequate stock of stores. Quest would increase the station's independence of visiting Shuttles because ISS crews would be able to perform spacewalks at any time, instead of waiting for the next Shuttle to arrive in order to employ its airlock. With the Quest airlock on the US segment, and a second airlock which was added to the Russian segment a few months later, station crews would be able to undertake EVA tasks that were deferred from the flight plans of visiting Shuttles. The installation of Quest required a coordinated effort using the Shuttle and ISS robotic manipulators plus support from three EVAs, with the last of these excursions being the first to be performed from the new Quest airlock. As events transpired, Quest would not be used operationally again until 2002.

This mission marked the end of the second phase of ISS assembly. During a period of just twelve months – from the docking of Zvezda to the attachment of Quest – over 31,292.5 lb. (69,000 kg) of hardware augmented the core station of Zarya-Unity. Since the docking of Zvezda, all the large components installed on the station (the Z1, PMA-3, the P6 with its solar arrays and radiator, Destiny, Canadarm2 and now Quest) were delivered by the Shuttle. This rapid expansion established a routine of crew exchange and logistics supply, and allowed the station to make a start on scientific research.

The next phase of assembly would focus on delivering the remaining truss elements to provide the planned power generating capacity, prior to completing the assembly by delivering the final modules and other facilities.

2001: The Joint Airlock Quest arrives on STS-104.

SHUTTLE MISSIONS TO EXPAND THE ISS (PHASE-III)

With the delivery of the Destiny laboratory, the SSRMS and the Quest airlock, the ISS was now at the cusp of supporting more extensive resident crew activities independent of visiting Shuttles. The remaining Shuttle flights of 2001, which kicked off Phase-III, would stock up the station in preparation for the expansion of the truss and solar array assemblies planned for the following year.

STS-105, Discovery: 11th ISS Assembly Flight 7A.1 (August 10–22, 2001): This logistics mission delivered over 6,615 lb (3,000 kg) of supplies. These included more internal experiment and equipment racks and additional external experiment platforms that were installed during two EVAs. The newly arrived ISS-3 crew took over from the retiring ISS-2 trio, who returned home aboard Discovery. The ISS-3 crew were aboard the station when the Russian Pirs ('Pier') Docking Compartment arrived in September. The airlock incorporated into this module provided alternative access for EVAs on the Russian segment that used Russian rather than American equipment.

STS-108, Endeavour: 12th ISS Assembly Flight UF1 (December 5–17, 2001): The final Shuttle visit of the year saw the exchange of the ISS-3 crew with the ISS-4 crew and the delivery of over 5,953.5 lb (2,700 kg) of supplies. The single EVA undertook routine maintenance tasks on the exterior of the station and several get-ahead tasks in preparation for more extensive spacewalking operations planned for the coming year.

2002: With the fourth resident crew on the ISS, the new year of Shuttle operations began in March with the 4th HST service mission (designated SM-3B). It was the only solo Shuttle flight on the manifest prior to the planned international scientific research mission of STS-107 with the Spacehab double module configuration for experiments to support future research aboard the ISS. The other four planned Shuttle missions for the year all focused on delivering truss and solar array elements, and stocking the station with more supplies. The year also saw the Shuttle complete the exchange of two new resident crews (ISS-5 and ISS-6) and it was also the year in which a new model of the Soyuz (TMA) was introduced, providing extra room for larger crewmembers (notably the Europeans and Americans) and allowing space tourists, space flight participants, and international cosmonaut researchers to make brief visits to the station.

STS-110, Atlantis: 13th ISS Assembly Flight 8A (April 8–20, 2002): The year's first Shuttle flight to the ISS featured a major expansion of the truss and solar array system. STS-110 installed the Starboard Zero (S0) truss element on a bracket atop the Destiny laboratory to serve as a platform to support other truss segments and their solar arrays. Four EVAs were made by four Shuttle astronauts working in pairs and alternating the assignments to give each pair a day off between excursions. One of their tasks was to release the launch clamps on the Mobile Transporter (MT) which was meant to enable future astronauts to travel along the truss on carts running on rails, thereby saving the spacewalkers time and energy. This 'railcar' system would be expanded across future missions to carry out inspection and maintenance tasks far away from the pressurized modules.

STS-111, Endeavour: 14th ISS Assembly Flight UF2 (June 5–19, 2002): The next Shuttle mission was another logistics flight. It delivered the ISS-5 crew and retrieved their predecessors. The fifth flight of an MPLM was the third for Leonardo, which on this trip carried over 7,938 lb (3,600 kg) of supplies. Another 992.25 lb (450 kg) was transferred from Endeavour's mid-deck lockers. Making the trip home would be over 4,630.5 lb (2,100 kg) of unwanted items, trash, equipment, and samples from science experiments.

STS-112, Atlantis: 15th ISS Assembly Flight 9A (October 7–18, 2002): The goal of this mission was to resume the expansion of the station's truss. It involved three EVAs. The Starboard 1 (S1) truss was connected to the starboard end of the S0 unit and Crew Equipment Translation Aid (CETA) Cart A was placed on S0 as the first of the pair of human-powered carts for the 'railcar' system.

STS-113, Endeavour: 16th ISS Assembly Flight 11A (November 21–December 7, 2002): The final visit of the year saw the installation of the Port 1 (P1) truss and CETA Cart B. In addition to delivering over 4,189.5 lb (1,900 kg) of logistics the mission saw the ISS-5 crew replaced by their ISS-6 counterparts. The new truss unit was attached to the port end of S0 by the crew across three EVAs. CETA B cart was attached to CETA A to form a double translation unit. Unknown at the time was that this would be the last Shuttle to dock with the station and the last successful landing for over two and a half years.

2003: In December 2002, the NASA Station Program Manager, William ('Bill') Gerstenmaier, said, "The year ahead will be the most complex so far in the history of ISS and its construction in orbit. The station literally becomes a new spacecraft with each assembly mission and that will be true next year, with dramatic changes in the configuration of its cooling and power systems as well as in its operations."[1] At the start of the year things looked good for ISS operations, with a new crew aboard and plans for major expansion of the truss system over five docking missions. Before that, however, there would be the solo flight of STS-107. It launched on January 16, 2003, and was an international science research flight designed to train scientists to develop their experiments and procedures for more extensive research on the ISS. The mission was originally created in order to bridge the gap between the final Spacelab mission (the Neurolab flown by STS-90 in

Table 2.1 Space Shuttle Missions To The International Space Station 1998–2002

International Designation	STS Flight	Station Mission	Orbiter Vehicle (OV)	Shuttle Crew Number	Launch Date	KSC	Docking Date	Date Undocked	Docked Duration DD:HH:MM	Landing Date	Land on Orbit	Shuttle Mission Duration DD:HH:MM:SS	Miles (m)	Km (m)
					A. SHUTTLE ISS ASSEMBLY AND RESUPPLY MISSIONS 1998–2002									
1998-069A	88	2A	105	6	1998 Dec 4	39A	1998 Dec 6	1998 Dec 13	6:20:38	1998 Dec 15	185	11:19:18:47	4.7	7.6
1999-030A	96	2A.1	103	7	1999 May 27	39B	1999 May 29	1999 Jun 3	5:18:17	1999 Jun 6	154	9:19:13:57	3.7	6.0
2000-027A	101	2A.2a	104	7	2000 May 19	39A	2000 May 20	2000 May 26	5:18:32	2000 May 29	155	9:21:10:10	4.1	6.6
2000-053A	106	2A.2b	104	7	2000 Sep 8	39B	2000 Sep 10	2000 Sep 17	7:21:54	2000 Sep 19	185	11:19:12:15	4.9	7.9
2000-062A	92	3A	103	7	2000 Oct 11	39A	2000 Oct 13	2000 Oct 20	6:21:23	2000 Oct 24	203	12:21:43:47	5.3	8.5
2000-078A	97	4A	105	5	2000 Nov 30	39B	2000 Dec 2	2000 Dec 9	6:23:13	2000 Dec 11	170	10:19:58:20	4.5	7.2
2001-006A	98	5A	104	5	2001 Feb 7	39A	2001 Feb 9	2001 Feb 16	6:21:15	2001 Feb 20	202	12:21:21:00	5.3	8.5
2001-010A	102	5A.1	103	7 U/7D*	2001 Mar 8	39B	2001 Mar 10	2001 Mar 18	8:21:54	2001 Mar 21	201	12:19:51:57	5.3	8.5
2001-016A	100	6A	105	7	2001 Apr 19	39A	2001 Apr 21	2001 Apr 29	8:03:35	2001 May 1	186	11:21:31:14	4.9	7.9
2001-028A	104	7A	104	5	2001 Jul 12	39B	2001 Jul 13	2001 Jul 22	8:04:46	2001 Jul 24	200	12:18:36:39	5.3	8.5
2001-035A	105	7A.1	103	7 U/7D*	2001 Aug 10	39A	2001 Aug 12	2001 Aug 20	7:20:10	2001 Aug 22	186	11:21:13:52	4.3	7.0
2001-054A	108	UF1	105	7 U/7D*	2001 Dec 5	39B	2001 Dec 7	2001 Dec 15	7:21:25	2001 Dec 17	185	11:19:36:45	4.8	7.7
2002-018A	110	8A	104	7	2002 Apr 8	39B	2002 Apr 10	2002 Apr 17	7:02:26	2002 Apr 20	171	10:19:42:44	4.5	7.2
2002-028A	111	UF2	105	7 U/7D*	2002 Jun 5	39A	2002 Jun 7	2002 Jun 15	7:21:05	2002 Jun 19	217	13:20:35:56	5.8	9.3
2002-047A	112	9A	104	6	2002 Oct 7	39B	2002 Oct 9	2002 Oct 16	6:22:56	2002 Oct 18	170	10:19:58:44	4.5	7.2
2002-052A	113	11A	105	7 U/7D*	2002 Nov 23	39A	2002 Nov 25	2002 Dec 2	7:22:11	2002 Dec 7	216	13:18:48:38	5.6	9.0

2003-2005 Shuttle flights to ISS suspended pending investigation following loss of Columbia and crew on February 1, 2003

Key:
(*)   Resident station crew member exchanged (See Table 5)
U/D   Total crew members Up / total crew members Down
A = American mission
E = European mission
J = Japanese mission
UF = Utilization Flight
LF = Logistics Flight
ULF = Utilization & Logistics Flight

April 1998) and the start of full scale research on the Destiny laboratory of ISS in 2001. Initially scheduled for launch in 2000, STS-107 was delayed no less than thirteen times before it finally lifted off. Once on-orbit, Columbia and its crew of seven performed admirably and there was talk of scheduling a second flight even before they wrapped up their studies, based upon their efforts and success. Tragically, however, while they worked in the Spacehab double module the fate of the crew had already been sealed by damage caused as debris struck the leading edge of one wing of the Orbiter during launch. A fortnight later, and just 16 min prior to the scheduled landing in Florida, the effects of that damage led Columbia to break up at hypersonic speed around 200,000 ft (60,960 m) above Texas.

For the second time, a Shuttle and seven astronauts had been lost. It was just a few days after the 17th anniversary of the loss of Challenger. This time the tragedy would have terminal implications for the Shuttle program. There would be an inquiry with a series of recommendations for changes and a Return-To-Flight program to qualify the Shuttle for further use, but this time it would be for the sole purpose of completing its obligations to the assembly of the ISS. Once this had been achieved, the fleet would be retired. At that time this was expected to be in 2010. As the recovery from Columbia's loss progressed the future of the entire US human space program, any replacement for the Shuttle, and indeed the continuation of the ISS were very much in doubt.

2004: This was only the second calendar year since 1981 that a Shuttle had not left the pad, and both times the cause was the loss of an Orbiter. During the post-Columbia inquiry the surviving ships – Discovery, Atlantis and Endeavour – remained in storage. Once the sequence of events that claimed Columbia and its crew were understood, a number of measures were devised to prevent a recurrence. As new Shuttle procedures, contingency actions and safeguards were being implemented, operations involving the ISS continued with a reduced two-person caretaker crew. The three-person ISS-6 crew who had arrived on STS-113 and were expected to leave on STS-114 had instead made the trip home in Soyuz TMA-1, having been relieved by the two-person ISS-7 crew who launched aboard TMA-2. The Russian Soyuz crew taxis and Progress resupply vessels became the savior of the ISS program at this stage. These veteran vehicles enabled the station to continue to be inhabited and supplied. The science programs were reduced until such time as the Shuttle could resume flying to deliver hardware and supplies to complete the ISS assembly.

2005: It took almost thirty months to prepare the Shuttle to fly again, with STS-114 as the Return-To-Flight mission. However, things did not go as well on that flight as hoped, and further work was required to understand the problems and come up with solutions. This led to the postponement of the second Return-To-Flight mission (STS-121), thus imposing a further, costly, twelve month delay. In the manifest of October 2000 there had been twenty-eight Shuttle flights assigned to assemble and supply the station through 2002. In February 2001 the FY2002 budget was announced. It had a serious reduction in Shuttle missions, reflecting questions which were already being raised about its longevity. After the loss of Columbia and the delays in implementing the Return-To-Flight, the manifest of June 2005, shortly prior to STS-114, listed only twenty-two Shuttle missions to the ISS, with the cancellation of some of the intended payloads and hardware. The grounded items included the US Habitation Module, the US Propulsion Module, and the X-38 Crew Rescue Vehicle. These deletions, coupled with the delays and uncertainty over the Shuttle program, had implications for the ISS partners in trying to plan their own long-term goals, fund programs, and develop new hardware. In 2005, the completion of the ISS and indeed its ultimate fate still hung in the balance. The key to the station's future, once again would depend on when, if ever, the Shuttle could resume operations.

RETURN TO FLIGHT, AGAIN

By the summer of 2005, the Shuttle was once again deemed ready to resume flying. It was now imperative to complete the expansion of the ISS truss so that the solar arrays could provide the power to support the remaining science laboratories and increase the resident crew to six persons.

However, prior to launching the truss elements, the next two Shuttle flights were to test the improvements to the Shuttle that had been developed in response to the loss of Columbia. The payloads for these missions were logistics and supplies to build up the stocks that had been depleted over the nearly two and a half years that the Shuttle fleet had been grounded. During that time it had been a challenge to sustain just a core crew of two on board the station. There was an urgency to resume the nominal three-person resident crew before implementing a rotational six-person presence and expanding the science program as the assembly phase drew to a close.

STS-114, Discovery: 17th ISS Assembly Flight LF1 (July 26–August 9, 2005): A mission originally designated ULF1 and planned for March 2003, it was intended to deliver both an extensive logistics payload and the three-person ISS-7 crew to replace the retiring ISS-6 crew. However, following the loss of Columbia this mission and all others planned at the time were grounded. As the program geared up to resume flying, the ISS launch manifest was revised and STS-114 was given the vital Return-To-Flight mission. Redesignated LF1 it carried the MPLM Raffaello loaded with over 3,748.5 lb (1,700kg) of urgent supplies to the station and returned with over 3.5 tons of unwanted material and trash that had accumulated in the intervening two and a half years.

In the post-Columbia plan, the ISS was to serve as a 'safe haven' in the event that a Shuttle was unable to return to Earth. Prior to docking, the crew used the RMS and the newly installed Orbiter Boom Sensor System (OBSS) to scan the surface of Discovery and downlink the results for ground analysis, as well as visually seeking any evidence of damage caused during launch. Then, as Discovery approached the ISS, Commander Eileen M. Collins executed the first Rendezvous Pitch Manoeuver when approximately 600 ft (183 m) away, back-flipping the spacecraft at a rate of 0.75° per second to allow the ISS crew to photo-document and visually inspect the underside and protective tiles. No significant damage was found on the Thermal Protection System (TPS), but further analysis of the images revealed protruding tile gap fillers that could potentially produce hot spots when re-entering the atmosphere. With the inevitable caution associated with this flight, the Shuttle crew performed an unplanned EVA to remove the fillers. There were three EVAs. The first involved working on deliberately damaged TPS tiles in the payload bay in order to evaluate various repair procedures and methods. In addition to miscellaneous long-awaited ISS maintenance work, the second EVA replaced a failed Control Moment Gyroscope (CMG) in the Z1 truss segment. The third EVA was added to tackle the issue discovered during the inspection of the thermal tiles. Extracting the protruding fillers by hand was simpler than predicted. Discovery returned home safely. Analysis of ground imagery taken at launch and imagery taken in flight indicated that foam insulation was still detaching from the ET and striking the Orbiter. The ensuing corrective work and tests of the improvements progressed so slowly that almost a year passed before the next Shuttle was ready to leave the pad.

2006: Back on the road again. After more than three difficult years, the Shuttle returned to operational status with the success of STS-121. Two other missions had flown by the end of the year, and the assembly of the ISS was once again underway.

STS-121, Discovery: 18th ISS Assembly Flight ULF1.1 (July 4–17, 2006): This was officially the second Return-To-Flight mission, the one that finally qualified the Shuttle for service and permitted assembly of the ISS to resume. The delivery of logistics to the station had fallen so far behind that this mission was added to the manifest in order both to certify the new safety procedures and to function as a utilization flight. It carried the Leonardo MPLM loaded with over 7,276.5 lb (3,300 kg) of logistics. Three EVAs were made when docked to undertake station maintenance, test improved EVA apparatus and procedures, and perform several delayed get-ahead tasks to prepare for extensive EVAs during later flights. One highlight was the transfer of Thomas Reiter, an ESA astronaut from Germany, to work with the ISS-13 and ISS-14 resident crews as Flight Engineer 2. He would return to Earth on a later Shuttle, which would deliver his replacement. This started a series of single-person transfers during assembly missions over the next three years. Three-person resident crews were no longer to be transported by Shuttle but this single exchange option brought an end to the two-person caretaker crews on the station after three years. It also meant that the third seat on each Soyuz carrying a new pair of residents to the ISS would be available for fee-paying participants who would return to Earth with the departing residents after about a week aboard the station.

STS-115, Atlantis: 19th ISS Assembly Flight 12A (September 9–21, 2006): Atlantis paid its first visit to the ISS in almost four years. It was carrying the P3 and P4 trusses, which were installed during the first few days, supported by three EVAs. The day after Atlantis undocked, Soyuz TMA-9 was launched. It was the first time since April 2001 that three crews were simultaneously in space on different craft. This activity reflected the complexity of ISS space operations during the initial years of the new millennium. At the homecoming celebration in Houston shortly after Atlantis landed on September 21, Mission Specialist Heidemarie Stefanyshyn-Piper collapsed twice and was taken to hospital for a checkup. The problem was diagnosed as being related to her personal re-adaptation to gravity after her first mission and twelve days in space, and there were no lasting effects.

STS-116, Discovery: 20th ISS Assembly Flight 12A.1 (December 9–22, 2006): As a reflection of the increased pace now that the Shuttle was again operational this mission flew less than three months after STS-115. Whilst mission safety remained paramount, the defined end to the Shuttle program left little time to spare in planning the remaining assembly and logistic missions to the ISS. With uncertainly remaining about what (if anything) would follow the Shuttle, pressure to complete the station was evident. Due to restrictions imposed by the loss of Columbia, STS-116 was the first night launch in four years. Its primary objective was to install the P5 truss section, a task supported by three

2006: High above New Zealand, the expansion of the station's truss and solar array system continues with STS-116 installing the P5 truss.

spacewalks. Some difficulties were encountered in fully retracting the P6 solar arrays to make room for the new arrays (the P6 unit was to be relocated from its initial position atop the Z1 truss to the end of the port truss by a mission in 2007) so a fourth EVA was added to finish this retraction process. The smaller mass and dimensions of the P5 truss (which would function as a spacer between P4 and P6 when the latter was relocated) meant that there was room in the payload bay for a Spacehab single module to carry supplies to stock up the station. While the EVAs were in progress, over 2 tons of logistics were simultaneously transferred internally. STS-116 also featured the first exchange of an ISS crew-member since STS-113 in 2003, with Thomas Reiter of ESA being replaced as FE2 by NASA's Sunita L. Williams, who would work in succession with the ISS-14 and ISS-15 resident crews.

2007: Despite the backlog, only three Shuttle missions were flown this year, due in part to the delay in launching STS-117 after a hail storm which caused damage to the thermal protection on both Atlantis and its External Tank. On the positive side though, the ISS continued to expand, with more of the truss structure being installed as well as the second pressurized node, named Harmony.

STS-117, Atlantis: 21st ISS Assembly Flight 13A (June 8–22, 2007): As the first mission to leave Pad 39A since STS-107 in January 2003, STS-117 delivered the S3 and S4 segments of the truss and several tons of logistics and supplies. At 42,671 lb (19,355.5 kg) this was the heaviest payload the Shuttle had lifted so far. Two pairs of astronauts working in the now familiar alternating system, made a total of four EVAs. After the recent spate of problems, this mission further demonstrated the flexibility of the Shuttle and its value to the ISS program. Following difficulties experienced by the computers on the Russian segment during the docked phase, the Orbiter's propulsion system was employed to assist in controlling the station's attitude in space while the Russian system was attended to. This mission also featured the next ISS crewmember exchange, with Clayton C. ('Clay') Anderson superseding Sunita Williams as FE2 to work sequentially with the ISS-15 and ISS-16 crews.

STS-118, Endeavour: 22nd ISS Assembly Flight 13A.12 (August 8–21, 2007): After a five-year program of major modifications, this was the first flight of Endeavour since it flew STS-113 in 2002. Four EVAs were carried out in support of the installation of the S5 truss section and an External Stowage Platform (ESP). The astronauts also replaced another failed CMG on the ISS. This mission also introduced the occasional change of EVA crewing where a serving American ISS resident participated in one or more of the planned spacewalks. In this case, ISS-16 FE Clay Anderson joined STS-118's Richard A. ('Rick') Mastracchio and Canadian Dafydd (David or 'Dave') R. Williams on the third and fourth EVAs respectively. STS-118 was the first mission to make use of the Station-to-Shuttle Power Transfer System (SSPTS) that permitted a docked Orbiter to draw additional power from the station so that flights could be extended, if needed. It was also the final flight of the Spacehab resupply module. One other highlight of this mission was the inclusion of Mission Specialist Barbara R. Morgan. The professional teacher from Idaho had served as back-up to fellow teacher Christa McAuliffe for the ill-fated STS-51 L Challenger mission. In 1998, Morgan had been selected for NASA astronaut training and during this mission she broadcast three educational events and answered many questions from school children, in addition to completing her normal Mission Specialist workload.

STS-120, Discovery: 23rd ISS Assembly Flight 10A (October 23–November 7, 2007): This very busy mission delivered the Node 2 (Harmony) module, relocated the P6 truss segment from atop Z1 to its permanent position on the port end of the main truss, and exchanged ISS crewmember Clay Anderson with Daniel M. Tani, who joined the ISS-16 crew as FE2. With the spacewalkers focusing upon the priority repair of the Solar Alpha Rotary Joint (SARJ) during four EVAs, a fifth planned EVA was reassigned to the ISS residents to undertake at a later date. Discovery also replenish the station with another logistics payload, and returned 2,020 lb (916.3 kg) of materials and scientific samples to Earth. The media tended to focus on the fact that, for the first time, female astronauts commanded both the Shuttle (Pamela A. Melroy) and the station (Peggy A. Whitson) and had therefore shaken hands in space.

2008. This was a very busy and significant year for ISS operations, with the delivery of several large elements for two of the remaining scientific laboratories, one European and the other Japanese. Their installation and outfitting paved the way to increase the permanent resident crew to six persons from 2009, and justified the assignment of both Japanese and European astronauts to resident crews, alongside those of Canada, who had supplied the station's main robotics systems.

2008: ESA's long-awaited Columbus laboratory arrives at the ISS via STS-122.

STS-122, Atlantis: 24th ISS Assembly Flight 1E (February 7–20, 2008): The ESA Columbus science laboratory, named after the historic 15th century Italian explorer Christopher Columbus, was a development of the earlier European Spacelab module that flew aboard several Shuttle missions between 1983 and 1998. It featured much-improved and expanded experiment racks and research facilities. This flight had been scheduled for launch in December 2007 but was delayed twice owing to problems in fueling the Shuttle's ET. The installation of Columbus was supported by three EVAs. Work on activating the internal systems and facilities was carried out by French ESA astronaut Leopold Eyharts, following his joining the ISS-16 crew as replacement for Dan Tani. For the first time since 2002, the Shuttle Orbiter was employed to re-boost the altitude of the ISS when the Orbital Maneuvering System (OMS) on Atlantis was fired to raise the orbit by 1.4 miles (2.25 km).

STS-123, Endeavour: 25th ISS Assembly Flight 1J/A (March 11–26, 2008): With Columbus added to the ISS, the focus turned to the Japanese laboratory called Kibo ('Hope'). The Japanese laboratory (like Columbus) had originally been intended for Space Station Freedom, and was formally named the Japanese Experiment Module (JEM). With a total mass exceeding 62,044 lb (28,138 kg), it was by far the biggest module delivered by the Shuttle and was too massive to be launched in one piece. Its installation was spread across three Shuttle missions, of which this was the first. STS-123 carried the Experiment Logistics Module Pressurized Section (ELM-PS) together with a new Canadian robotic arm that had the impressive name of the Special Purpose Dexterous Manipulator and was more affectionately known as 'Dextre.' The ELM-PS was temporarily attached to the Harmony Node while awaiting the delivery of the next segment of the Japanese laboratory on the next Shuttle. Five EVAs were conducted (a record for a Shuttle assembly flight) totaling 33 hr 28 min. Much of this duration was spent on various tasks associated with the attachment of Kibo and Dextre, as well as get-ahead tasks in preparation for later spacewalks. In addition, this mission also saw the exchange of Leopold Eyharts after just one month on the ISS-16 crew by NASA astronaut Garrett E. Reisman who would also work sequentially with the ISS-16 and ISS-17 crews.

2008: STS-123 delivers the Special Purpose Dexterous Manipulator known as 'Dextre.'

STS-124, Discovery: 26th ISS Assembly Flight 1J (May 31–June 14, 2008): The second mission associated with the delivery of Kibo carried the Pressurized Module (JEM-PM) as its primary payload. Since this was so large, there was no capacity to carry the OBSS inspection device as well. That had been temporarily stowed on the station during STS-123 and would be brought home at the end of STS-124 once the Japanese laboratory had been offloaded. The ELM-PS was relocated to its intended position on top of the JEM-PM. The installation operations were supported by three spacewalks. Garrett Reisman was replaced by Gregory E. ('Greg') Chamitoff, who served as FE2 with first the ISS-17 crew and then the ISS-18 crew.

STS-126, Endeavour: 27th ISS Assembly Flight ULF2 (November 14–30, 2008): The final Shuttle flight of the year saw 6.5 tons of logistics delivered using MPLM Leonardo. Sandra H. Magnus replaced Greg Chamitoff as FE2 on the ISS-18 crew. This mission was primarily "home improvements and maintenance," with the cargo featuring two large water recycling racks, a new 'kitchen' unit, a second toilet, two additional sleep stations, more exercise apparatus, and a multitude of other essential supplies. In addition to performing three EVAs, the Shuttle crew assisted the station crew with the installation of a new Water Recovery System designed to make waste water drinkable and maximize the supply of potable water aboard the ISS. This was another vital step toward the kind of closed-cycle life support system that would be essential for establishing an outpost on the Moon or making exploratory missions to asteroids or to Mars.

2009. The final element of the huge truss structure was delivered this year, more than eight years after the installation of the first piece. The four Shuttle flights to the ISS were overshadowed by STS-125's daring and highly successful final HST service mission (SM-4). Originally deleted from the manifest following the loss of Columbia, but reinstated to fly in May between STS-119 and STS-127, this became the final solo Shuttle mission and supported the last EVAs directly from the Shuttle airlock, a series which began with STS-6 in April 1983. The year also saw the size of the resident crew aboard the ISS increased to six for the first time, including representatives from ESA, Canada, and Japan. As a result, the availability of third seats on the Soyuz flights for fee-paying participants was greatly reduced. The year also saw the final resident crew transfers via the Shuttle.

STS-119, Discovery: 28th ISS Assembly Flight 15A (March 15–28, 2009): The task of this mission was to deliver and install the S6 truss segment with the fourth and final set of solar arrays in order to increase the overall power generation capacity of the ISS to the specified 120 kW. In addition to doubling the power available for the research laboratories from 15 kW to 30 kW, there would still be sufficient power to support a permanent crew of six and, for short periods, up to nine. The first of the three EVAs conducted on this flight was in support of installing the S6 truss. The other two were associated with maintenance and get-ahead tasks. Sandra Magnus was superseded by JAXA astronaut Koichi Wakata on the ISS-18 crew, making him the first Japanese to serve on a resident crew.[*] He would work with both the ISS-19 and ISS-20 crews.

STS-127, Endeavour: 29th ISS Assembly Flight 2J/A (July 15–31, 2009): Just two months after the highly successful final HST service mission, attention returned to the ISS with STS-127 completing the Kibo laboratory by adding its Exposed Facility (EF). Five EVAs supported this work, as well as conducting various other tasks including the addition of the Integrated Cargo Carrier – Vertical Light Deployable (ICC-VLD) to the port side of the Mobile Base System. By this point, all of the major science facilities of the US Segment were in place, but the planned, and much-delayed, remaining Russian science modules were still in the factory. Astronaut Timothy L. Kopra replaced Koichi Wakata on the ISS-20 crew.

STS-128, Discovery: 30th ISS Assembly Flight 17A (August 28–September 12, 2009): This was the first of two missions in this year primarily devoted to stocking up the ISS with supplies and logistics. STS-128 also featured the final exchange of an ISS resident crewmember by Shuttle, with Nicole M. P. Stott superseding Tim Kopra on the ISS-20 crew and later working with the ISS-21 crew. The three EVAs of the mission involved relocating apparatus and the installation of new experiment platforms, EVA aids and a range of hardware for future replacement. NASA later stated that this mission marked the transition of the ISS from an assembly site to a long-term research facility. MPLM Leonardo carried 7.5 tons of supplies that included new experiment racks and sample freezers for Destiny. The MPLM was packed with 5,223 lb (2,400 kg) of experiment results, trash, and unwanted hardware for return to Earth. A further 861 lb (390.5 kg) was stored in Discovery's mid-deck lockers.

---

[*]The **J**apanese **A**erospace e**X**ploration **A**gency was formed on October 1, 2003, by merging the former Institute of Space and Astronautical Sciences (ISAS), the National Aerospace Laboratory of Japan (NAL) and the National Space Development Agency of Japan (NASDA).

2008–2009 saw a series of three Shuttle missions deliver elements of the Japanese Experiment Module Kibo ('Hope').

STS-129, Atlantis: 31st ISS Assembly Flight ULF3 (November 16–27, 2009): The final Shuttle mission of the year ferried over a ton of supplies to the ISS and returned with a similar mass of unwanted materials and trash. The robotic arms on the Shuttle and station installed a pair of Express Logistics Carriers (ELC-1 and ELC-2) on the Earth-facing (nadir) side of the port truss. The three EVAs on this flight once again supported a wide variety of activities, tasks, and chores across the exterior. Another closure occurred when Nicole Stott, having served with the ISS-20 and ISS-21 crews, transferred across to the Orbiter without being superseded. She was the last of eleven resident crewmembers exchanged by Shuttles between July 2006 and November 2009: eight Americans, two Europeans (one from Germany and the other from France), and one Japanese. For the foreseeable future, *all* station crewmembers would arrive and depart via Soyuz spacecraft.

2010. The penultimate year of Shuttle operations saw a series of flights designed to fill the station with sufficient supplies to last the six-person resident crews at least one year after the Shuttle was retired. These flights would also remove much of the clutter and unwanted material from inside the station, and stock up as many spare parts and Orbital Replacement Units (ORU) as possible inside and outside of the facility.

STS-130, Endeavour: 32nd ISS Assembly Flight 20A (February 8–21, 2010): This mission delivered two major components. After the cancellation of the US Habitation Module, the newly delivered Node 3, named Tranquility, carried facilities to improve living conditions for the six-person crew. The second major item was the Cupola, the seven-paned panoramic viewing module which would ultimately serve as the station's central robotic control facility. These were the last major habitable modules of the US segment of the ISS. Their installation was supported by three EVAs. In addition to the much-needed living quarters Tranquility provided air revitalization, oxygen regulation and waste handling systems. Eventually almost all of the US segment's waste and crew hygiene equipment would be relocated to Tranquility. The recently installed Combined Operational Load Bearing External Resistance Treadmill with the contrived acronym of COLBERT[2] was later moved to Tranquility, along with other exercise apparatus, to free up working volume elsewhere. The Cupola, which had been awaiting flight assignment for several years, provided a 360° view of the world, the station structure, and the starry background. It would also come to serve as a quiet place for crewmembers to reflect on their unique experiences and the job they were doing.

2010: STS-130 delivers the station's Cupola.

STS-131, Discovery: 33rd ISS Assembly Flight 19A (April 5–20, 2010): When the Shuttle was retired the opportunity to deliver substantial cargoes to the ISS and return bulky hardware to Earth would be lost, probably for many years. Other vehicles were operating, or would soon become available, but their payloads were smaller and they had limited (or zero) ability to return items to Earth. With STS-130 the ISS assembly was 98% complete, therefore STS-131 was to deliver as many supplies, consumables and spares as possible and retrieve unwanted items. With less clutter to deal with, the residents would be able to focus upon routine maintenance and scientific experiments. The Leonardo MPLM delivered 17,999 lb (7,711.2 kg) of cargo, and once it had been emptied it was refilled with bags of unwanted material, broken equipment and various discarded items for the trip home. The tasks undertaken during the three EVAs of this mission included replacing an empty 1,800lb (816.5 kg) ammonia coolant tank with a new unit.

STS-132, Atlantis: 34th ISS Assembly Flight ULF4 (May 14–26, 2010): An all-veteran crew flew Atlantis on this mission, completing the Shuttle launch manifest for the year. An Integrated Cargo Carrier (ICC) was transferred to the exterior of the ISS. However, most of the interest focused upon the installation of the Russian-built Mini Research Module 1 called Rassvet ('Dawn'). This was the first major item of Russian hardware to be carried by Shuttle since the Docking Module augmented Mir in 1995. Rassvet was permanently attached to the lower (nadir) port of Zarya. The ICC carried spares and supplies to support the ISS through to at least 2020. Its payload included a spare K_u-band antenna and truss support, six nickel-hydrogen batteries, and a range of spares for the Dextre manipulator. The flight once again included three EVAs, totaling 21 hr 20 min. These supported the installation of Rassvet and the ICC and performed a number of get-ahead tasks. In total, over 2,800lb (1,300 kg) of cargo was moved from Atlantis to the station, and over 822 lb (3,730 kg) was moved the other way. STS-132 was initially to have been the final flight of Atlantis, but its retirement was delayed to allow the Orbiter to be prepared as a Launch-On-Need (LON) rescue ship (designated STS-335) for the final manifested Shuttle flight (STS-134). But then it was decided to prepare Atlantis for one more flight (designated STS-135) to the ISS.

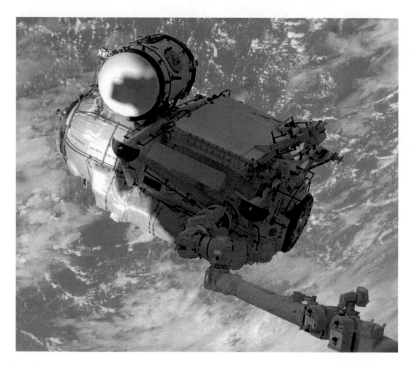

2010: Installation of the Russian Rassvet ('Dawn') Mini Research Module 1 on the nadir port of Zarya by STS-132.

2011. The year that saw the 50th anniversary of Yuri Gagarin's historic first manned orbital flight also witnessed the completion of the first flight a new Soyuz variant, Soyuz TMA-M, after 160 days in space. In addition there was the launch of the first unmanned Chinese space laboratory (Tiangong-1), the automated docking with that laboratory by the unmanned Shenzhou 8, and a second redocking 11 days later. Alongside these new programs, 2011 also witnessed the ending of two eras. The main assembly phase of the ISS was finally completed, as were the final three flights of the Shuttle before the fleet was retired.

STS-133, Discovery: 35th ISS Assembly Flight ULF5 (February 24–March 9, 2011): With ISS assembly essentially complete, this was the first of three missions to deliver supplies. In the payload bay was a newly modified Leonardo Multi-Purpose Logistics Module, redesignated the Permanent Multipurpose Module (PMM). The MPLM had been designed for temporary berthing on the ISS during a Shuttle visit to allow direct transfer of materials without having to pass through the narrow hatches and tunnels of the Orbiter Docking System. It had been long recognized that *space* (volume) aboard the station was at a premium, so one of the three MPLMs had been modified so that it could be permanently affixed to the ISS to provide additional storage. Discovery also carried miscellaneous apparatus and spares on Express Logistics Carrier 4 for external transfer, and a humanoid robotic device (Robonaut R2) that was to be assessed inside the ISS, both as a potential aid for automated repair tasks and an assistant for complex EVAs. This investigation could lead to advanced forms of robot capable of supporting more demanding work on the Moon, on asteroids, or on Mars. The two EVAs for the mission achieved a range of maintenance tasks on the exterior of the station, including placing ELC-4 on the truss. This final flight by Discovery was its thirteenth mission to the ISS and its thirty-ninth mission overall.

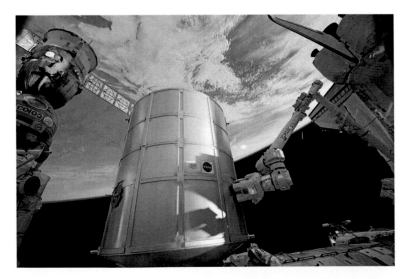

2011: The attachment of the Permanent Multipurpose Module that was the modified MPLM Leonardo.

STS-134, Endeavour: 36th ISS Assembly Flight ULF6 (May 16–June 1, 2011): The penultimate Shuttle flight of the program was also the twenty-fifth and final flight for Endeavour, which was on its twelfth visit to the ISS. Once Endeavour was docked, it provided a unique photo-opportunity, since this was the first and only time that all of the current types of resupply craft were present simultaneously: the American Shuttle OV-105 Endeavour, the Russian Progress M-10M and Soyuz TMA-21, the European ATV-2 Johannes Kepler, and the Japanese HTV-2 Kounotori. The historic images of this configuration were captured by ESA astronaut Paolo Nespoli while departing the station aboard Soyuz TMA-20. In addition to the Express Logistics Carrier 3 that was loaded with spares, Endeavour delivered an Alpha Magnetic Spectrometer, a particle physics experiment that was an improved version of one evaluated during the STS-91 mission to Mir in 1998. By observing cosmic rays, it was to investigate the nature of matter in the universe. Four EVAs were included in the flight plan, the last of which was the final EVA by members of a Shuttle crew (although via Quest rather than the Orbiter's airlock). This drew to a close the long sequence of Shuttle spacewalks that began in April 1983. STS-134 also saw the permanent relocation of the OBSS to the ISS to extend the reach of Canadarm2. The placing of AMS-02 on the main truss on May 19 officially marked the completion of the US segment of ISS. Several planned Russian modules had yet to be finished and a few outstanding large experiments and smaller items of hardware remained to be delivered but now the ISS was officially no longer an assembly site, it was a fully-fledged laboratory. Indeed, NASA declared the ISS to be a National Resource Laboratory for the benefit of the American people. This was in addition to it serving as an International Resource Laboratory as a result of the presence of the Russian segment and the European and Japanese facilities.

STS-135, Atlantis: 37th and final ISS Assembly Flight ULF7 (July 8–21, 2011): The end of the Shuttle era came thirty years and three months after the launch of Columbia on STS-1. It carried a crew of just four. The remaining operational MPLM, Raffaello, had over 9,000 lb (4,880 kg) of supplies, including sufficient food (2,677 lb or 1,214.3 kg) to sustain the ISS for a year, as well as other items and spares that would be of use in the future. This cargo was crammed into a total of seventeen racks – including eight Resupply Stowage Platforms (RSP), six Resupply Stowage Racks (RSR), and one Zero Stowage Rack. There was another 2,228 lb (1,010.6 kg) of materials on the mid-deck. The empty MPLM was then refilled with 5,666 lb (2,570.1 kg) of junk. With so much cargo to move, a review of the flight plan led to another day being added to the mission to finish off the operation. Also transferred across to the ISS was the Robotic Refueling Mission (RMM) experiment hardware, which was later to be evaluated by the residents to investigate the techniques, hardware, and procedures that were being developed for robotic satellite refueling and supporting EVA during servicing, even though the target spacecraft for this process had yet to be developed. The single planned EVA during the docked phase of was made by station crewmembers, mainly because the relatively late decision to add this mission to the manifest had provided little time for the Shuttle crew to train up. On 19 July, Atlantis undocked for the final time. The mission ended with an uneventful landing at the Cape two days later. The International Space Station had been assembled and the Shuttle passed into retirement.

Table 2.2 Space Shuttle Missions To The International Space Station 2005–2011

International Designation	STS Flight	Station Mission	Orbiter Vehicle (OV)	Shuttle Crew Number	Launch Date	KSC	Docking Date	Date Undocked	Docked Duration DD:HH:MM	Landing Date	Land on Orbit	Shuttle Mission Duration DD:HH:MM:SS	Miles (m)	Km (m)
\multicolumn{15} SHUTTLE ISS ASSEMBLY AND RESUPPLY MISSIONS 2005-2011														
2005-026A	114	LF1	103	7	2005 Jul 26	39B	2005 Jul 28	2005 Aug 6	8:19:54	2005 Aug 9	219	13:21:32:48	5.8	9.3
2006-028A	121	ULF1.1	103	7 U/6D*	2006 Jul 4	39B	2006 Jul 6	2006 Jul 15	8:19:06	2006 Jul 17	202	12:18:37:54	5.3	8.5
2006-036A	115	12A	104	6	2006 Sep 9	39B	2006 Sep 11	2006 Sep 17	6:02:04	2006 Sep 21	187	11:19:07:24	4.8	7.8
2006-055A	116	12A.1	103	7 U/7D*	2006 Dec 9	39B	2006 Dec 12	2006 Dec 19	7:23:58	2006 Dec 22	204	12:20:45:16	5.3	8.5
2007-024A	117	13A	104	7 U/7D*	2007 Jun 8	39A	2006 Jun 10	2006 Jun 19	8:19:06	2007 Jun 22	219	13:20:12:44	5.8	9.3
2007-035A	118	13A.1	105	7	2007 Aug 8	39A	2007 Aug 10	2007 Aug 19	8:17:54	2007 Aug 21	201	12:17:55:34	5.2	8.4
2007-050A	120	10A	103	7 U/7D*	2007 Oct 23	39A	2007 Oct 25	2007 Nov 5	9:21:52	2007 Nov 7	238	15:02:24:02	6.2	10.0
2008-005A	122	1E	104	7 U/7D*	2008 Feb 7	39A	2008 Feb 9	2007 Feb 18	8:16:07	2008 Feb 20	202	12:18:21:50	5.3	8.5
2008-009A	123	1 J/A	105	7 U/7D*	2008 Mar 11	39A	2008 Mar 12	2008 Mar 26	11:20:36	2008 Mar 26	250	15:18:10:54	6.5	10.5
2008-027A	124	1 J	103	7 U/7D*	2008 May 31	39A	2008 Jun 2	2008 Jun 11	8:17:39	2008 Jun 14	217	13:18:13:07	5.7	9.2
2008-059A	126	ULF2	105	7 U/7D*	2008 Nov 14	39A	2008 Nov 16	2008 Nov 28	11:16:46	2008 Nov 30	251	15:20:29:27	6.6	10.6
2009-012A	119	15A	103	7 U/7D*	2009 Mar 15	39A	2009 Mar 17	2009 Mar 24	7:22:33	2009 Mar 28	202	12:19:29:33	5.3	8.5
2009-038A	127	2 J/A	105	7 U/7D*	2009 Jul 15	39A	2009 Jul 17	2009 Jul 28	10:23:41	2009 Jul 31	248	15:16:44:57	6.5	10.5
2009-045A	128	17A	103	7 U/7D*	2009 Aug 28	39A	2009 Aug 30	2008 Sep 8	8:18:32	2009 Sep12	219	13:20:53:43	5.7	9.2
2009-062A	129	ULF3	104	6 U/7D*	2009 Nov 16	39A	2009 Nov 18	2009 Nov 25	6:17:02	2009 Nov 27	171	10:19:16:13	4.9	7.2
2010-004A	130	20A	105	6	2010 Feb 8	39A	2010 Feb 9	2010 Feb 20	9:19:48	2010 Feb 21	217	13:18:06:22	5.7	9.2
2010-012A	131	19A	103	7	2010 Apr 5	39A	2010 Apr 7	2010 Apr 17	10:05:08	2010 Apr 20	238	15:02:47:10	6.2	10.0
2010-019A	132	ULF4	104	6	2010 May 14	39A	2010 May 16	2010 May 23	7:00:54	2010 May 26	186	11:18:29:09	4.8	7.8
2011-008A	133	ULF5	103	6	2011 Feb 24	39A	2011 Feb 26	2011 Mar 6	7:23:55	2011 Mar 9	202	12:19:03:51	5.3	8.5
2011-020A	134	ULF6	105	6	2011 May 16	39A	2011 May 18	2011 May 29	11:17:41	2011 Jun 1	249	15:17:38:22	6.5	10.4
2011-031A	135	ULF7	104	4	2011 Jul 8	39A	2011 Jul 10	2011 Jul 19	8:15:21	2011 Jul 21	200	12:18:27:52	5.2	8.5

(continued)

Table 2.2 (continued)

At conclusion of STS-135 on July 21, 2011 the Shuttle fleet was retired.

Key:

(*) Resident station crew member exchanged (See Table 3.2)

U/D Total crew members Up / total crew members Down

A = American mission

E = European mission

J = Japanese mission

UF = Utilization Flight

LF = Logistics Flight

ULF = Utilization & Logistics Flight

TOTAL SPACE SHUTTLE DOCKINGS WITH SPACE STATIONS (MIR & ISS)

103 Discovery [2 Mir missions (1 rendezvous 1 docking); 13 ISS dockings= 15 missions, 1 rendezvous and 14 dockings]

104 Atlantis [7 Mir dockings; 12 ISS dockings = 19 missions, 19 dockings]

105 Endeavour [1 Mir docking; 12 ISS dockings = 13 missions, 13 dockings]

Total of 47 space station missions [10 to Mir, 1 rendezvous and 9 dockings; and 37 dockings to ISS]

2011: This historic picture shows STS-135 Atlantis (right) docked to the ISS with the US segment complete. In the foreground is the recently attached AMS-02 experiment.

Shuttle Station Assembly: A Summary

The thirty-seven missions flown between December 1998 and July 2011 were the core of the effort to assemble the ISS. Remarkably, each mission flew without encountering major failures or requiring re-flights. This was a truly stunning achievement. Three of the four-Orbiter fleet undertook the work, with Discovery flying thirteen missions and Endeavour and Atlantis each making twelve flights. Columbia, being heavier, was not involved. The three Orbiters could have kept flying to resupply the station and perhaps to expand it with additional modules and hardware, but this was not to be.

Twelve years and seven months after Endeavour roared off to initiate the assembly process, Atlantis glided to its final landing with that task accomplished.

SHUTTLE AT STATION

If the ten Shuttle-Mir missions are added to the Shuttle-ISS missions, then fully one-third (34.81%) of the total of 135 launches over the thirty-year Shuttle program were associated with space stations. However, it was by no means apparent at the time that this focus would be practicable. Although the assembly of a space station was one of the tasks intended for the Shuttle when it was first conceived, it was unable to pursue this challenge until the second half of its operational lifetime.

But the story of the Shuttle at the ISS, and to a similar degree at Mir, is not simply about dockings by Atlantis, Discovery, and Endeavour. It is the story of hundreds of workers on the ground who prepared the hardware and controlled the flights, and the astronauts who applied the lessons learned during Shuttle-Mir to ensure that the ISS dockings were

2A (STS-88)   2A.1 (STS-96)   2A.2a (STS-101)   2A.2b (STS-106)

3A (STS-92)   4A (STS-97)   5A (STS-98)   5A.1 (STS-102)

6A (STS-100)   7A (STS-104)   7A.1 (STS-105)   UF1 (STS-108)

8A (STS-110)   UF2 (STS-111)   9A (STS-112)   11A (STS-113)

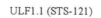

LF1 (STS-114)   ULF1.1 (STS-121)   12A (STS-115)   12A.1 (STS-116)

13A (STS-117) 13A.1 (STS-118) 10A (STS-120) 1E (STS-122)

1J/A (STS-123) 1J (STS-124) ULF2 (STS-126) 15A (STS-119)

2J/A (STS-127) 17A (STS-128) ULF3 (STS-129) 20A (STS-130)

19A (STS-131) ULF4 (STS-132) ULF5 (STS-133) ULF6 (STS-134)

ULF7 (STS-135)

SPACE SHUTTLE ISS PAYLOAD EMBLEMS

Each Shuttle flight had its own mission emblem which is widely known. However, for ISS each mission also had a distinctive payload emblem which is reproduced here and are less well known. All images courtesy of Spacefacts.de from NASA originals achieved without incident and that the tons of logistics were relocated safely and efficiently. The story also involves new developments in robotics and EVA, with a coordinated effort between the two skills to greatly expand and secure the final ISS configuration and hopefully facilitate years of research.

We are now far beyond the point at which the surviving Orbiters of the Shuttle fleet were retired, yet the ISS still operates around the clock thanks to the efforts of various players, including privately developed and operated space vehicles. However, without the Shuttle to bear the initial burden of the assembly process the ISS wouldn't exist as we know it today, and *that* must be one of the greatest legacies of the Shuttle program.

Notes

1. NASA News Release 02-256, December 20, 2002
2. Named after the American comedian and TV personality Stephen Colbert.

3

The Human Element

> *I have an extremely excited crew that is very well trained,*
> *and we are anxious to go... We've reached that point*
> *[that] this is going to happen, the crew is*
> *really getting focused and psyched*
> *to go do this, and we're ready.*
>
> Robert D. Cabana, Commander STS-88 (ISS Assembly Mission 2A)
> during a pre-flight interview

When Bob Cabana made that statement some months prior to the launch of the first ISS assembly mission, it came after a long gestation of not only his crew's mission training profile but also familiarity with both their own payload and Zarya, the Russian element they were to link up with in space. The crew had been named in August 1996 and their mission was delayed by almost a year. There comes a point in the training cycle where it becomes more difficult to maintain the 'keen edge' required to fly the mission before re-training becomes necessary. Each flight crew hopes that their training reaches a peak just prior to their embarking on the mission for which they have trained. In some cases the time between selection and flight can be short, but just occasionally it can be drawn out, often by several months but sometimes by years. Finding the balance in the training of a crew to maintain a high level of readiness amidst delays, setbacks and cancellations can be tricky. In the face of protracted delays, decisions have to be made about whether a crew should be kept intact, or split up and its individual members reassigned.

BECOMING VERY WELL TRAINED

The companion volume *Linking the Space Shuttle and Space Stations: Early Docking Techniques from Concept to Implementation* includes an overview of Shuttle training, crew positions and roles. Depending on the mission profile and the preparation of the hardware for the flight, training for a Shuttle mission took about one year. In a major

D.J. Shayler, *Assembling and Supplying the ISS*, Springer Praxis Books,
DOI 10.1007/978-3-319-40443-1_3

program like assembling a space station, the preparation of one mission can also require the on-time actions by earlier missions.

A Shuttle crew consists of a Commander, a Pilot and several Mission Specialists, the number of the latter being fixed by the mission plan and usually not exceeding five. The Flight Deck Crew comprised the Commander and Pilot who 'flew' the Orbiter, handled the rendezvous, docking, undocking and fly-around activities, and Mission Specialist 2 acting as ascent and re-entry Flight Engineer in support of the two pilots in their duties and actions to control and operate the Orbiter. Other Mission Specialist roles supported the robotics objectives or EVA activities, and on station docking missions managed the flow of logistics, transferable payloads and experiments. There was some cross-training in which the Commander and Pilot occasionally took the role of handling the robotics, supported the EVA preparations, and participated in the transfer of materials under the direction of the Mission Specialist who was the loadmaster. Whatever their objectives, the crew on each mission were fully occupied with assigned tasks, some of which were rather less glamorous or likely to make the headlines than others. This was the routine of space flight.

The flight crew. The STS-118 (ISS-13A.1) crew poses for a traditional pre-flight crew image: [l-r] Richard Mastracchio, Barbara Morgan, Charles Hobaugh, Scott Kelly, Tracy Caldwell, Canadian Dave Williams, and Alvin Drew.

Between 2001 and 2002 and again from 2006 to 2009 some resident crew members were launch or returned as a member of a Shuttle crew. Such ISS residents were given abbreviated MS training for launch and re-entry prior to the 2003 Columbia accident. After that tragedy the main resident crews from ISS-7 onwards received Soyuz training for launch and re-entry. Following the resumption of Shuttle flights in 2005, a series of single-person resident crew transfers were completed by the Shuttle between 2006 and 2009, and those crewmembers received nominal ascent and re-entry Shuttle training.

There was a change in training for the final Shuttle mission. In March 2011 the four-person crew of STS-135 went to Russia to be issued Sokol pressure suits and undergo a pressure chamber test. The suits would be carried on the Shuttle as a precaution, in case Atlantis received damage that rendered it unable to safely execute re-entry and landing. With no further Shuttle missions available to potentially rescue the STS-135 astronauts, that crew needed to rely on a ride home on a Soyuz, so the qualification in a Sokol suit was essential in case they had to remain on the ISS for up to a year pending the arrival of Soyuz rescue ships. If they did live on the ISS for that long, they would have had to undertake a fitness regime on board to maintain their condition pending their return to Earth. While in Russia therefore, each of the four astronauts was also measured for the Penguin load-bearing suit that would help to counter the effects of an unplanned long-duration stay in space.

As the complexity of the ISS increased, so did the requirement for crews to conduct a certain amount of 'international training.' This mainly focused on 'increment' crews but included some familiarization training for the crews of Shuttle assembly flights in Europe, Canada and Japan as well as Russia.

Canada: The Operations Engineering Training Facility (OETF) of the CSA at the John H. Chapman Science Center, Longueil, just outside Montreal, Canada, is where the astronauts trained to operate the Shuttle RMS and, later, the ISS Canadarm2 from the robotic workstation located in the Destiny laboratory.

Europe: In the early days of the Shuttle program, crews visited Europe to inspect the Spacelab module, train on European equipment and experiments, unpressurized pallets, and other hardware. For the ISS, the STS-122 crew underwent familiarization training on the European Columbus laboratory module, and this was also useful for the transfer of logistics into that module.

Japan: Familiarization training in Japan on the elements of the Kibo laboratory was useful in supporting delivery of this hardware by the STS-123, -124 and -127 missions.

Specialist Training

As well as proficiency in flying the Orbiter, space station missions required skills in a number of other areas. Some of these had been addressed on non-station missions (in particular the extensive EVAs required for the Hubble service calls) but such training peaked between 1995 and 2011 for the series of Shuttle-Mir and Shuttle-ISS missions.

This additional training included learning the intricacies of rendezvous, docking and proximity operations ('prox ops'), skills for which American astronauts had not trained in depth since the Apollo era. A significant amount of work would be completed by the Canadian-supplied robotic manipulators during station missions, so members of Shuttle crews trained for robotic activities on the Orbiter and on the ISS. One specific focus of this training was the support role that these manipulators would play in the substantial program of EVAs required to build and outfit the ISS (as well as two missions to Mir). The fact that each Shuttle docking mission to Mir and the ISS succeeded was testament to this additional training program, as was the extremely skilled execution of the EVAs.

In preparation for assignment to a crew, astronauts assume a number of support roles for other missions. Here Ellen Ochoa is a Capcom for STS-92, just over a year prior to flying as MS on STS-108.

Looking back now – with the assembly of the ISS completed – it is evident that the whole process from selection to training, then flight assignment, was absolutely right.

The success of the International Space Station is a credit to the skills of those who planned, trained for, and executed those missions.

FIRST ISS ASSEMBLY CREW NAMED

In the midst of announcements of crews being assigned to Mir missions, or in some cases losing their seats, there was news on August 16, 1996, that flight operations for the ISS were advancing. The first ISS assembly crew (2A) was named to the mission manifested as STS-88.[1] Commanding this historic crew would be Robert ('Bob') D. Cabana. His Pilot would be rookie Frederick ('Rick') W. Sturckow, while the three Mission Specialists would be veterans Nancy J. Currie (Sherlock), James ('Jim') H. Newman and Jerry L. Ross. The flight was originally scheduled for late 1997, but it slipped into 1998 and finally launched in December of that year. Of this crew, only Ross had flown to Mir. The robotic arm would be used to grapple the Russian FGB, called Zarya, and mate it with the PMA on the end of the Unity Node. Once the two modules were fully linked, the assemblage would be released as the nascent station. Such an important mission required a long lead time to master the techniques.

The first ISS Assembly crew (STS-88) shown in the Unity node during the mission, December 1998: [l-r] Rick Sturckow, Jerry Ross, Jim Newman, Nancy Currie, Bob Cabana and cosmonaut Sergei Krikalev.

Change Of ISS-1 Crewing

In September 1996, Anatoli Solovyov was appointed by the Russian training group to the ISS-1 crew that would launch on a Soyuz TM and return on a Shuttle. This would require him to undergo the appropriate ascent and re-entry training for both vehicles. But by November he had asked to swap places with Gidzenko, who was then receiving Mir training. Solovyov had been assigned as Soyuz Commander for ascent and docking but had stood down from EO-1 when it became clear that Bill Shepherd was to serve as the first *Station* Commander. Given the great difference in their space flight experience (over 650 days, compared to just 18 for Shepherd) Solovyov was, understandably, a bit put out, and opted instead for a fifth mission to Mir.

Expanding ISS Preparations

Apart from the near calamity of a collision at Mir, June 1997 also brought new crewing announcements for the early ISS assembly flights. On June 2, Japanese astronaut Koichi Wakata was named as primary RMS operator for STS-92 (3A), the mission that was to deliver the Zenith truss segment (Z0) and the third Pressurized Mating Adapter (PMA-3).[2] In the statement, NASA Administrator Daniel S. Goldin said, "NASA is honored to have Mr. Wakata participate in such an early and significant space station assembly mission. His participation on this flight is symbolic of the close bond that has developed between the American and Japanese space programs, and the extent to which we rely upon each other to meet our mutual objectives in space."

A week later, on June 9, fourteen Shuttle astronauts were assigned to an intensive training program to prepare for the major series of EVAs which ISS assembly would require.[3] Jerry Ross and Jim Newman had already been named to STS-88, which was the first assembly mission, at that time expected to launch in July 1998, but the other twelve were new assignments which reflected the long lead time necessary to prepare the EVA crews for their complicated tasks at the station. These astronauts were Leroy Chiao, P. Jeffrey ('Jeff') K. Wisoff, Michael E. Lopez-Alegria and William ('Bill') S. McArthur Jr., for STS-92 (3A) planned for January 1999; Joseph ('Joe') R. Tanner and Carlos I. Noriega for STS-97 (4A) in March 1999; Mark C. Lee and Thomas D. Jones for STS-98 (5A) in May 1999; Robert L. Curbeam Jr., and the CSA astronaut Chris A. Hadfield for STS-99 (6A, and later re-manifested as STS-100) planned for June 1999; and finally Michael Gernhardt and Jim Reilly for STS-100 (7A, later re-manifested as STS-104) in August 1999. The plan was to fly six very challenging missions during a period of thirteen months, but the plan relied on the crucial launches of the initial two Russian modules.

Another assignment was announced on June 22, this time of the eighth astronaut to take the duties of the Director of Operations in Russia (DOR) at Star City in Moscow.[4] Brent W. Jett was to relieve Michael Lopez-Alegria, who would return to JSC to start training for STS-92 and the extensive program of EVAs assigned to that flight.

First Four ISS Resident Crews Named

On November 17, 1997, in a joint press release by NASA and its Russian counterpart, twelve crewmembers were named as the first four three-person resident crews for the ISS.[5] The first crew would launch on a Soyuz and dock at the Zvezda Service Module. They would be relieved by the second trio, who would be delivered by Shuttle. There would always be a Soyuz available as a rescue vessel when no Shuttle was present. As occurred with Mir, the Soyuz spacecraft were to be exchanged during a succession of short-term visiting 'taxi' flights. This timing was due to their designed orbital lifetime limits of six months or about 180 days. These flights would provide the cash-strapped Russian program the opportunity to offer 'seats for sale' to passengers, dubbed 'space tourists' in the press. This particular idea raised some tensions between the Russians and the Americans, who initially did not officially allow tourists into the US segment due to their not having received NASA safety and familiarization training. For a three-person resident crew, a launch aboard Soyuz would mean a landing on the Shuttle and vice versa. Consequently,

In this age of computers and digital information the amount of hard documentation for each Shuttle crew still filled a small library. Here the STS-135 crew receives updates to the Flight Data File…and (bottom) there were still more handouts to collect.

each member of a resident crew would undertake a certain amount of either launch or re-entry training and get safety briefings about both Soyuz and Shuttle, as appropriate to their specific mission requirements.

The first ISS crew was to launch early in 1999 on a Soyuz TM and be retrieved five months later by Atlantis. This crew was confirmed as William M. Shepherd as ISS-1

A mission Commander is responsible for ensuring that a crew keeps up to speed with the latest developments. Here Chris Ferguson updates his notes for STS-135 (top)...while Eileen Collins briefs her STS-114 crew.

Commander, Yuri P. Gidzenko as Soyuz Commander and Sergei K. Krikalev as Flight Engineer. They would be backed up by the ISS-3 crew, consisting of Kenneth ('Ken') D. Bowersox (ISS-3 Commander), Vladimir N. Dezhurov (Soyuz Commander) and Mikhail V. Tyurin (Flight Engineer).

The second ISS crew would launch in the summer of 1999 on the same Shuttle that retrieved their predecessors. This crew comprised Yuri V. Usachev, who was both the ISS-2 Commander and Soyuz Commander, with Jim Voss (having just backed up two

Personal fitness has always been the responsibility of each individual astronaut. Here Japanese astronaut Soichi Noguchi works out in the Astronaut Gym at JSC.

American Mir astronauts) and Susan J. Helms as Flight Engineers. They would return in the Soyuz TM that delivered the third crew. The ISS-4 crew of Yuri I. Onufriyenko (ISS-4 and Soyuz Commander) and Flight Engineers Carl E. Walz and Daniel ('Dan') W. Bursch were to start out by backing up the second crew. This sequence was to see the ISS-3 crew launch in a Soyuz TM in late 1999 for a two month residency and then return aboard Discovery when it arrived with the ISS-4 crew in early 2000. That crew would remain aboard the station for four months and return in the Soyuz.

This system of rotation would continue to feature alternating command of the station between the Russians and Americans. The crews announced reflected the joint training that had been undertaken by the two nations. Krikalev and Dezhurov had experience of

All mission preparations include bench reviews of the crew equipment.

STS-101 MS Mary Ellen Weber practices deployment of the Foot Restraint Equipment Device (FRED) during a series of induced parabolic flight profiles in the NASA DC-9 aircraft.

serving on a Shuttle crew, and Gidzenko, Usachev and Onufriyenko were aboard Mir at times when American astronauts were visiting or resident there. The Soyuz taxi flights would be commanded by experienced Russian cosmonauts. Although there would be at least one seat for sale (or barter) with other ISS partners, there would be no places for Canadian, ESA or Japanese astronauts on long-duration crews. At that time there were no such astronauts in training to serve on resident crews, and it would be over a decade before they could serve on resident crews aboard the station.

January 22–31, 1998: STS-89, the eighth, penultimate Shuttle-Mir docking mission delivered Andrew Thomas and returned with Dave Wolf.

June 2–12, 1998: STS-91, the ninth and final Shuttle-Mir docking mission drew to a conclusion Phase-I by bringing Andrew Thomas home.

Gearing Up For ISS

Alongside the preparations for the final Shuttle-Mir flights, announcements of future ISS assembly crews appeared regularly during 1998, starting on February 2 with the remaining STS-92 crewmembers.[6] Brian Duffy (CDR) and Pamela A. Melroy (PLT) joined the already named RMS operator Koichi Wakata, along with the four Mission Specialists assigned to complete the EVA program. These four, who would conduct the spacewalks in pairs, were Jeff Wisoff, Leroy Chiao, Bill McArthur and Michael Lopez-Alegria. Interestingly, Wakata, Chiao, McArthur and Lopez-Alegria would later return to the ISS as members of resident crews. Initially planned as the third assembly flight, STS-92 would eventually fly as the fifth in the series, carrying the first element of the Integrated Truss Structure.

At the end of July 1998, veteran cosmonaut Sergei Krikalev (already in training as one of the members of the first ISS resident crew) was named to the STS-88 crew that was due to launch in December and mate the Russian-built Zarya Control Module with the US-built Unity Node.[7] With over fifteen months of space flight experience from a number of Mir missions and on STS-60, Krikalev would bring a wealth of knowledge to STS-88 and would gain an early insight into condition of the station on-orbit.

The final crewing announcement prior to the start of ISS assembly was released on August 4, 1998, when an additional thirteen astronauts were named to support Shuttle missions.[8] Seven of them formed the STS-96 crew (2A.1) for a logistics and resupply flight that was planned for mid-May 1999. Kent V. Rominger was CDR, with Rick D. Husband as Pilot. The Mission Specialists were Ellen L. Ochoa, Tamara ('Tammy') E. Jernigan and Daniel ('Dan') T. Barry, along with Canadian Julie Payette and Russian Yuri I. Malenchenko. The plan was to fly the mission after the Zvezda Service Module had been permanently docked at the rear of the Zarya module. This was scheduled for April 1999. The assignment of Malenchenko was very significant because his training was to work on the internal configuration of Zvezda and to unload an already docked Progress resupply ship. Over seven days of docked activities, at least one EVA would be undertaken and supplies and logistics would be transferred to continue preparations for the arrival of the first resident crew, whose launch was at that point manifested for July 1999.

The other six astronauts identified on August 4 were to join previously assigned EVA crews for two missions to the station once the first residents were aboard. The STS-97

(4A) mission scheduled for August 1999 was to deliver the first set of solar arrays, batteries, and truss components and associated cooling radiators. Joining the previously named EVA crew of Joe Tanner and Carlos Noriega for this mission were Brent W. Jett (CDR), Michael J. Bloomfield (PLT) and Canadian Marc Garneau as an additional Mission Specialist. The STS-98 (5A) mission scheduled for October 1999 was to deliver the US Destiny laboratory module. The EVA crew of Mark C. Lee and Thomas D. Jones were joined by Kenneth Cockrell (CDR), Mark L. Polansky (PLT) and Mission Specialist Marsha Ivins.

Virtual Reality was extensively used during Shuttle-ISS assembly missions. Here STS-88 MS Nancy Currie uses VR for RMS training.

ISS Assembly Commences

Just four days prior to the launch of the first component of the ISS, the five astronauts assigned to the STS-101 (2A.2a) logistics mission were named as James D. Halsell Jr. (CDR), Scott D. Horowitz (PLT) and Mission Specialists Mary E. Weber, Jeffrey N. Williams and Edward T. Lu.

VR simulations can reproduce full size images of the payload and offer realistic views out of the aft and overhead flight deck windows, in this case during training by the STS-132 crew.

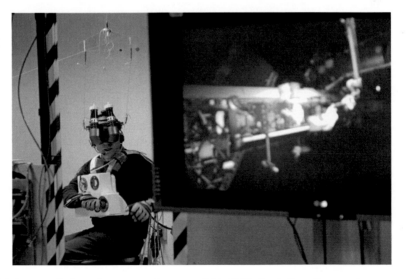

VR helmets and gloves have proved very useful in EVA simulations for over twenty years.

November 20, 1998: Russia launched the unmanned Zarya Control Module on a Proton rocket as the first launch of the ISS program (designated as assembly mission 1A/R because, although the module was built by Russia it was paid for and owned by America).

December 4–15, 1998: STS-88 (2A) Endeavour saw the Unity Node mated with the Zarya Control Module to create the embryonic International Space Station.

On February 9, 1999, ESA named Italy's Umberto Guidoni as the first European astronaut on the ISS.[9] His assignment was to STS-102, a logistics flight planned for April 2000 which was later postponed by slippage in launching the Russian Zvezda Service Module. His assignment was in the context of a bilateral agreement between NASA and the Italian Space Agency (ASI) involving the fabrication and delivery of three Multi-Purpose Logistics Modules (MPLM). The first of these (Leonardo) was to fly on STS-102. In the re-shuffle of flights and assignments prompted by the delay in launching Zvezda, Guidoni would later be reassigned to the STS-100 mission.

Just three days later, on February 12, it was reported that alterations to the Shuttle manifest would necessitate changes to crewing for the planned missions.[10] Owing to delays with hardware and the resultant changes to the launch schedule, three Russian cosmonauts were reassigned. Having received specialized training, Yuri Malenchenko would be replaced on STS-96 by Valeri I. Tokarev. Malenchenko moved to STS-101, which would fly *before* Zvezda was launched, not afterwards as initially planned. This decision was taken because of the need to further stock up and prepare Zarya ahead of the docking by Zvezda. Dr. Boris V. Morukov, whom the Russians had unsuccessfully nominated to fly on the STS-88 mission, would join Malenchenko on STS-101.

May 27–June 6, 1999: STS-96 (2A.1) Discovery flew the first logistics mission to the ISS in preparation for the launch of the Zvezda Service Module, which at this point was still expected later in the year.

Originally scheduled for launch in mid-1999 and then later in that year, the Russian Zvezda Service Module had endured a number of delays that meant it would not reach orbit before July 2000. This delay raised worries about the operation of the embryonic station. In particular, the chemical storage batteries of the Zarya Control Module were not meant to power the complex for the length of time which it now seemed would be necessary pending the arrival of the Service Module. The next logistics mission, STS-101 (2A.2), was originally to have visited *after* the docking of Zvezda but *prior to* the arrival of the resident crew. The earlier delays to Zvezda had revised this to a flight in advance of adding the Service Module. By the end of 1999, however, the Zarya-Unity stack had been in space for a year and been visited by only two Shuttle crews.

The delay to Zvezda was not the only problem facing the program. Additional issues found during electrical wiring inspections on the Shuttle fleet imposed further delays on the launch manifest, both for the ISS and independent missions. The problem was acute because the Hubble Space Telescope was experiencing difficulties with the gyroscopes that pointed it to make celestial observations. The delays also slipped the Shuttle Radar Topographical Mission (SRTM) into 2000. These pressures on the Shuttle manifest led to the decision to split the tasks for STS-101 across two flights. Initially, some thought was given to flying STS-101 as planned ahead of the launch of Zvezda, then recycling the Orbiter and crew to fly a second mission (informally designated STS-101A) *after* the launch and docking of the Service Module. This, it was said, would fully apply the crew's experience and training. Upon reflection, however, this plan was judged overly complicated and a new flight was added to the manifest instead. The same remedy was applied to the tasks that had been planned for the third HST service mission.[11]

The revised plan was published on February 18, 2000.[12] STS-101 became 2A.2a while the additional mission, 2A.2b, would be flown by a new crew as STS-106, this being the next available number on the manifest. These additional delays now meant that Malenchenko and Morukov, who had trained extensively to activate the Zvezda module and to unload the docked Progress, were moved across to the new crew. Since Ed Lu had trained for an EVA with Malenchenko, it seemed logical to keep that team together and save training time, so Lu also joined STS-106 and the EVA intended for STS-101 was moved to the added mission. The other members of this new crew were Terrance ('Terry') W. Wilcutt (CDR) and Scott D. Altman (PLT), along with Mission Specialists Daniel ('Dan') C. Burbank and Richard ('Rick') A. Mastracchio. And Lu, Malenchenko and Morukov were replaced on the original STS-101 crew by the three members of the ISS-2 crew, Usachev, Voss and Helms. It was reasoned that because the latter trio were already in training to operate ISS systems, their addition to the STS-101 crew would assist with on board preparations ahead of the arrival of the first residents. It would also give them valuable hands-on experience aboard the station a year prior to their own scheduled residency. The idea was that experience aboard the nascent station would facilitate smooth handovers between the first three resident crews.

On May 9, 2000, still awaiting the arrival of Zvezda, NASA named the crew to the eighth Shuttle mission to the ISS, STS-102.[13] The crew for assembly mission 5A.1 in 2001 would be Jim Wetherbee (CDR), James M. Kelly (PLT) and Mission Specialists Andy Thomas and Paul W. Richards. The plan included two EVAs and the supply of additional equipment for the US Destiny laboratory. The amount of cargo meant this mission carried the first of the Italian-built Multi-Purpose Logistics Modules (MPLM). This Leonardo module would be hoisted out of the payload bay by the RMS and mated to the ISS for unloading, then be filled with unwanted material and trash and restored to the Orbiter for return to Earth. This mission would also deliver the three ISS-2 crew to replace the ISS-1 crew that would launch by Soyuz and return by Shuttle.

May 19–29, 2000: STS-101 (2A.2a) Atlantis flew the second logistics mission to the ISS.

July 26, 2000: The long-delayed Zvezda Service Module finally docked at the aft port of the Zarya Control Module. This meant that plans could at last be finalized to launch the first resident crew to the station.

On September 7, NASA reported Mark Lee had been withdrawn from STS-98 "for undisclosed reasons" and was replaced by Bob Curbeam. According to spokesman Ed Campion it was "an internal Astronaut Office matter." No further details were given.[14] Curbeam had been in training for STS-99 (later re-manifested as STS-100) and his seat on that mission was reassigned to Scott Parazynski.

September 8–20, 2000: STS-106 (2A.2b) Atlantis delivered more supplies to the ISS in preparation for the launch of the first resident crew.

By September, the increasing pace of Shuttle training was evident with the news that a cadre of twenty astronauts and a single Russian cosmonaut had been selected for four missions planned for 2001.[15] Two of these were ISS assembly missions. The third, STS-107, was a stand-alone science research mission with the Spacehab module. The fourth, STS-109, was manifested as the fourth scheduled Hubble servicing mission, designated SM-3B. This mixture of ISS and other flights was becoming more unusual, particularly now that its assembly process began to dominate the manifest. Very few solo missions remained on the manifest or would be added in the future. Now that NASA was finally able to pursue its dream of assembling a space station, this became the raison d'être for the Shuttle fleet.

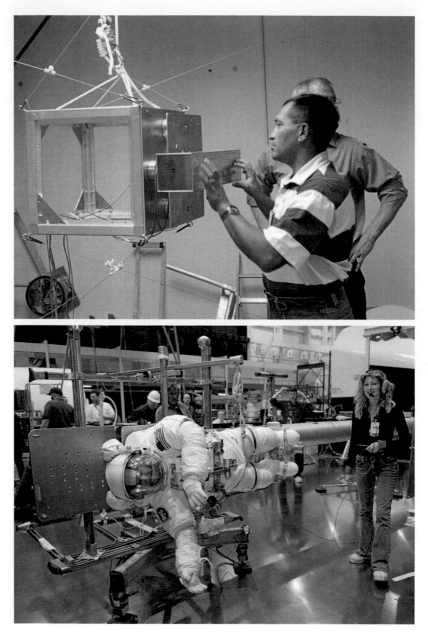

Despite all of the technology, most hands-on EVA training sessions start with basic one-g simulations of equipment and hardware (top), then progressed to one-g fully-suited sims on the air-bearing floor in Building 9A at JSC (below).

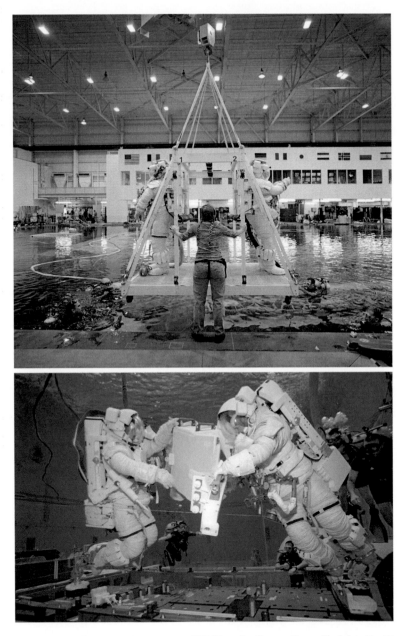

In 1997 the Neutral Buoyancy Laboratory (NBL) at the Sonny Carter Training Facility near JSC became the primary water tank for all underwater ISS assembly EVA training (top). Supported by scuba divers and a pool-side team, each EVA crew undertook a choreographed simulation of every task on each of their assigned spacewalks (below).

The two new ISS assembly flights were STS-100 (6A) and STS-104 (7A). The first of these was to deliver the Canadian-built Space Station Remote Manipulator System (SSRMS) nicknamed Canadarm2. It would also feature three EVAs, as well as further logistics transfers, this time using MPLM Raffaello. The crew was Kent V. Rominger (CDR), Jeffrey ('Jeff') S. Ashby (PLT) and Mission Specialists John L. Phillips, Scott E. Parazynski and Russian cosmonaut Yuri V. Lonchakov. Already named to the flight were Mission Specialists Chris A. Hadfield of Canada and Umberto Guidoni, an Italian flying for ESA. The second ISS flight, STS-104, featured three EVAs in support of the delivery and installation of the ISS airlock, known as Quest. This crew was Steven W. Lindsey (CDR), Charles O. Hobaugh (PLT) and Mission Specialists Janet L. Kavandi, Michael L. Gernhardt and James F. Reilly.

October 11–24, 2000: STS-92 (3A) Discovery installed the Zenith (Z1) segment of the truss and PMA-3 on the ISS.l

November 2, 2000: Soyuz TM-31(1S), docked with the three-person ISS-1 crew of Gidzenko, Krikalev and Shepherd for the ISS. This was the first of the permanent crews aboard the station, able to remain without the presence of a docked Shuttle. They had launched from the Baikonur Cosmodrome on October 31.

On December 1, 2000, NASA announced that the crew for STS-105 (7A.1) would be Scott J. Horowitz (CDR), Frederick W. Sturckow (PLT) and Mission Specialists Daniel T. Barry and Patrick G. Forrester.[16] This was manifested as a logistics supply mission for June 2001. Also aboard the Orbiter for ascent would be the ISS-3 crew of Frank L. Culbertson, Vladimir N. Dezhurov and Mikhail Tyurin, whilst returning would be the ISS-2 of Yuri Usachev, Jim Voss and Susan Helms.

November 30–December 11, 2000: STS-97 (4A) Endeavour delivered the P6 truss and solar arrays to the ISS.

One of the less demanding tasks in preparing for a Shuttle mission was the selection of each crewmember's personal menu.

A REAL SPACE ODYSSEY IN 2001

In 1968, as Apollo began to gear up to land Americans on the Moon, Stanley Kubrick released his movie *2001: A Space Odyssey*. One early scene showed a Pan-Am 'space liner' arriving at a large Earth-orbiting station. The movie imagined a future in which flights to Earth orbit and the Moon were common by the turn of the millennium. Sadly the reality of the year 2001 fell far short of this vision. There had been stations in orbit for several decades, most recently the case of Mir, but none were as impressive as that of Kubrick's movie.

February 7–20, 2001: STS-98 (5A) Atlantis delivered the US Destiny laboratory to the ISS, significantly expanding its capabilities.

March 8–21, 2001: STS-102 (5A.1) Discovery delivered the ISS-2 crew and further logistics in the Leonardo MPLM. The mission also saw the first exchange of command on board the station, from ISS-1 to ISS-2. The retirees returned to Earth as part of the STS-102 crew.

On March 26, five days after STS-102 landed, NASA issued the names of fourteen astronauts who would start training for long-duration ISS crews.[17] Many of these had previous experience on Shuttle-Mir and ISS assembly missions. They would be joined by as yet unnamed Russian crewmembers. On April 11, NASA appointed the crew for STS-110 (8A).[18] Mike Bloomfield (CDR) was a veteran of both Mir and ISS missions. Stephen Frick (PLT) and Ellen Ochoa (FE-MS2) rounded out the flight crew. The four Mission Specialists were Jerry Ross, who was assigned to his seventh mission, Steven Smith, Lee Morin and Rex Walheim. They were to undertake an extensive program of spacewalks with the support of Ochoa operating the RMS.

The next announcement on July 3 named four astronauts as the core of the crew for STS-111 (UF2).[19] These were Ken Cockrell (CDR), Paul S. Lockhart (PLT), Mission Specialist Franklin R. Chang-Diaz (who, like Ross, was being assigned to his seventh flight) and French CNES astronaut Philippe Perrin as MS2-FE. Chang-Diaz and Perrin would carry out the planned EVAs during a mission that included delivering the ISS-5 crew of Valeri G. Korzun, Peggy A. Whitson and Sergei Y. Treschev and returning the ISS-4 crew of Onufriyenko, Walz and Bursch.

The August 17 announcement, on the other hand, named no fewer than twenty-three astronauts and cosmonauts for three ISS assembly flights scheduled for 2002, namely STS-112 (9A) in July, STS-113 (11A) in August, and STS-114 (ULF1) in November, with the latter two flights carrying resident crews to and fro.[20] The STS-112 crew was to be Jeffrey S. Ashby (CDR), Pamela A. Melroy (PLT) and Mission Specialists David A. Wolf and rookies Piers J. Sellers, Sandra Magnus and Russian cosmonaut Fyodor N. Yurchikhin. The flight deck crew for STS-113 was Jim Wetherbee (CDR), Christopher J. Loria (PLT) and Mission Specialists Michael Lopez-Alegria and John B. Herrington. On the mid-deck during the ascent would be the ISS-6 crew of Kenneth D. Bowersox, Donald A. Thomas and Nikolai M. Budarin. On the return to Earth their seats would be taken by the ISS-5 crew of Korzun, Whitson and Treschev. These two missions were to expand the Integrated Truss Structure.

The STS-114 utilization and logistics mission had been designated as STS-113 in earlier planning manifests and press announcements. It would be flown by Eileen M. Collins

(CDR) and James M. Kelly (PLT), along with Mission Specialists Stephen K. Robinson and Japanese NASDA astronaut Soichi Noguchi. The ISS-7 crew of Yuri I. Malenchenko, Sergei I. Moschenko and Edward T. Lu would be aboard for the ascent and the ISS-6 trio of Bowersox, Budarin and Pettit would make the return trip.

April 19–May 1, 2001: STS-100 (6A) Endeavour delivered the Space Station RMS system (SSRMS, known as Canadarm2) and the second MPLM Raffaello packed with logistics and supplies.

April 28, 2001: Soyuz TM-32 (2S) arrived at the ISS to perform the first short-term visiting crew mission. The replacement of a Soyuz demonstrated the capability of the Russian vessel to serve as a crew return and rescue vehicle whilst the Shuttle was not present. In fact, it was this mission that signaled the beginning of the end of using the Shuttle as a resident crew transport vehicle.

July 12–24, 2001: STS-104 (7A) Atlantis delivered the Joint Airlock Module named Quest.

August 10–22, 2001: STS-105 (7A.1) Discovery was the second flight for MPLM Leonardo, packed with supplies. The mission delivered the ISS-3 crew of Culbertson, Dezhurov and Tyurin to the station and retrieved the ISS-2 crew of Usachev, J. Voss and Helms.

December 5–17 2001: STS-108 (UF1) Endeavour was the second flight of MPLM Raffaello. The mission delivered the ISS-4 crew of Onufriyenko, Bursch and Walz and retrieved the ISS-3 crew of Culbertson, Dezhurov and Tyurin.

Success Before Setback, 2002

On February 26, 2002, two new crews were named by NASA.[21] The STS-115 (11A) crew was to be Brent W. Jett Jr. (CDR), Christopher ('Chris') J. Ferguson (PLT) and Mission Specialists Joe Tanner, Dan Burbank, Heidemarie M. Stefanyshyn-Piper and Canadian Steven ('Steve') G. MacLean. It was manifested to deliver the second port-side segment of the ITS (designated P3/P4, because the P2 segment of Space Station Freedom had been deleted from the reduced ISS configuration). STS-116 (9A.1) was assigned another four-person core crew of Terry Wilcutt (CDR), William ('Bill') A. Oefelein (PLT) and Mission Specialists Bob Curbeam and A. Christer Fuglesang, the latter being a Swedish ESA astronaut. The mission was manifested to install the third port-side segment of the ITS (designated P5), deliver logistics, deliver the ISS-8 crew of C. Michel Foale, William S. McArthur and Valeri I. Tokarev, and return the ISS-7 crew of Malenchenko, Moschenko and Lu.

April 8–20, 2002: STS-110 (8A) Atlantis delivered the S0 segment of the ITS along with the Mobile Transporter.

June 5–19, 2002: STS-111 (UF2) Endeavour completed a logistics and resident crew exchange mission, replacing the ISS-4 crew of Onufriyenko, Bursch and Walz with the ISS-5 crew of Korzun, Whitson and Treschev. It also featured the fifth MPLM flight, the third for Leonardo.

A crewmember substitution was announced on July 26. Donald ('Don') A. Thomas was stepping down in favor of Donald ('Don') R. Pettit on the ISS-6 crew that was in training for launch later in the year.[22] The official explanation was "a medical issue that affected Thomas' long-duration space flight qualification." Astronaut Charles Precourt explained

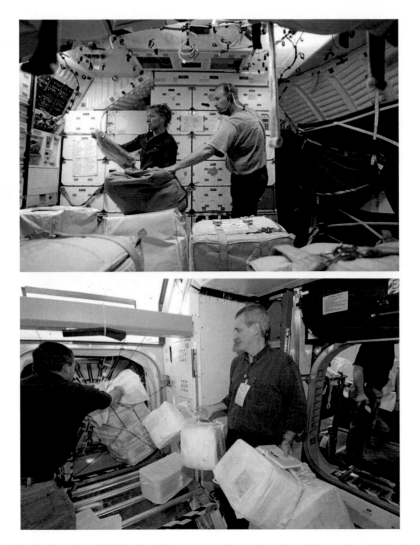

Learning the location of all the logistics, crew equipment, and supplies on the mid-deck takes a while (top). And organizing the transfer of the logistics to and from the Orbiter occupies further training time (bottom).

that because activities on the ISS were still in the early stages, it was prudent to adopt a conservative approach to the demanding nature of long missions. He pointed out that such changes had been made in the past. This action vindicated the decision to resume back-up crew assignments for ISS crews, something NASA had not done since the back-up system ended in 1982. Pettit was to join Bowersox and Budarin, launch on STS-113 to the ISS and be retrieved by STS-114.

On August 15, NASA announced several assignments to the ISS assembly flights.[23] The six-person crew for STS-117 (13A) in the fall of 2003 was Rick Sturckow (CDR), Mark

L. Polansky (PLT), and Mission Specialists James F. Reilly II, Richard ('Rick') A. Mastracchio, Joan E. Higginbotham and Patrick G. Forrester. It was also revealed that Paul S. Lockhart would replace the previously named Christopher J. Loria as PLT for STS-113. As Lockhart had recently flown in that role on STS-111, that experience would be a benefit to his abbreviated training cycle for the STS-113 crew, particularly as the mission to expand the ITS would be similar to his last one. Loria had requested reassignment following an "injury sustained at home." Having two herniated discs in his lower back, Loria had withdrawn from training the previous month and transferred to administrative roles. In the fall of 2004 his injures were deemed inoperable. Having been disqualified from making a space flight he returned to the USMC.

October 7–18 2002: STS-112 (9A) Atlantis delivered the S1 truss and the first of two Crew Equipment Translation Aids (CETA).

November 23–December 7, 2002: STS-113 (11A) Endeavour delivered the P1 truss and CETA B, and replaced the ISS-5 crew of Korzun, Treschev and Whitson with the ISS-6 crew of Bowersox, Pettit and Budarin.

As 2002 closed, the scheduled expansion of the station in the new year prompted a raft of crewing assignments. On December 12, two announcements together reflected the continuing progress of station assembly and the rotation of expedition crews using Shuttle missions.

The first release revealed that STS-118 (12A) was planned for November 2003 with a crew of six to further expand the truss.[24] STS-119 would then deliver the ISS-9 crew and retrieve the ISS-8 crew. By installing Node 2 in early 2004, STS-120 would mark the completion of the US core assembly phase. The crew for STS-118 was revealed to be Scott J. Kelly (CDR), Charles O. Hobaugh (PLT) and Mission Specialists Scott E. Parazynski, Canadian Dafydd ('Dave') R. Williams, Lisa M. Nowak and Barbara R. Morgan. Morgan's assignment as the first Educator Mission Specialist was explained in a follow-up announcement.[25] Earlier in the year NASA Administrator Sean C. O'Keefe had confirmed his commitment to Morgan's flight as the first of what was expected to be "many flights of the new Educator Astronaut program." He said that Morgan's crew assignment was, "[One] part of NASA's responsibility to cultivate a new generation of scientists and engineers."*

The crew announced for STS-119 (12A.1) was Steven W. Lindsey (CDR), Mark E. Kelly (PLT) and Mission Specialists Michael L. Gernhardt and Carol L. Noriega. The flight was to deliver the ISS-9 crew of Gennady Padalka, E. Michael Fincke and Oleg D. Kononenko and retrieve the ISS-8 crew of C. Michael Foale, William S. McArthur and Valeri I. Tokarev. The crew for STS-120 would be James D. Halsell (CDR), Alan G. Poindexter (PLT) and Mission Specialists Wendy B. Lawrence, Piers J. Sellers, Stephanie D. Wilson and Michael J. Foreman.

---

*Barbara Morgan had originally been selected for the Teacher-In-Space program in 1985 and served as back-up to Christa McAuliffe, who died aboard Challenger on January 28, 1986. Morgan resumed her teaching career later that year, but retained connections with NASA to support its educational effort. In January 1998 Morgan was selected to train as a fully-fledged NASA Mission Specialist and prior to receiving her flight assignment she served as a Capcom in Mission Control.

The whole crew also perform a number of stand-alone or integrated simulations with trainers and flight controllers in the mid-deck and flight deck simulators (top). They also practice getting into and out of the compartments wearing their orange launch and entry 'pumpkin' suits (bottom).

February 1, 2003: Columbia and its crew of seven were lost as the vehicle broke up at an altitude of 200,000 ft (60,960 m) over Texas near the end of what had been a very successful sixteen day international science research flight. Although STS-107 was an independent mission with a double Spacehab module in its payload bay, the work was intended to assist with the broader ISS science program.

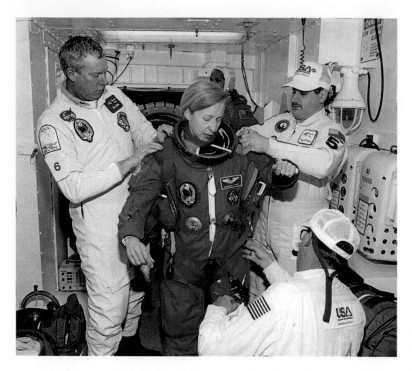

Training continues at the Cape during the Terminal Countdown Demonstration Test. Here STS-98 MS Marsha Ivins has the straps of her launch and entry suit adjusted in the White Room at the top of the launch tower.

RECOVERY

After the loss of Columbia, the other Orbiters were grounded pending an investigation and the implementation of actions to prevent a recurrence of such a tragedy.

For the foreseeable future, ISS crews would be reduced to caretaker pairings of one American astronaut and one Russian cosmonaut whose main job would be to maintain the station until the Shuttle could resume assembly operations. The three-person ISS-6 crew that was in space would be replaced at the end of their tour by the first caretaker crew, comprised of two members of the planned three-person ISS-7 crew. The retiring trio were to return to Earth in Soyuz TMA-1, which had been docked since November, rather than aboard STS-114 as originally planned. That Shuttle would become the first Return-To-Flight mission.

Reducing ISS crews to two people meant that the third seat on the Soyuz could now be occupied by representatives from partner agencies, or by fee-paying space tourists, who would visit the station for up to 10 days and then return to Earth with the retiring residents.

With the caretaker crew process in place for the ISS, NASA's focus was to prepare for the Return-To-Flight mission. On November 7, 2003 it announced a revision of the designation of STS-114 from ULF1 to LF1.[26] It also revised the crew. In 2001, Eileen Collins,

James Kelly, Stephen Robinson, and Japanese astronaut Soichi Noguchi had been named to the mission. Originally planned to ferry a three-person resident crew to the ISS and collect the ISS-6 crew, the revised crew now included Mission Specialists Andrew Thomas, Wendy Lawrence and Charles J. Camarda. The requirement to re-qualify the Shuttle system following the loss of Columbia had altered STS-114 from a logistics flight to one that would evaluate new safety procedures, including inspecting the TPS tiles and using a payload bay rig to evaluate various repair techniques. There would therefore be fewer station-related tasks.

Emergency training includes using a 'sky genie' harness to exit the Orbiter via the overhead flight deck windows (left). This training continues in the Neutral Buoyancy Laboratory (NBL) where water egress training is performed (right). In this case one of the STS-104 crew drops out of the simulated Orbiter side hatch into the pool. Note the support safety divers already in the water.

Explaining the new crew assignments for STS-114, Bob Cabana, Director of Flight Crew Operations at JSC, said, "This is a demanding mission and the addition of Andy, Wendy and Charlie to this already well-qualified crew ensures they have all the skills necessary to meet the challenge of the Return-To-Flight, and the resumption of Shuttle support of the International Space Station." Thomas, Cabana pointed out, had a wealth of experience in space, including a long visit to Mir in 1998 and making an EVA at the ISS in 2001. Lawrence, a veteran of two Shuttle flights to Mir, was "a superb robotics operator." Camarda had completed technical assignments involving thermal protection repair activities and had trained as a back-up ISS crewmember prior to this assignment.

On December 2, there came the surprise announcement of a second Return-To-Flight mission, STS-121 (ULF1.1), in order to "accommodate the growing list of requirements

After the loss of Columbia each crew was trained to perform limited EVA repairs to the thermal protection system of the Orbiter. Here the STS-135 crew receive a briefing for such a contingency (top). Although STS-135 did not have an EVA assigned, the crew still had to train for contingency spacewalks in the event of an emergency. Suspended in the Partial (ZerO) Gravity SimulatOr (POGO) MS Sandra Magnus uses a power ratchet tool during EVA simulation training (bottom).

originally assigned to the [STS-114] Return-To-Flight mission."[27] This additional flight would continue to supply the ISS with much-needed logistics and other hardware. The core crew was Steve W. Lindsey (CDR), Mark E. Kelly (PLT) and Mission Specialists Carlos I. Noriega and Michael E. Fossum. The remaining seats would be announced in due course, as would those assigned to STS-119 to replace Lindsey, Kelly and Noriega after their reassignment to this new mission. The first such announcement was made on November 18, 2004, reporting that Stephanie Wilson from STS-120 and Lisa Nowak from STS-118 had both been reassigned to STS-121 simply because "when STS-121 was added to the schedule, [their] training and experience provided an opportunity to move them up to an earlier flight."[28]

The next crew announcement came from ESA on April 28, 2005.[29] It reported that Thomas Reiter of Germany was to be the first European astronaut on a long-duration ISS crew. He would launch on STS-121 and land on STS-116. The press release also reported Frenchman Leopold Eyharts as Reiter's back-up, and said their assignments were part of a set of bilateral understandings between Roscosmos, NASA and ESA to fly members of ISS partner nations. According to Daniel Sacotte, the ESA Director of Human Spaceflight, Microgravity and Exploration, the agreement, which had recently been signed with the Russian Federal Space Agency, "covers the ESA astronaut's flight in a crew position which was originally planned for a Russian cosmonaut. [Reiter] will perform all the tasks originally allocated to the second Russian cosmonaut on board the ISS and, in addition, an ESA experimental program."

Resuming Training

As the recovery from the loss of Columbia gathered pace, further announcements were made with the eventual goal of resuming full time, three-person crewing of the ISS. On February 9, 2005, NASA announced that two crews were to resume training for Shuttle missions.[30] After STS-116 (12A.1) had installed the P5 segment of the truss, STS-117 (13A.1) would deliver the S3/S4 segment. The new Commander of STS-116 was Mark Polansky, replacing Terry Wilcutt, who had moved into managerial roles following the loss of Columbia. William Oefelein was named as PLT. The Mission Specialists were Robert Curbeam, Joan E. Higginbotham, Nicholas J. M. Patrick and Christer Fuglesang – the latter being a Swedish ESA astronaut. Oefelein, Curbeam and Fuglesang had been named to the crew way back in February 2002, but Patrick and Higginbotham were new assignments.

A crew for STS-117 had been named in August 2002 but, like other assigned crews, this was stood down from training after the loss of Columbia. The new crew would be a mixture of old and new assignments. Rick Sturckow retained his position as CDR, but Lee J. Archambault became PLT instead of Polansky, who had become STS-116 CDR. Jim Reilly, Rick Mastracchio and Patrick Forrester remained as Mission Specialists and were joined by Steven Swanson, who replaced Joan Higginbotham when she moved to STS-116.

Medical Condition

On July 15, NASA reported that Piers J. Sellers was to replace Carlos I. Noriega as a Mission Specialist on STS-121 owing to "a temporary medical condition that affected his qualification for the flight."[31] The name of the astronaut who would replace Sellers on the STS-120 crew would be revealed in due course. Noriega performed a series of

The STS-102 crew participate in simulated emergency slide-wire training at LC-39.

management positions at JSC without receiving another flight assignment and then he retired from the agency in August 2011.

July 26–August 9, 2005: STS-114 (LF1) Discovery's Return-To-Flight mission carried over 3,748.5 lb (1,700 kg) of supplies and cargo to the ISS and tested new procedures and safety requirements following the loss of Colombia over two years earlier. It also marked the sixth use of the MPLM, and the third flight of Raffaello.

A press release on May 2, 2006, named Michael Lopez-Alegria, Sunita Williams and Mikhail Tyurin as the ISS-14 resident crew.[32] This was the first three-person crew since ISS-6. Lopez-Alegria and Tyurin would launch by Soyuz. Williams would be delivered by STS-116 to replace Thomas Reiter, who would return with the Shuttle crew. Clayton Anderson was to back-up Williams. Several days later, on May 5, a joint announcement by NASA and JAXA assigned Japanese astronaut Takao Doi to the 1J/A mission which would deliver the first element of that nation's Kibo laboratory.[33] Doi would undertake the initial configuration of this facility.

A fortnight later, on May 17, NASA finalized the crew assignments for two Shuttle missions scheduled in 2007.[34] It confirmed that John D. Olivas would join the crew of STS-117 (13A.1, later redesignated 13A) and that Tracy Caldwell would join STS-118 (which became the new 13A.1). A further shuffle of seats was announced with Richard Mastracchio, previously assigned to STS-117, transferring to STS-118 to replace Scott Parazynski, who was said to be "preparing for a new flight assignment with a different crew." STS-117 was to deliver the S3/S4 segment of the truss. STS-118 would add the S5 segment and an external stowage platform and also ferry supplies to the station in a Spacehab single module.

The STS-104 crew meet the press to explain the objectives of their mission: [l to r] J. Reilly (MS2), J. Kavandi (MS3), M. Gernhardt (MS1), C. Hobaugh (PLT) and S. Lindsey (CDR).

The STS-117 crew observe the traditional end-of-training cake cutting ceremony.

The next month, Parazynski's reassignment was revealed in an announcement that reflected NASA's confidence in a resumption of ISS assembly. The crew for STS-120 (10A) was named on 19 June as Pamela A. Melroy (CDR), George D. Zamka (PLT), with Parazynski as a Mission Specialist together with Douglas H. Wheelock, Michael J. Foreman and Italian ESA astronaut Paolo A. Nespoli.[35] Parazynski had been reassigned from STS-118 because of his spacewalking experience. The manifested payload for this mission would be Node 2. This announcement posed the question of what happened in the intervening time to the crew that had been assigned to STS-120 in December 2002. Though it was not stated in the release, James D. Halsell, the original CDR, had taken on management roles pending his retirement in November. The original PLT, Alan G. Poindexter, was soon to be reassigned to STS-122. Mission Specialists Piers J. Sellers and Stephanie D. Wilson had been reassigned to STS-121 (and she would be re-assigned back to STS-120 in November 2006, after flying STS-121). Meanwhile Mission Specialist Wendy B. Lawrence had retired from the agency soon after flying on STS-114. Only Michael J. Foreman remained on the STS-120 crew from the original announcement.

On 20 July another assignment was revealed. The crew for STS-122 (1E) would be Stephen N. Frick (CDR) alongside reassigned PLT Alan G. Poindexter. The Mission Specialists were Rex J. Walheim, Stanley G. Love, Leland Melvin and German ESA astronaut Hans Schlegel.[36] This mission would deliver the ESA Columbus laboratory.

July 4–17, 2006: STS-121 (ULF1.1) Discovery performed this second Return-To-Flight mission and restored the Shuttle fleet to full operational status (as of the next mission). It featured logistics and resupply activities and delivered the German ESA astronaut Thomas Reiter to join the ISS-13 crew as a flight engineer.

September 9–21, 2006: STS-115 (12A) Atlantis resumed ISS assembly by installing the P3/P4 truss.

October 18 saw the announcement of further ISS crewmembers.[37] NASA and the Russian Federal Space Agency named the main residents for the ISS-15 and ISS-16 crews. Clayton C. Anderson and Daniel M. Tani were both to be Flight Engineers on ISS-15, flying in sequence. When STS-118 delivered Anderson in June 2007 he would replace Sunita Williams, who would return aboard that Shuttle. When Tani arrived on STS-122 in October 2007, Anderson would return to Earth.

December 9–22, 2006: STS-116 (12A.1) Discovery delivered the P5 truss. It also delivered logistics and cargo in a Spacehab single module. Sunita Williams replaced Thomas Reiter on the resident crew, with Reiter returning on the Shuttle. This was the first of a series of partial crew exchanges using the Shuttle in advance of the start of permanent crews of six, which would be ferried to and from the station on Soyuz taxis from 2009.

Expansion

With the solar panels on the almost complete ITS boosting the power supply, the next phase of ISS assembly would see scientific modules from Japan and Europe attached. The astronauts assigned to these missions were announced in early 2007. A release on January 29 reported the crew for the STS-123 (1J/A) mission, scheduled for December 2007.[38] This was to deliver the first part of the Japanese Kibo laboratory, known as the Experiment Logistics Module-Pressurized Section (ELM-PS), as well as the Canadian Dextre robotics system. Joining the already named Japanese Mission Specialist Takao Doi were Richard M. Linnehan, Robert L. Behnken and Michael J. Foreman. The latter had transferred from STS-120 and his place on that mission was filled by Stephanie D. Wilson, who had rejoined the crew in November 2006. Dominic L. Gorie was the CDR for the mission and Gregory H. Johnson was the PLT. It was also to undertake a partial crew transfer by delivering and returning as yet unnamed ISS resident crewmembers.

Just over two weeks later, on February 13, and again in a joint release by NASA and its international partners, the names of crewmembers from four agencies were reported who would live on the ISS over the next two years as part of three expeditions.[39] In the fall of 2007 the ISS-16 crew would launch by Soyuz to replace the ISS-15 crew, which included Dan Tani. He would remain on the station working with ISS-16 until STS-122 delivered his successor, French ESA astronaut Leopold Eyharts, in the fall of 2007. He, in turn, would return to Earth when STS-123 delivered Garrett E. Reisman. This cycle was to continue throughout the summer of 2008 with the arrival of Sandra Magnus on STS-119. She would be replaced by Japanese astronaut Koichi Wakata when STS-126 arrived in the autumn. He, in turn, would hand over to Gregory ('Greg') E. Chamitoff, but it had yet to be decided whether Wakata would return aboard a Shuttle or a Soyuz. The back-up slots were: Belgian ESA astronaut Frank DeWinne for Eyharts; Timothy L. ('Tim') Kopra for Reisman; Nicole Stott for Magnus; Soichi Noguchi for Wakata, and Timothy J. ('TJ') Creamer for Chamitoff.

On March 22 the crew for STS-124, the 1J mission that would add the Pressurized Module of the Kibo laboratory and Japanese robotic arm system, was named as Mark E. Kelly (CDR), Kenneth T. Ham (PLT) and Mission Specialists Karen L. Nyberg, Ronald J. Garan Jr., Michael E. Fossum, Stephen G. Bowen and Akihiko Hoshide of Japan.[40]

A release on April 26 announced an amendment to the crew rotation between Sunita Williams and Clayton Anderson.[41] The exchange was initially assigned to STS-118 in June but that mission was postponed to August after the ET suffered hailstorm damage. The revised plan was for Williams to come home on STS-117, which was to launch on June 8 with Anderson. His assignment to ISS-15 was revised to have him return aboard STS-120, which had an October 20 launch date. His move from launching on STS-118 to STS-117 left a vacant seat on the former flight. On May 3 it was announced that this position had been assigned to Mission Specialist Benjamin A. ('Alvin') Drew.[42] An in-depth review indicated that these changes would not seriously affect either space station operations or future Shuttle objectives.

June 8–22, 2007: STS-117 (13A) Atlantis installed the S3/S4 trusses on the ISS and delivered Clayton Anderson to replace Sunita Williams, who returned on the Shuttle.

August 8–21, 2007: STS-118 (13A.1) Endeavour delivered the S5 truss, together with External Stowage Platform #3 and additional supplies transported in a Spacehab single module.

The assignment for STS-126 (ULF2) was "to deliver equipment to that ISS that will enable larger crews to reside aboard the complex." This equipment included additional crew living quarters, a second treadmill, and apparatus for the regenerative life support system. On October 1, 2007, the crew was announced as Christopher Ferguson (CDR), Eric A. Boe (PLT) and Mission Specialists Stephen G. Bowen, Joan E. Higginbotham, Robert S. Kimbrough and Heidemarie M. Stefanyshyn-Piper.[43] Back in March, Bowen had been appointed to STS-124 so his transfer to STS-126 opened the way for another, as yet unidentified Mission Specialist on that mission.

October also saw the resolution of the crewing for the much-delayed STS-119 (15A) mission. The original crew, assigned in 2002, was Steven W. Lindsey (CDR), Mark E. Kelly (PLT) and Mission Specialists Michael L. Gernhardt and Carlos L. Noriega but a series of changes and delays meant these astronauts were transferred to earlier flights.* On October 19 the revised manifest scheduled this mission in the fall of 2008, flown by Lee J. Archambault (CDR), Dominic Antonelli (PLT) and Mission Specialists Richard Arnold, Joseph Acaba, John Phillips and Steven R. Swanson.[44] By adding the S6 truss segment to the station, STS-119 would complete its 361 ft (110 m) long backbone.

October 23–November 7, 2007: STS-120 (10A) Discovery delivered the Harmony Node to the ISS and performed the partial resident crew exchange between Dan Tani and Clayton Anderson, bringing the latter home.

Two weeks after STS-120 landed, NASA revealed on November 21, 2007, that Joan E. Higginbotham had withdrawn from the STS-126 crew and resigned from the agency in order to accept a position in the private sector.[45] With the mission not due for launch until September 2008, her replacement, Donald R. Pettit, would have plenty of time to train for the flight.

With the Shuttle fleet's retirement expected in 2010, the year 2008 was intended to see a succession of flights that would deliver the last major ISS hardware elements as rapidly as possible in order to gain flexibility during the limited time in which the fleet would be available.

On January 11, 2008, NASA announced updates to ISS expedition crews owing to changes in the schedule for the Shuttle missions that would ferry residents to and from the station.[46] Now Garrett Reisman, who would serve on the ISS-16 and ISS-17 crews, would return aboard STS-124, which was targeted for launch on April 24, 2008, rather than on STS-126. The flight that would deliver him, STS-123 in March, would remain the same. Greg Chamitoff would launch on STS-124 to replace Riesman on the ISS-17 crew, and return to Earth on STS-126, which was expected to launch on September 18. Both of these assignments were backed up by Timothy Kopra. STS-126 was to deliver Sandra Magnus to replace Chamitoff. After working on the ISS-17 and ISS-18 crews, she would return to Earth on STS-119 in the fall of 2008. Her back-up was to be Nicole Stott. Koichi Wakata would arrive on STS-119 to replace Magnus on the ISS-18 crew, then return to Earth on STS-127. Soichi Noguchi was named as his back-up.

On February 12, 2008, NASA named the crew for STS-127 (2J/A), the mission that was manifested to deliver the final elements of the Kibo Japanese laboratory (the JEM Exposed Facility and Experiment Logistics Module Exposed Section).[47] Joining Mark L. Polansky (CDR) and Douglas G. Hurley (PLT) were Mission Specialists Christopher J. Cassidy, Thomas H. Marshburn, David A. Wolf and Canadian Julie Payette. It would also deliver Tim Kopra to replace Wakata on the ISS-18 crew. The release also pointed out that although Nicole Stott would arrive on STS-128 to replace Kopra, the plan was for her to return by Soyuz, not by Shuttle. In May 2009, Yuri V. Lonchakov, Frank De Winne and Robert B. Thirsk would launch by Soyuz to join the ISS-19 crew, thereby expanding the crew to six people for the first time. The combined teams were to form the ISS-20 crew. Stott was to return to Earth along with Lonchakov and DeWinne by Soyuz at the end of their residency. Thirsk would be collected by STS-129. The back-up assignments were Creamer for Kopra, Coleman for Stott, and Hadfield for Thirsk. They would all receive

---

*All except Carlos Noriega who had left the Astronaut Office to perform managerial roles within NASA.

Shuttle launch and re-entry training as part of their residency assignments but this would be the final round of resident crew training on the Shuttle system. When Lonchakov later stood down as Soyuz Commander he was replaced by Roman Y. Romanenko.

February 7–20, 2008: STS-122 (1E) Atlantis delivered the ESA Columbus laboratory to the ISS, along with ESA astronaut Leopold Eyharts from France to replace Dan Tani on the station. The retrieval of Tani had been delayed two months due to difficulties in getting a Shuttle off the launch pad. Eyharts' primary responsibility was to support the integration and activation of Columbus into the station.

March 11–26, 2008: STS-123 (1J/A) Endeavour delivered the Experiment Logistics Module, Pressurized Section (ELM-PS) of the Japanese Kibo laboratory along with the Special Purpose Dexterous Manipulator ('Dextre') supplied by Canada. It conducted a partial resident crew exchange between Garrett Reisman and Leopold Eyharts, with the latter returning home at the end of the flight.

May 31–June 14, 2008: STS-124 (1J) Discovery delivered the Pressurized Module (JEM-PM) of the Japanese Kibo laboratory, plus the associated robotic manipulator system. In the partial resident crew exchange Gregory E. Chamitoff replaced Garrett Reisman, with the latter returning home at the end of the flight.

Before a crew could launch, another Shuttle tradition at the Cape was for the CDR to use up their bad luck in a game of poker. Here Jerry Ross deals cards to the final Shuttle crew before they leave for the pad.

Launch Dates Set For The Remaining Shuttle Missions

On July 7, 2008, NASA announced an integrated assessment of the remaining Shuttle flights prior to ending the program at the close of Fiscal Year 2010 on September 30.[48]

One of the two remaining missions for 2008 was to be Atlantis's launch on October 8 as STS-125, the final HST service mission (designated SM-4). The other, scheduled for November 10, was to be Endeavour's STS-126 ULF2 mission to supply the station and service the Solar Alpha Rotary Joints on the port and starboard ends of the ITS.

As a precaution, NASA warned that this date was provisional, being "based on processing and other launch vehicle schedules." Nevertheless, NASA was committed to finishing the assembly of the ISS prior to retiring the Shuttle. Follow-on work would have to be assigned to new launch vehicles and spacecraft.

The remaining flights, excluding STS-125, were listed as follows:

2009 schedule

- Launch planned for February 12 (Discovery STS-119) ISS/5A. Delivery of the final pair of arrays on the starboard end of the ITS. The fourteen day flight was to include four EVAs and see Wakata replacing Magus on the resident crew.
- Launch planned for May 15 (Endeavour STS-127) 2J/A. Deliver the Japanese Kibo Exposed Facility and Experiment Logistics Module Exposed Section. It was to last fifteen days, and deliver six new batteries for the P6 truss, a spare drive for the MT, and a spare boom assembly for the K_u-band antenna. An ISS assembly flight record of five spacewalks would be attempted. Kopra would replace Wakata on the resident crew.
- Launch on July 30 (Atlantis STS-128) 17A. This eleven day logistics flight would carry an MPLM, perform three EVAs, and see Stott replace Kopra on the station.
- Launch on October 15 (Discovery STS-129) ULF3. This mission would last fifteen days and undertake "at least three EVAs to install two large external carriers." It would return with Canadian Robert Thirsk, thereby finishing the Shuttle's involvement in delivering or returning ISS resident crewmembers.
- Launch on December 10 (Endeavour STS-130) 20A. An eleven day mission that would deliver Node 3 and the Cupola with three EVAs.

2010 schedule

- Due for launch on February 11 (Atlantis STS-131) 19A. Manifested for eleven days, this flight would carry logistics in an MPLM and undertake three EVAs.
- Due for launch on April 8 (Discovery STS-132) ULF4. The main cargo would be the second Russian Mini Research Module ('Rassvet') with miscellaneous apparatus attached to its exterior, including at the time the European Robotic Arm which was then planned to be installed on the Multi-Purpose Laboratory Module ('Nauka') after that was launched by Russia.
- Due for launch on May 31 (Endeavour STS-133) ULF5. Lasting fifteen days, this 35th and final Shuttle-ISS flight would deliver miscellaneous spares and conduct three EVAs.

On July 16, 2008, NASA named the crew for the STS-128 (17A) mission that was to launch on July 30, 2009, with an MPLM of science and storage racks for the ISS: Rick Sturckow (CDR), Kevin A. Ford (PLT) and Mission Specialists John D. Olivas, Patrick G. Forrester, Jose M. Hernandez and Swedish ESA astronaut Christen Fuglesang.[49] The mission would also deliver station crewmember Nicole Stott and retrieve Tim Kopra.

The next announcement, on September 30, revealed the assignments to the STS-129 (ULF3) flight on which Discovery would deliver two experiment racks, two gyroscopes and a cargo of other spares.[50] Charles O. Hobaugh (CDR) and Barry E. Wilmore (PLT) were to fly with Mission Specialists Robert L. Satcher Jr., Michael Foreman, Randolph

('Randy') J. Bresnik and Leland D. Melvin. By retrieving Canadian Robert Thirsk, who was to serve on the ISS-20 and ISS-21 crews, this would end the Shuttle's involvement in carrying station personnel begun in 2001.

November 14–30, 2008: STS-126 (ULF2) Endeavour completed a logistics and outfitting mission to permit six-person ISS crewing from 2009. Sandra Magnus took over resident crew status on ISS-18 from Greg Chamitoff, with the latter returning home at the end of the flight.

On November 21, while Endeavour was docked at the ISS, NASA announced the joint planning for ISS crewmembers through to 2010, together with the identification scheme that would be applied with the six-person crewing.[51] The new scheme was to start with the docking of the Soyuz in May 2009 that would start full crewing. "From that point, Expeditions will end with the undocking of a Soyuz [and the] numbers will change every two to four months, as new crewmembers arrive and depart." Each six-person crew would live and work as a cohesive team, but with members of that team changing regularly. A typical period of residency by a specific crewmember would be about six months and would span two Expeditions. Once a given Soyuz departed, the command of the station would transfer to the new Expedition.

One part of the release highlighted a significant change in the operation of the ISS. "In addition to the Russian Soyuz, the Space Shuttle will continue to provide transport for station crewmembers through STS-129, targeted for the fall of 2009." Placed in the middle of the release, this clearly signaled the end of ISS crewmember ferry flights by the Shuttle, a far cry from the promotion of the spacecraft earlier in the program as the alternative to Soyuz for crew rotations to and from the station. Its imminent retirement put pressure on the remaining manifest to ensure that the assembly of the ISS could be completed. The remaining Shuttle missions would also have to deliver as much food, consumables, equipment and spares as possible, as well as prevent a buildup of trash, unwanted material, redundant equipment and miscellaneous waste.

Crewing announcements made by this same press release observed that, for ISS-20, Tim Kopra would launch on STS-127 and return on STS-128, Canadian Robert Thirsk would launch by Soyuz but return on STS-129, and Nicole Stott would launch on STS-128 but land by Soyuz. The remaining twenty-one crewmembers who were in training for the ISS-20 through ISS-26 crews would all launch and return by Soyuz. Following the retirement of the Shuttle, the opportunity for American astronauts to launch on an American vehicle from an American launch site and land on American territory would be lost. Flights for *all* ISS crewmembers from every partner nation would be reliant on the Russian Soyuz system for the foreseeable future.

On December 5, NASA reported crew assignments for two of the remaining Shuttle flights, reflecting the urgency to deliver some of the final elements for the station that could only be launched on the Shuttle.[52] STS-130 (20A) would deliver Node 3 and the Cupola. It would be flown by George D. Zamka (CDR), Terry W. Virts Jr. (PLT) and Mission Specialists Robert ('Bob') L. Behnken, Nicholas J. M. Patrick, Kathryn P. Hire and Stephen K. Robinson. STS-131 (19A) would be flown by Alan G. Poindexter (CDR), James P. Dutton Jr. (PLT) and Mission Specialists Rick Mastracchio, Clayton Anderson, Dorothy Metcalf-Lindenburger, Stephanie Wilson and Naoko Yamazaki of JAXA. It was to deliver research and science apparatus, a new sleeping compartment, and further supplies. The years leading to the retirement of the Shuttle were certainly going to be busy.

The traditional walkout from the crew quarters to the astro-van. Here the STS-110 crew are on their way to the launch pad.

2009 And All-Change On The ISS

Early in the new year, further changes were made to the Shuttle manifest. The prospect of delays to the launch of future missions, particularly STS-129, would impact resident crew durations and the decision was made to amend how two future ISS crewmembers would return home. This was announced on March 3, 2009.[53] Nicole Stott and Robert Thirsk would now swap seats on the Shuttle and Soyuz for their ride home. Thirsk was to launch on a Soyuz in May, as part of the ISS-20 and ISS-21 crews, then land aboard that Soyuz instead of on Atlantis as part of the STS-129 crew. Stott was now to launch on STS-128 and return on STS-129 in November, instead of by Soyuz. In essence, this meant that any postponement of STS-129 would cause Thirsk's mission to be extended beyond the preferred six months duration.

March 15–28, 2009: STS-119 (15A) Discovery delivered the final truss segment S6. During the partial exchange of resident crewmembers Koichi Wakata replaced Sandra Magnus.

March 28, 2009: On the day that STS-119 landed, Soyuz TMA-14 docked at the ISS to deliver the ISS-19 and ISS-20 resident crew, joining Wakata. He was later replaced by Tim Kopra and finally by Nicole Stott. This expedition saw the resident crew of the ISS increased to six, formed of two overlapping three-person crews flying to and from the station on Soyuz vehicles. Henceforth, when three new crewmembers arrived they would serve as Flight Engineers in support of the current crew, prior to becoming the primary team on the next expedition with one of the trio as Commander. They in turn would be supported by the next three crewmembers, who would take over for the next expedition. This eliminated the chore of delivering resident crewmembers on board the few remaining Shuttle flights, thereby allowing those missions to focus on transferring hardware and supplies to the station before the fleet was retired. The additional science facilities (and extra crew) aboard the station would also greatly expand the scientific research of the ISS. Crews of six finally provided opportunities for representatives of the other partners (Canada, ESA and Japan) to serve as part of the resident crew but curtailed the availability of Soyuz seats for non-resident fee-paying visitors known as Space Flight Participants. These Soyuz-only crew rotations would begin towards the end of 2009.

NASA named the crew for STS-132 (ULF4) on May 14, 2009, as Ken Ham (CDR), Dominic A. ('Tony') Antonelli (PLT), and Mission Specialists Steve Bowen, Karen L. Nyberg, Garrett Reisman and Piers Sellers.[54] The primary payload would be Russia's Mini Research Module (MRM1).

May 29, 2009: Soyuz TMA-15 delivered Roman Y. Romanenko, Frank De Winne and Robert B. Thirsk. Upon joining the ISS-19 residents Gennady Padalka, Michael Barratt and Wakata they created the new six-person ISS-20 crew. For the first time there were representatives of all five leading agencies aboard the ISS as a residents: Padalka and Romanenko (Russia), Barratt (USA), Wakata (Japan), Thirsk (Canada) and De Winne (Belgium, ESA).

July 15–31, 2009: STS-127 (2J/A) Endeavour installed the Japanese Experiment Module Exposed Facility (JEM-EF). It retrieved Wakata, whose place on the ISS-20 crew was taken by Timothy L. Kopra. *

It had been intended to retire the Shuttle with the STS-133 mission, but after several ISS components were canceled to allow the program to meet its termination date there was a strong argument to fly at least one further mission to install science instruments on the ISS, in particular an improved Alpha Magnetic Spectrometer (a state-of-the-art cosmic-ray detector to investigate fundamental issues about matter and the structure of the universe) and stores on an ExPRESS Logistics Carrier. In June 2008, the House of Representatives authorized funding to fly STS-134. There was initial opposition from the outgoing Administration of George W. Bush because this would divert funds from Constellation, its own program to develop vehicles to supersede the Shuttle, but in the spring of 2009 the new Administration of Barack H. Obama approved the mission.

On August 11, 2009, NASA named the crew for STS-134 (ULF6). Planned for May 2010, it was to install the Alpha Magnetic Spectrometer on the ITS of the station. The crew was Mark E. Kelly (CDR), Gregory H. Johnson (PLT) and Mission Specialists Michael

---

* The previous flight, STS-125 in early May, was the final Hubble servicing mission and was therefore independent of ISS operations.

Fincke, Greg Chamitoff, Andrew Feustel, and Italian ESA astronaut Roberto Vittori. It was also reported that Michael Good was to replace Karen Nyberg on the STS-132 crew. Nyberg had been grounded by a medical condition. She was assigned technical duties while recovering, and then in December 2010 was named to the prime crew for the ISS-36 and ISS-37 residency.

August 28–September 12, 2009: STS-128 (17A) Discovery carried an MPLM with supplies, stores, and equipment for the ISS. The final partial resident crew exchange saw Nicole Stott replace Tim Kopra.

CREW FOR THE FINAL SCHEDULED SHUTTLE MISSION

On September 18, 2009, NASA announced the crew for STS-133 (ULF5) as Steven W. Lindsey (CDR), Eric A. Boe (PLT), and four Mission Specialists: Benjamin A. Drew, Michael R. Barratt, Timothy Kopra and Nicole Stott. At that time Lindsey was serving as Chief of the Astronaut Office so in October 2009 he was replaced in that position by Peggy Whitson to enable him to begin his mission training.[55]

October 2–11, 2009: Soyuz TMA-16 docked at the ISS on October 2 with ISS-20 and ISS-21 crewmembers Jeffrey N. Williams and Maxim V. Surayev and the last scheduled Space Flight Participant, Canadian Guy Laliberté, who returned with the outgoing ISS-20 cosmonauts Padalka and Barrett in TMA-14 on October 11, thus leaving De Winne, Romanenko, Thirsk, Stott, Williams and Surayev as the ISS-21 crew.

November 16–27, 2009: STS-129 (ULF3) Atlantis delivered Express Logistics Carriers 1 and 2 and returned with Nicole Stott at the end of her residency on the ISS, leaving a five-person ISS-21 crew for a short time. Five days later the TMA-15 trio of Romanenko Thirsk and De Winne also left, leaving Williams and Surayev as the two-person core of the ISS-22 crew. Three other ISS-22 and ISS-23 crewmembers arrived on TMA-17 at the end of December, resuming five-person crewing through to the next crew exchange in April 2010. From that point, six-person resident crews continued to be the core of ISS expeditions.

2010 Towards Assembly Completion

On July 1, 2010, NASA updated its target dates for the final two missions prior to the scheduled retirement of the fleet, with Discovery launching as STS-133 on November 1, 2010 and Endeavour flying as STS-134 on February 26, 2011.[56] These target dates had altered because the critical payload manifested for STS-133 wouldn't be ready in time for the planned date of September 16, necessitating its move to November and a corresponding delay to STS-134.

February 8–21, 2010: STS-130 (20A) Endeavour delivered the Cupola module and Node 3 (called Tranquility).

April 5–20, 2010: STS-131 (19A) Discovery carried the MPLM Leonardo, which was making its seventh flight. This was the tenth and final round trip for an MPLM.

May 14–26, 2010: STS-132 (ULF4) Atlantis delivered the Mini Research Module (MRM1) called Rassvet ('Dawn') for the Russian segment of the ISS, and additional supplies and spares to stock-up the station.

Table 3.1 Shuttle-ISS Assembly Mission Crewing 1998–2011

STS	Mission	Commander	Pilot	MS-1	MS-2	MS-3	MS-4	MS-5
88	2A	Cabana	Sturckow	Ross	Currie	Newman	Krikalev[1]	-
96	2A.1	Rominger	Husband	Jernigan	Ochoa	Barry	Payette[2]	Tokarev[1]
101	2A.2a	Halsell	Horowitz	Weber	William J.	Voss J.S.	Helms	Usachev[1]
106	2A.2b	Wilcutt	Altman	Lu	Mastracchio	Burbank	Malenchenko[1]	Morukov[1]
92	3A	Duffy	Melroy	Chiao	McArthur W.	Wisoff	Lopez-Alegria	Wakata[5]
97	4A	Jett	Bloomfield	Tanner	Garneau[2]	Noriega	-	-
98	5A	Cockrell	Polansky	Curbeam	Ivins	Jones T.	-	-
102	5A.1	Wetherbee	Kelly J.	Thomas A.	Richards P.	*Voss J.S. [U]* / *Krikalev[1] [D]*	*Helms [U]* / *Shepherd W. [D]*	*Usachev[1] [U]* / *Gidzenko[1] [D]*
100	6A	Rominger	Ashby	Hadfield[2]	Phillips	Parazynski	Guidoni[3]	Lonchakov[1]
104	7A	Lindsey	Hobaugh	Gernhardt	Reilly	Kavandi	-	-
105	7A.1	Horowitz	Sturckow	Forrester	Barry	*Culbertson [U]* / *Voss J. S. [D]*	*Tyurin[1] [U]* / *Helms [D]*	*Dezhurov[1] [U]* / *Usachev[1] [D]*
108	UF1	Gorie	Kelly M.	Godwin	Tani	*Onufriyenko[1] [U]* / *Culbertson [D]*	*Bursch [U]* / *Tyurin[1] [D]*	*Walz [U]* / *Dezhurov[1] [D]*
110	8A	Bloomfield	Frick	Walheim	Ochoa	Morin	Ross	Smith S.
111	UF2	Cockrell	Lockhart	Chang-Diaz	Perrin[4]	*Korzun[1] [U]* / *Onufriyenko[1] [D]*	*Whitson [U]* / *Bursch [D]*	*Treschev[1] [U]* / *Walz [D]*
112	9A	Ashby	Melroy	Wolf	Magnus	Sellers	Yurchikhin[1]	-
113	11A	Wetherbee	Lockhart	Lopez-Alegria	Herrington	*Bowersox [U]* / *Treschev [D]*	*Budarin [U]* / *Korzun [D]*	*Pettit [U]* / *Whitson [D]*
114	LF1	Collins E.	Kelly J.	Noguchi[5]	Robinson	Thomas A.	Lawrence	Camarda
121	ULF1.1	Lindsey	Kelly M.	Fossum	Nowak	Wilson	Sellers	*Reiter[3] [U]*
115	12A	Jett	Ferguson	Tanner	Burbank	Stefanyshyn-Piper	Maclean[2]	-
116	12A.1	Polansky	Oefelein	Patrick	Curbeam	Fuglesang[3]	Higginbotham	*Williams S. [U]* / *Reiter[3] [D]*
117	13A	Sturckow	Archambault	Forrester	Swanson	Olivas	Reilly	*Anderson C. [U]* / *Williams S. [D]*
118	13A.1	Kelly S.	Hobaugh	Caldwell	Mastracchio	Williams D.[2]	Morgan	Drew

(continued)

Table 3.1 (continued)

STS	Mission	Commander	Pilot	MS-1	MS-2	MS-3	MS-4	MS-5
120	10A	Melroy	Zamka	Parazynski	Wilson	Wheelock	Nespoli[3]	*Tani [U]* / *Anderson C. [D]*
122	1E	Frick	Poindexter	Melvin	Walheim	Schlegel[3]	Love	*Eyharts[3][U]* / *Tani [D]*
123	1 J/A	Gorie	Johnson	Behnken	Foreman	Doi[5]	Linnehan	*Reisman [U]* / *Eyharts[3][D]*
124	1 J	Kelly M.	Ham	Nyberg	Garan	Fossum	Hoshide[5]	*Chamitoff [U]* / *Reisman [D]*
126	ULF2	Furguson	Boe	Pettit	Bowen	Stefanyshyn-Piper	Kimbrough	*Magnus [U]* / *Chamitoff [D]*
119	15A	Archambault	Antonelli	Acaba	Swanson	Arnold	Phillips	*Wakata[5][U]* / *Magnus [D]*
127	2 J/A	Polansky	Hurley	Wolf	Cassidy	Payette[2]	Marshburn	*Kopra [U]* / *Wakata[5] [D]*
128	17A	Sturckow	Ford	Forrester	Hernandez	Olivas[3]	Fuglesang[3]	*Stott [U]* / *Kopra [D]*
129	ULF3	Hobaugh	Wilmore	Melvin	Bresnik	Foreman	Satcher	*Stott [D]*
130	20A	Zamka	Virts	Hire	Robinson	Patrick	Behnken	-
131	19A	Poindexter	Dutton	Mastracchio	Metcalf-Lindenburger	Wilson	Yamazaki[5]	Anderson
132	ULF4	Ham	Antonelli	Reisman	Good	Bowen	Sellers	-
133	ULF5	Lindsey	Boe	Drew	Bowen	Barratt	Stott	-
134	ULF6	Kelly M.	Johnson G.H.	Fincke	Vittori[3]	Feustel	Chamitoff	-
135	ULF7	Furguson	Hurley	Magnus	Walheim	-	-	-

NOTES:

MS = Mission Specialist.

[1] RSA = Russian Space Agency; [2] CSA = Canadian Space Agency; [3] ESA = European Space Agency; [4] CNES = French Space Agency; [5] JAXA = Japanese Space Agency.

Names in *italic's* denote station resident crewmember; U = Up [Ascent only]; D = Down [Descent only].

Table 3.2 Space Station (ISS) Resident Crewmembers Who Launched And/Or Landed On The Space Shuttle [in order of first launch or landing]

ISS resident crewmember	Nationality	Agency	STS launch date	STS flight up	STS MS up	Station Expedition	STS flight down	STS MS down	STS landing date
Voss, James S.	American	NASA	2001 Mar 8	102	3	ISS-2	105	3	2001 Aug 22
Helms, Susan J.	American	NASA	2001 Mar 8	102	4	ISS-2	105	4	2001 Aug 22
Usachev, Yuri V.	Russian	RSA	2001 Mar 8	102	5	ISS-2	105	5	2001 Aug 22
Krikalev, Sergei K.	Russian	RSA	N/A[1]	N/A	N/A	ISS-1	102	3	2001 Mar 21
Shepherd, William M.	American	NASA	N/A[1]	N/A	N/A	ISS-1	102	4	2001 Mar 21
Gidzenko, Yuri P.	Russian	RSA	N/A[1]	N/A	N/A	ISS-1	102	5	2001 Mar 21
Culbertson, Frank L.	American	NASA	2001 Aug 10	105	3	ISS-3	108	3	2001 Dec 17
Tyurin, Mikhail V.	Russian	RSA	2001 Aug 10	105	4	ISS-3	108	4	2001 Dec 17
Dezhurov, Vladimir N.	Russian	RSA	2001 Aug 10	105	5	ISS-3	108	5	2001 Dec 17
Onufriyenko, Yuri I.	Russian	RSA	2001 Dec 5	108	3	ISS-4	111	3	2002 Jun 19
Bursch, Daniel W.	American	NASA	2001 Dec 5	108	4	ISS-4	111	4	2002 Jun 19
Walz, Carl E.	American	NASA	2001 Dec 5	108	5	ISS-4	111	5	2002 Jun 19
Korzun, Valery N.	Russian	RSA	2002 Jun 5	111	3	ISS-5	113	4	2002 Dec 7
Whitson, Peggy A.	American	NASA	2002 Jun 5	111	4	ISS-5	113	5	2002 Dec 7
Treschev, Sergei V.	Russian	RSA	2002 Jun 5	111	5	ISS-5	113	3	2002 Dec 7
Bowersox, Kenneth D.	American	NASA	2002 Nov 23	113	3	ISS-6	N/A	N/A	N/A[2]
Budarin, Nikolai M.	Russian	RSA	2002 Nov 23	113	4	ISS-6	N/A	N/A	N/A[2]
Pettit, Donald R.	American	NASA	2002 Nov 23	113	5	ISS-6	N/A	N/A	N/A[2]
Reiter, Thomas	German	ESA	2006 Jul 4	121	5	ISS-13/14	116	5	2006 Dec 22
Williams, Sunita L.	American	NASA	2006 Dec 9	116	5	ISS-14/15	117	5	2007 Jun 22
Anderson, Clayton C.	American	NASA	2007 Jun 8	117	5	ISS-15/16	120	5	2007 Nov 7
Tani, Daniel M.	American	NASA	2007 Oct 23	120	5	ISS-16	122	5	2008 Feb 20
Eyharts, Leopold	French	ESA	2008 Feb 7	122	5	ISS-16	123	5	2008 Mar 26
Reisman, Garrett E.	American	NASA	2008 Mar 11	123	5	ISS-16/17	124	5	2008 Jun 14
Chamitoff, Gregory E.	American	NASA	2008 May 31	124	5	ISS-17/18	126	5	2008 Nov 30
Magnus, Sandra H.	American	NASA	2008 Nov 14	126	5	ISS-18	119	5	2009 Mar 28
Wakata, Koichi	Japanese	JAXA	2009 Mar 15	119	5	ISS-18/19/20	127	5	2009 Jul 31

(continued)

Table 3.2 (continued)

ISS resident crewmember	Nationality	Agency	STS launch date	STS flight up	STS MS up	Station Expedition	STS flight down	STS MS down	STS landing date
Kopra. Timothy L.	American	NASA	2009 Jul 15	127	5	ISS-20	128	5	2009 Sep 12
Stott. Nicole M.P.	American	NASA	2009 Aug 28	128	5	ISS-20/21	129	5	2009 Nov 27

NOTES.

[1] ISS-1 resident crewmembers Krikalev Shepherd, and Gidzenko were launched on October 31, 2000 onboard Soyuz TM-31.

[2] ISS-6 resident crewmembers Bowersox, Budarin and Pettit landed on March 5, 2003 onboard Soyuz TMA-1, a result of grounding the Shuttle fleet following the loss of Columbia on February 1, 2003.

Launch-On-Need (LON) Crew Named

On September 14, 2010, NASA released the names of four astronauts assigned to the flight designated STS-335.[57] This was a rescue flight that would be launched only if STS-134, the last planned Shuttle flight, was unable to return home. This decision was based on recommendations made following the loss of Columbia in 2003, after which crew training included a Launch-on-Need crew, ready to launch if irreparable damage prevented the crew of an Orbiter either from returning safely to Earth or reaching the 'safe haven' of the ISS. With STS-134's launch set for February 26, 2011, the rescue flight would be available from that date until the conclusion of the mission. There was also the possibility that another mission could be added to the manifest, with STS-335 being redesignated STS-135 and scheduled for launch in June. The crew for the LON mission were Christopher J. Ferguson (CDR), Douglas ('Doug') G. Hurley (PLT) and two Mission Specialists: Sandra H. Magnus and Rex J. Walheim. Their training would include the possibility of going on to fly an extra mission to the ISS.

As William ('Bill') H. Gerstenmaier, Associate Administrator for NASA's Space Operations Mission Directorate in Washington DC, explained, "These astronauts will start training immediately as a rescue crew as well as in the baseline requirements that would be needed to fly an additional Shuttle flight. The normal training template for a Shuttle crew is about one year…so we need to start training now in order to maintain the flexibility of flying a rescue if needed, or alter course and fly an additional Shuttle mission if that decision is made."

STS-135 Added To The Manifest

The addition of STS-135 into the Shuttle manifest had an interesting background. The requirement to provide a rescue mission for STS-134 meant that there was a complete stack of Shuttle hardware certified to fly and authorized by the agency's management but funded *only* to undertake a rescue. The formal agreement to include the mission in the manifest came on October 11, 2010 as part of the space agency's Fiscal Year 2011 authorization, but the funding would not come until the next Bill. The main contractor for the ground support of Shuttle operations, United Space Alliance, had agreed a six month extension to the contract into 2011 to continue to support flight operations for both STS-134 and a possible STS-135 mission.

In November 2010, NASA Administrator Charles F. Bolden noted that likely delays in the development of commercial space vehicles implied a requirement to fly STS-135. These commercial vehicles were intended to deliver cargo to the ISS but they were not yet available, and with limitations to the number of European ATV and Japanese HTV automated resupply freighters, it was imperative to stock up the ISS before the Shuttle retired. The Russian Progress ships would still continue to supply the station, but their cargo capacity was limited and they did not provide the return-to-Earth capacity of the new spacecraft being developed by SpaceX. By January 2011 there was optimism that STS-135 would fly to further stock the station but no certainty. Regardless of funding, processing towards the mission had to get underway, so as to be ready to proceed with the flight if the formal agreement was forthcoming.

On January 20, Shuttle management altered the designation on the manifest from STS-335 LON to STS-135, allowing the training documentation and other mission-specific material and procedures to begin, whilst retaining the option to rescue STS-134. On February 13 the workforce were told the Shuttle would fly regardless of the funding issue. There was a growing interest in the mission and an expectation that it would indeed launch, but the uncertainty continued for another two months. The US federal budget approved in April included $5.5 billion for Shuttle and ISS operations through to September 30, 2011. This confirmed that the STS-135 mission would fly.

2011: THE FINAL COUNTDOWN(S)

With the retirement of the Shuttle in the summer of 2011 confirmed, a hectic final few months of vehicle processing, crew training and mission preparations continued for the final three flights, starting with STS-133 in February, STS-134 in April, and the newly manifested STS-135 in July. This last spurt of activity by a dedicated workforce would see not only the end of ISS assembly and resupply using the Shuttle, but the conclusion of the thirty-year Shuttle program itself.

A tragic event occurred just over a week into the new year that would influence so much more than flying the final missions. On January 8, Congresswoman Gabrielle D. ('Gabby') Gifford, the wife of STS-134 CDR Mark Kelly, was critically wounded after being shot in Tucson, Arizona. In the wake of the incident, and at a very difficult time for his family, Kelly recommended to NASA management that steps be taken to fulfill mission preparations in his absence. Kelly remained hopeful that he would still be able to command the mission. As Chief of the Astronaut Office, Peggy Whitson, observed, "[Mark] is facing many uncertainties now as he supports Gabrielle, and our goal is to allow him to keep his undistracted attention on his family, while allowing preparation for the mission to progress."

On January 13, NASA reported that Rick Sturckow, currently Deputy Chief of the Astronaut Office, would serve as back-up CDR for STS-134 "to facilitate continued training for the crew and support teams" during the absence of Mark Kelly.[58] Peggy Whitson explained, "Designating a back-up allows the crew and the support team to continue training, and enables Mark to focus upon his wife's care." Sturckow began training with the other members of the crew the next week. Kelly started his personal leave of absence on January 8. Gifford's condition stabilized and her prospects were encouraging, but she would have a long road to recovery.

On February 4, NASA announced that Kelly, who had managed to continue some basic proficiency training during his absence, would resume training for STS-134 on February 7, when Sturckow would resume his role as CB Deputy Chief.[59] Despite the obvious concerns for his wife and the stress that the situation had placed on Kelly, he was grateful for the confidence that his astronaut colleagues and NASA management had shown in him. Whitson assured him that everyone was glad to have him back and that, as a veteran Shuttle Commander, the space agency was confident in his ability to lead a successful mission.

In the meantime, there had been a further change to one of the other Shuttle crews. On January 19 it had been announced that Steven Bowen would replace Tim Kopra on STS-133.[60] Kopra had been injured in a bicycle accident the previous weekend. Whilst his injuries were not serious, they were sufficient to prevent him flying the mission the following month. Had the launch for STS-133 slipped to later in the year, Kopra might have been able to re-join the crew, but this did not happen. As a result, Bowen became the only astronaut ever to fly back to back Shuttle missions. His recent experience on STS-132 ensured a smooth integration into the STS-133 crew and their training. There were minor adjustments to duties assigned to crewmembers, but nothing to impact the mission objectives or launch window.

As the astro-van arrives at LC-39, ahead lies the Shuttle that will take the crew into orbit.

Strapped in on the mid-deck and ready to launch.

February 24–March 9, 2011: STS-133 (ULF5) This logistics and resupply mission was the final flight of Discovery. It delivered the refurbished Leonardo MPLM which was berthed at the Earth-facing port of Unity as the Permanent Multipurpose Module (PMM), essentially serving the role of a rather large storage cupboard.

May 16–June 1, 2011: STS-134 (ULF6) This final flight of Endeavour delivered the Alpha Magnetic Spectrometer-2 (AMS-02) and ELC-3. The mission also saw the final EVA of an impressive twenty-eight year program by Shuttle crewmembers. The 50 ft (15.24 m) Orbiter Boom Sensor System (OBSS) was permanently transferred across to the starboard truss of the ISS as an inspection aid and to provide further reach to the station's robotic arm. Italian ESA astronaut Robert Vittori became the final European to fly on the Shuttle, ending a series that started in 1983 when German ESA astronaut Ulf Merbold flew on STS-9 Spacelab 1.

July 8–21 2011: STS-135 (ULF7) The final flight of the Shuttle era came just over thirty years after the first, with Atlantis delivering additional supplies to stock-up the ISS.

The End Of An Era

The first Shuttle crews had been assigned to the Approach and Landing Test flights in 1976. The first assignments for the Orbital Flight Tests were announced in 1978. Now after thirty-five years, the final Shuttle crews had flown. This ended an era of *at least* 150 proposed Shuttle mission manifests, of which only 135 had made it off the launch pad.

BY THE NUMBERS

The statistics for crewing and flight assignments involving space stations were (in part) as follows.

Shuttle-ISS

The flight opportunities available on the thirty-seven Shuttle missions to the ISS totaled 246, and these were made up of 216 Orbiter crew seats and thirty ISS resident seats. Of the residents, twenty-seven flew both ascent and return aboard Shuttles. The remaining three seats were occupied for ascent on STS-113 by the ISS-6 crew of Ken Bowersox, Don Pettit and Nikolai Budarin (who returned to Earth on Soyuz TMA-1 after the loss of Columbia) and for descent on STS-102 by the ISS-1 crew of Bill Shepherd, Sergei Krikalev and Yuri Gidzenko (who launched to the ISS on Soyuz TM-31). In terms of crewmembers there were eighteen NASA astronaut allocations with fifteen astronauts making the return trip (Shepherd completing a descent only and Bowersox and Pettit just the ascent). Added to this were nine Russians, with six making the trip both ways and three flying either an ascent (Budarin) or decent (Gidzenko and Krikalev). There were also two-way journeys on Shuttle flights for a French, a German and a Japanese ISS resident. Individual residents were seventeen American (five female), ten Russian (all male) and one each (all male) from France, Germany and Japan.

Summarizing the Orbiter crew statistics, 144 individuals (twenty-four female) flew missions to the ISS. Of these 121 were American (twenty-two female), seven Russian (all male), five from Canada (one female), five from Japan (one female), three from Italy (all male) and one each (all male) from France, Germany and Sweden. Of those who made Shuttle flights, no fewer than thirty (five female) would go on to serve in residencies, while nine (one female) later flew on a Shuttle mission after a period of residency or visiting crew mission.

Regarding the Shuttle crewing of the 144 individuals who made one Orbiter flight, fifty-seven went on to fly two Shuttle missions to the ISS, thirteen made three visits, and one (Sturckow) made four visits. Finally, there were twenty members of Shuttle-Mir crews (five female) who also made Shuttle-ISS flights, with all except one of the males (Canadian Hadfield) being American.

This was a remarkable accumulation of experience and breadth of training, with many people either moving on to senior roles within their space agencies, or retiring from the program. The record of achievement, the data archived from these missions, and the personal achievements are written in the history books. Many of these people have remained in the aerospace industry in order to pass on their skills. Others make presentations and deliver speeches in order to share their knowledge and experiences from the Shuttle, Mir and the ISS with audiences across the globe. This sharing is an important part of the legacy of the Space Shuttle.

Notes

1. NASA News Release 96-169
2. NASA News (JSC) 97-117, June 2, 1997
3. NASA News (JSC) 97-126 June 9, 1997
4. NASA News (JSC) 97-142, June 24, 1997
5. NASA News (JSC) 97-269, November 17, 1997
6. NASA News (Headquarters) H98-21
7. NASA News H98-137 (Headquarters) July 30, 1998
8. NASA News H98-143 (Headquarters) August 4, 1998
9. ESA Press Release 7-1999
10. Release 99-19, February 12, 1999
11. *Enhancing Hubble's Vision* pp. 58–60
12. Release 00-29, February 18, 2000
13. Release H00-78, May 9, 2000
14. *Florida Today*, September 8, 1999 p. 1A
15. Release H00-154, September 28, 2000
16. Release J00-82
17. Release H01-45a, March 26, 2001
18. Release 01-70, April 11, 2001
19. Release 01-135, July 3, 2001
20. Release H01-167, August 17, 2001
21. Release H02-35

22. Release H02-139, July 26, 2002
23. Release H02-155, August 15, 2002
24. Release H02-249
25. Release H02-250
26. Release H03-360
27. Release H03-385
28. Release 04-378, November 18, 2004
29. ESA News, 23-2005
30. Release 05-040, February 9, 2005
31. Release H04-224, July 15, 2004
32. Release J06-053, May 2, 2006
33. Release J06-054, May 5, 2006
34. Release J06-059, May 17, 2006
35. Release J06-070, 19 June, 2006
36. Release J06-078, 20 July, 2006
37. Release H06-337, October 18, 2006
38. Release H07-016, January 29, 2007
39. Release H07-039, February 13, 2007
40. Release H07-073, March 22, 2007
41. Release H07-093, April 26, 2007
42. Release H07-099, May 3, 2007
43. Release H07-217, October 1, 2007
44. Release H07-229, October 19, 2007
45. Release M07-259, November 21, 2007
46. Release H08-005, January 11, 2008
47. Release H08-052, February 12, 2008
48. Release H08-167, July 7, 2008
49. Release H08-176, July 16, 2008
50. Release H08-250, September 30, 2008
51. Release H08-306, November 21, 2008
52. Release H08-321, December 5, 2008
53. Release M09-34c, March 3, 2009
54. Release H09-105, May 14, 2009
55. Release H09-218
56. Release H10-157, July 1, 2010
57. Reference 10-222, September 14, 2010
58. Release 11-015, January 13, 2011
59. Release 11-036, February 4, 2011
60. Release 11-023, January 19, 2011

4

Putting It All Together

STS-135 Closeout Crew's Farewell
Friday, 8 July 2011 03:20:04 PM GMT
Before leaving the White Room for what might be
the final time, each member of the Closeout Crew
held up signs that bore the following message:

*On behalf of all who have designed and built,
serviced and loaded, launched and controlled,
operated and flown these magnificent space vehicles,
thank-you for 30 years with our nation's Space Shuttles!
Godspeed Atlantis! God bless America!*

And after some final words from
Closeout Crew Chief Travis Thompson,
the team departed the pad.[*]

As astronauts were preparing to fly a mission, there was feverish activity underway at the Kennedy Space Center in Florida to ensure that all the hardware components were checked and ready to go. Teams of dedicated engineers worked hard on a complicated processing system designed to bring together the Orbiter, twin Solid Rocket Boosters and External Tank on schedule. In parallel the payload, logistics, experiments, and the personal equipment and supplies of the astronauts, underwent separate and joint tests and checks. In addition, there was the training of the flight control teams and support staff.

It can be surprising to learn about the paths to orbit that a space explorer follows from the classroom to the launch pad. Their journeys are many and varied. For those with prior experience of working at KSC earlier in their careers, their first link to the ISS could occur many years before they were chosen for astronaut training.

---

[*]http://www.nasa.gov/mission_pages/Shuttle/Shuttlemissions/sts135/launch/launch_blog.html; also http://www.youtube.com/watch?v=ldlphfRuk1Q ("Shuttle Closeout Crew Says Goodbye").

© Springer International Publishing Switzerland 2017
D.J. Shayler, *Assembling and Supplying the ISS*, Springer Praxis Books,
DOI 10.1007/978-3-319-40443-1_4

ESTABLISHING AN EFFICIENT TIMELINE

A decade prior to being selected as a NASA astronaut in 1994, Janet Kavandi was an employee of the Boeing Aerospace Company in Seattle, Washington. While there she worked on power systems for space and defense contracts, including for Space Station Freedom. As she says, "We worked on power systems for the space station, the actual energy storage devices, batteries and solar arrays, and power system for the station, the power for any experiments… Those kind of things."[1] Starting in 1996, after ASCAN training, Kavandi received a technical role in the payloads and habitation branch of the Astronaut Office. This included integrating the payloads for the ISS. "I was primarily responsible for doing the crew evaluation of the payload. As the payloads came in, the flight documentation came to us. We verified that the procedures were correct, that the payloads matched the labeling on the payload, and that by following the described procedures the operations on the payloads followed exactly what was described. For example, sometimes the procedures were accurate, the drawings were accurate, but the labeling wasn't quite accurate. So we'd go through that and do the final run through so that, when it reached orbit, everything would work properly and there wouldn't be any hiccups. That was the intention. We also checked out the prescribed tools for using the hardware to make sure that the interfaces were accurate. There were lots of issues with the checklists to make sure that everything was ready to go for the crew."

This was all prior to signing off the flight data file by which the crew executed their procedures and timeline for each of the assigned payloads. Kavandi explained that her work during this time was with the documentation. "It was the version before the final version. We helped determine the timeline by actually running through the procedures to find out how much time it actually took to do it. We tried to save time and make the procedure more efficient by eliminating unnecessary steps or perhaps doing things in a different order. This could simply be fixing the labels of things that could be confusing or it could be trying to use fewer tools, such as using one tool instead of five. When we could, we tried to make those inputs. It just made operations a lot more efficient."

This is only one example of how the members of the Astronaut Office can become involved in the process of preparing for programs, missions or activities many months, even years, prior to hardware or procedures being assigned to a flight. Even if an item of hardware or detailed timeline doesn't make it to flight, it mustn't be seen as wasted time and effort. On the contrary this type of work continues to contribute to an overall appreciation of how to prepare items of hardware, develop an experiment, or sequence the steps in a procedure to make the actual mission proceed efficiently, and above all safely. It also highlights the fact that the principal role of an astronaut (if not actually flying a mission) is to test, re-test and test again. There are hundreds of hours spent on the ground performing such technical support tasks long before an astronaut receives a flight assignment, and the tasks are generally run again in between missions to further enhance efficiency in space.

MISSION MANAGEMENT TEAM

Another major participant in the preparation and support of a Shuttle mission was the Mission Management Team. As recalled by former astronaut and Deputy Director of Flight Crew Operations at JSC, Steve Hawley, their responsibilities were not directly related to planning for future flights.[2] "The way the Shuttle Program was organized, was there was the Program, run by the Program Manager. The Program consisted of many Projects. Some were 'real' projects like the SRB Project or the Orbiter Project, and there were other functions that were treated like projects. One of those was Flight Crew Operations. Our project responsibilities were to provide crews ready to fly the mission (in collaboration with Mission Operations), provide specific support personnel (Capcoms, ASPs, TALCOMS, etc.), and family support. And we were also responsible for the aircraft, including readiness to ferry the Orbiter back to the Cape should that be necessary. The projects were often referred to as Level III. The Program was Level II, and headquarters was Level I. We would conduct a Level III Flight Readiness Review and report our readiness for flight to the Program and to the Level I Flight Readiness Review.

"The way the Program was run, there would be a daily meeting of the project reps with the program. That would normally be for dealing with issues that had come up in preparation for the next couple of missions, or where some anomaly required prompt decisions. There was also a weekly Program Requirements Control Board that dealt with higher level issues, including establishing the manifest and planning for future missions. As its name implies, the PRCB was responsible for managing the Program requirements. I normally represented Flight Crew Operations at these meetings, and sometimes Jim Wetherbee would represent FCOD when he was Director. As I said, I was accountable for the things for which FCOD was responsible. I also had a vote as a member of the team on any decision that had to be made.

"The Mission Management Team consisted of the same players and was convened specifically to oversee flight activities. We would initially meet a day or two prior to launch to deal with issues that might have come up during the countdown. We would also be a part of the final count, representing the Program as necessary, should issues arise. The MMT would give the Go for launch; I had a vote. We would then normally meet once per day during the mission to oversee the execution. Real-time issues were the responsibility of Mission Control and the Flight Director, as was the overall conduct of the mission within the scope of the flight rules. We would approve any changes to flight rules, decide which landing site would be prime and whether to activate a back-up site, authorize changes to the mission plan including extending or shortening the flight, put together any teams needed to investigate anomalies and things like that – anything that would be outside of the original plan. All of the activities I have described worked throughout the Shuttle Program, whether we were doing Shuttle-Mir, HST servicing or ISS assembly.

"The NASA Engineering and Safety Council (NESC) was established after the Columbia accident to provide an independent technical capability in support of HQ Safety, Reliability & Quality Assurance (SR&QA). It was staffed with engineers on loan from the various Centers with expertise in subjects such as materials, software, propulsion, mechanisms, etc. I felt that they should have someone with operations experience, so I suggested this to Ralph Roe, who was the first Director of NESC. I either volunteered or he asked me to join the NESC. I carried that out as a collateral assignment. We'd be tasked by HQ to look into various technical issues. I think we also could decide on our own to investigate something. The idea was to give HQ an expert technical opinion independent of what the Program was thinking. We also had the capability to contract with discipline experts in industry or academia."

PREPARATIONS AT THE CAPE

As well as the administrative and planning organization of the program, there came a time when actual hardware had to be prepared and brought together to fly the mission. The processing of hardware intended to assemble and supply the ISS focused mostly upon existing facilities at the Kennedy Space Center in Florida, supported by various other NASA field centers across the country.

KSC Launch Directors

Directing launch operations for the Shuttle-ISS assembly flights was the responsibility of three men over the thirteen-year period of that task. David ('Dave') A. King served as Launch Director at the end of the Shuttle-Mir program and the start of Shuttle-ISS (1998–2000) with missions STS-89 and -91 to Mir, then STS-101 and -106 to the ISS. In parallel, Ralph R. Roe Jr. (1998–1999) dealt with STS-88 and -96. From STS-92 in 2000 through to the retirement of the Shuttle with STS-135 eleven years later, the role was handled by Michael ('Mike') D. Leinbach.

Shuttle-ISS Mission Processing

Fully detailing the long and convoluted program of ground processing throughout this schedule is beyond the scope of this current work, but the components of each Shuttle stack are presented in Table 4.1. Details of the ground facilities at the Cape for Shuttle-Mir were described in *Linking the Space Shuttle and Space Station: Early Docking Technologies from Concept to Implementation.* They were changed very little for the ISS program, but ground preparations for the Shuttle were rarely straightforward and the program was plagued by difficulties throughout its operational lifetime. Although many of these problems were easily rectified, others were more complicated and had the capability to temporarily ground the fleet.

Some of the issues which affected the smooth flow of Shuttle processing during the ISS assembly period included:

- 1999–2000: The delayed launch by Russia of the Zvezda Service Module; electrical wiring and fuel line problems on the Shuttle fleet; the non-ISS flights of STS-93, -103, and -99 delayed and revised the launch manifests for the ISS missions.

Table 4.1 Orbiter Ground Processing Summary

STS	OV	Total OPF Work Days	Total VAB Work Days	PAD LC/39	Total Pad Work Days	Total Flow Work Days	MLP
Ground Processing For Shuttle ISS Assembly Missions							
88	105	187	5	Pad A	37	229	3
96	103	122	12	Pad B	30	164	2
101	104	333	8	Pad A	50	391	1
106	104	66	5	Pad B	22	93	2
92	103	197	10	Pad A	21	238	3
97	105	203	5	Pad B	26	234	1
98	104	70	30	Pad A	28	128	2
102	103	84	8	Pad B	24	113	3
100	105	82	3	Pad A	23	110	1
104	104	82	11	Pad B	21	114	2
105	103	79	8	Pad A	31	118	3
108	105	142	6	Pad B	34	182	1
110	104	132	6	Pad B	28	166	3
111	105	92	7	Pad A	33	132	1
112	104	106	6	Pad B	25	139	3
113	105	79	9	Pad A	35	123	2
114	103	994	25	Pad B	85	1,104	3
STS-114 extended time in OPF & on pad due to post Columbia (STS-107) recovery & Return-to-Flight efforts							
121	103	264	7	Pad B	41	312	1
115	104	264	7	Pad B	41	312	2
116	103	105	8	Pad B	28	141	1
117	104	125	80	Pad A	42	247	2
STS-117 Includes rollback to VAB (+72 days) and return to pad (+25 days)							
118	105	1477	9	Pad A	25	1,511	1
STS-118 OPF processing (1332+64+63+18 days) occurred over a total time period of 1665 days							
120	103	234	7	Pad A	23	264	2
STS-120 OPF processing occurred over a total time period of 273 days							
122	104	121	7	Pad A	76	204	1
STS-122 add 1 day holiday @ OPF add 10 holiday & 4 contingency days @ pad							
123	105	159	7	Pad A	23	189	3
STS-123 add 14 days holiday @ OPF							
124	103	157	7	Pad A	29	193	3
STS-123 add 13 days holiday @ OPF							
126	105	162	7	Pad B	19	206	3
		-	-	Pad A	18		
STS-126 add 3 holidays and 3 weathers days @ OPF; 1 weather day @ VAB; 15 contingency days @ Pad B and 5 contingency days @ Pad A							
119	103	191	6	Pad A	47	244	1

(continued)

Table 4.1 (continued)

STS	OV	Total OPF Work Days	Total VAB Work Days	PAD LC/39	Total Pad Work Days	Total Flow Work Days	MLP
STS-119 add 13 holiday and 3 weather days @ OPF;							
14 contingency days @ Pad							
127	105	109	7	Pad B	32	190	3
		-	-	Pad A	42		
STS-127 processing @ OPF occurred over a total time period of 118 days; add 9 holiday days							
@ OPF;10 contingency,1 scrub & 1 crew rest day @ Pad B (supporting STS-125); add 3							
contingency & 1 holiday day @ Pad A							
128	103	117	9	Pad A	25	151	2
STS-128 processing @ OPF occurred over a period of 119 days							
(117 & 2 holiday days)							
129	104	113	7	Pad A	32	152	3
STS-129 processing @OPF occurred over a total time period of 124 days; add @ OPF 10 non							
work days and 1 holiday day; add 1 contingency day @ VAB; add 2 contingency days @ Pad							
130	105	130	9	Pad A	31	170	2
STS-130 processing @ OPF occurred over a total time period of 133 days; add 3 holiday days							
@ OPF; add 5 contingency and 11 holiday days @ VAB; add 3 contingency days @ Pad							
131	103	142	9	Pad A	32	183	3
STS-131 processing @ OPF occurred over a total time period of 153 days; add 11 holidays @							
OPF; add 2 contingency days at Pad							
132	104	127	7	Pad A	22	156	2
STS-132 processing @OPF occurred over a total time period of 136 days; add 9 holiday days @							
OPF; add 2 weather days @ VAB;							
add 1 contingency day @ Pad							
133	103	138	10	Pad A	82	284	3
		-	35	Pad A	19		
STS-133 processing @ OPF occurred over a total time period of 141 days; add 3 holiday days;							
during Flow Cycle A add 2 days contingency @ VAB; add 8 days contingency & 2 holiday days							
@ Pad; during Flow Cycle B add 5 holiday days @ VAB and 5 contingency days at Pad							
134	105	263	9	Pad A	53	325	2
STS-134 processing @ OPF occurred over a total time period of 371 days; add 89 non work							
days,17 holiday days and 2 safety days @ OPF; add 1 contingency and 1 weather day @ VAB;							
add 14 contingency days @ Pad							
135	104	242	9	Pad A	35	286	3
STS-135 processing @ OPF occurred over a total period of 355 days; add 96 non work and 17							
holiday days @ OPF; add 4 contingency and 1 holiday days @ VAB; add 2 contingency and 1							
holiday days @ Pad							

Key: OV Orbiter Vehicle (Shuttle)
OPF Orbiter Processing Facility (Bay)
VAB Vehicle Assembly Building
HB High Bay (within the VAB)
LC Launch Complex
MLP Mobile Launcher Platform

Reference: Space Shuttle Mission Summary, NASA/TM-20111-216142, compiled by Robert D. Legler and Floyd V. Bennett, Mission Operations, Johnson Space Center, Houston, Texas, September 2011

- May 8, 1999: Discovery (STS-96) was damaged by a hail storm and returned to the VAB on May 16; technicians evaluated the damage to the vehicle; the inspections revealed 648 divots in the outer foam of the ET; 189 were judged acceptable for flight, but blending or sanding work was required for 211 hits and 248 divots had to be patched with new foam; the vehicle was rolled back to the pad on May 20.
- January 2, 2001: Atlantis (STS-98) left the VAB heading for Pad 39A but was halted on the crawler way an hour later when a computer processor failed on Crawler Transporter #1; the stack was returned to the VAB later the same day.
- June 20, 2001: Atlantis (STS-105) was rolled to Pad 39B; then rolled back to the VAB as a precaution against the possibility of lightning in the vicinity of the pad.
- 2001–2003: The SM-3B Hubble servicing and STS-107 Spacehab research missions, both of which were to be flown by Columbia, had to be taken into account when planning the ISS assembly flights during this period.
- 2003–2006: The loss of Columbia and the protracted Return-To-Flight had to be factored into the ISS assembly, together with the plan to retire the fleet in 2010.
- February 26, 2007: The thermal protection on both the Orbiter Atlantis and its ET (STS-117) was damaged by a hail storm while it stood on Pad 39A.
- October 23, 2008: Following the slippage of STS-125 (HST SM-4) into 2009, the STS-126 (Endeavour) stack which was no longer needed for the potential Launch-on-Need rescue flight was relocated from Pad 39B to the now vacant Pad 39A.
- 2009: The final HST service mission and the Launch-on-Need requirements had to be taken into account in planning the final few Shuttle-ISS assembly missions.
- May 16, 1999 (STS-96 Discovery): Hail storm damage to the ET.
- January 2, 2001 (STS-98 Atlantis): A failed computer aboard the Crawler Transporter left the vehicle on the crawler way for an hour while engineers worked to resolve the issue; when this proved unsuccessful, the stack was returned to the VAB using a secondary computer processor.
- January 19, 2001 (STS-98 Atlantis): The second roll back for this mission stemmed from uncertainties over the integrity of SRB cables.
- May 26, 2005 (STS-114 Discovery): Roll back to install a new modified ET during the preparations for the Return-To-Flight mission.
- August 29, 2006 (STS-115 Atlantis): This was the only partial roll back of the series, initiated by the threat of Tropical Storm Ernesto. Throughout the day, weather updates revealed that the storm was weakening, so it was decided to halt the roll back and instead to restore the stack to the pad.
- December 21, 2010 (STS-133 Discovery): This roll back was to make tests on the Inter-tank of the ET; workers removed 89 sensors from the aluminum skin and applied insulating foam to those areas. They also scanned below the foam insulation surrounding the Inter-tank section, seeking any indication of cracks or other issues.

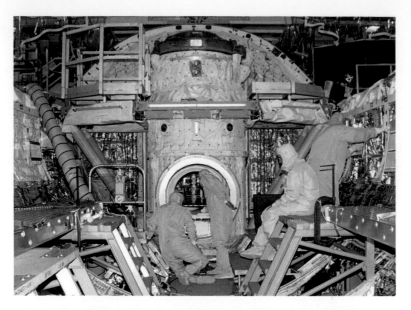

In the OPF workers check paperwork concerning the wiring required to support the first use of the OBSS on STS-114.

Atlantis is hoisted high in the VAB prior to mating with the ET and SRBs for STS-101.

A Crawler Transporter delivers an MLP bearing STS-116 to Pad 39B.

Table 4.2 STS 'Stack' Components For International Space Station Missions

STS	Orbiter	Orbital Vehicle	External Tank	SRB Set	RSRM Set	SSME #1	SSME #2	SSME #3
STS Launch Vehicle 'Stack' Components For ISS Assembly Missions								
88	Endeavour	OV-105	ET-097	BI-095	67	2050	2044	2041
96	Discovery	OV-103	ET-100	BI-098	70	2047	2051	2049
101	Atlantis	OV-104	ET-102	BI-101	74	2043	2054	2049
106	Atlantis	OV-104	ET-103	BI-102	75	2052	2044	2047
92	Discovery	OV-103	ET-104	BI-104	76	2045	2053	2048
97	Endeavour	OV-105	ET-105	BI-103	72	2054	2043	2049
98	Atlantis	OV-104	ET-106	BI-105	77	2052	2044	2047
102	Discovery	OV-103	ET-107	BI-106	78	2048	2053	2045
100	Endeavour	OV-105	ET-108	BI-107	79	2054	2043	2049
104	Atlantis	OV-104	ET-109	BI-108	80	2056	2051	2047
105	Discovery	OV-103	ET-110	BI-109	81	2052	2044	2045
108	Endeavour	OV-105	ET-111	BI-110	82	2049	2043	2050
110	Atlantis	OV-104	ET-114	BI-112	85	2048	2051	2045
111	Endeavour	OV-105	ET-113	BI-113	84	2050	2044	2054
112	Atlantis	OV-104	ET-115	BI-115	87	2048	2051	2047
113	Endeavour	OV-105	ET-116	BI-114	86	2050	2044	2045
114	Discovery	OV-103	ET-121	BI-125	92	2057	2054	2056
121	Discovery	OV-103	ET-119	BI-126	93	2045	2051	2056
115	Atlantis	OV-104	ET-118	BI-127	94	2044	2048	2047
116	Discovery	OV-103	ET-123	BI-128	95	2050	2054	2058
117	Atlantis	OV-104	ET-124	BI-129	96	2059	2052	2057
118	Endeavour	OV-105	ET-117	BI-130	97	2047	2051	2045
120	Discovery	OV-103	ET-120	BI-131	98	2050	2048	2058
122	Atlantis	OV-104	ET-125	BI-132	99	2059	2052	2057

(continued)

Table 4.2 (continued)

STS	Orbiter	Orbital Vehicle	External Tank	SRB Set	RSRM Set	SSME #1	SSME #2	SSME #3
STS Launch Vehicle 'Stack' Components For ISS Assembly Missions								
123	Endeavour	OV-105	ET-126	BI-133	101	2047	2044	2054
124	Discovery	OV-103	ET-128	BI-134	102	2051	2048	2058
126	Endeavour	OV-105	ET-129	BI-136	104	2047	2052	2054
119	Discovery	OV-103	ET-127	BI-135	103	2048	2051	2058
127	Endeavour	OV-105	ET-131	BI-138	106	2045	2060	2054
128	Discovery	OV-103	ET-132	BI-139	107	2052	2051	2047
129	Atlantis	OV-104	ET-133	BI-140	108	2048	2044	2058
130	Endeavour	OV-105	ET-134	BI-141	109	2059	2061	2057
131	Discovery	OV-103	ET-135	BI-142	110	2045	2060	2054
132	Atlantis	OV-104	ET-136	BI-143	111	2052	2051	2047
133	Discovery	OV-103	ET-137	BI-144	112	2044	2048	2058
134	Endeavour	OV-105	ET-122	BI-145	113	2059	2061	2057
135	Atlantis	OV-104	ET-138	BI-146	114	2047	2060	2045

KEY: BI   Booster Integration (set of 2)
          ET   External Tank
          OV   Orbital Vehicle
          RSRM Redesigned Solid Rocket Motor (set of 2)
          SRB   Solid Rocket Booster
          SSME   Space Shuttle Main Engine
             (Position looking forward: # 1 (top), #2 (lower left) or #3 (lower right)
          STS   Space Transportation System

Reference: Space Shuttle Mission Summary, NASA/TM-20111-216142, compiled by Robert D. Legler and Floyd V. Bennett, Mission Operations, Johnson Space Center, Houston, Texas, September 2011.

SPACE STATION HARDWARE PROCESSING

The primary objective of preparing Orbiters, ETs, SRBs and associated apparatus for missions to the ISS was to carry out the assembly program, but towards the end of the program the emphasis switched to stocking the facility with sufficient stores and spare parts to enable it to continue operating after the retirement of the Shuttle fleet by using only smaller logistics carriers.

Most of the payloads carried on Shuttle-Mir missions were logistics, but during the ISS assembly much of the payload would be physical hardware that was to be attached or assembled on-orbit. For some of the largest elements ever launched on the Shuttle, NASA built a new facility at KSC to receive, check, test, and prepare each element of station hardware prior to its installation in the Orbiter payload bay during preparations for launch.

Creating A Fail-Safe System

Boeing was the prime contractor to NASA for the design and engineering of all the elements of the ISS fabricated in the United States. The aerospace company was also responsible for processing the elements supplied by international partners that were to be launched by Shuttle. Every element destined for the Shuttle had first to complete processing in the Space Station Processing Facility (SSPF) at KSC. Here, a team of Boeing engineers and technicians would spend months carrying out various checks, inspections and tests, before finally integrating the components into the package of experiments, flight hardware, and other payloads that made up each manifested ISS assembly payload. Before a payload was accepted, representatives of both KSC and JSC performed a walk-down quality inspection to ensure all safety and engineering specifications had been satisfied. After processing, the hardware was transported in Payload Canisters either to the OPF or to Launch Complex 39 for insertion into the Orbiter payload bay.

While each payload or piece of hardware was being processed, the database of processing requirements would be regularly updated prior to any component being certified for dispatch to the station. These requirements were based on a document called the Operations and Maintenance Requirements and Specifications Document (OMRSD), which was essentially the textbook source for the necessary operations, maintenance, data and analysis requirements and specifications used to maintain or verify all of the hardware for flight and operational readiness. These standards were applied both to launch by the Shuttle and installation on the ISS. As part of the risk mitigation process for a given flight, it was essential to keep track of every tool and piece of apparatus utilized in preparing the payload and Orbiter stack. Every Shuttle launch was risky enough without the prospect of a loose strap or lost tool interfering with vehicle systems or payloads. To that end, every tool, access platform, restraint device, and test apparatus was logged while being used within the Orbiter or on any element of ISS hardware. Any failure of apparatus or procedural mishap generated a report that specified the item or action, its location, any damage, and the searches or investigations that would be required to prevent a further risk or hazard to the given payload, flight safety, or worker safety.

The Environmental Control and Life Support System (ECLSS) test facility at MSFC tests the life support technologies intended for the ISS and troubleshoots problems encountered in space.

The US Destiny laboratory, which was constructed at MSFC by Boeing between 1995 and 1998, is prepared for delivery to KSC by a Supper Guppy aircraft.

The ESA Columbus laboratory manufactured in Germany is unloaded from a Beluga Airbus at KSC on May 30, 2006.

Furthermore, because much of the payload came from the international partners or other outside vendors, JSC developed the Launch Site Support Plan (LSSP). This was essentially an agreement between KSC and the ISS customer which related to ground processing for any launch package. It detailed the KSC ground processing services for each element supplied, and covered what would be required of the customer or vendor to achieve that processing. The LSSP addressed coordination and planning prior to the arrival of the hardware at the SSPF, as well as transportation inside the KSC area, pre-Shuttle integration activities, physical integration, and any post-landing activities to be performed at KSC or any other landing site at which the mission might land.

Specific documents were also created to address any operational requirements for each element of every payload launched on every mission. These documents formed only part of the huge 'library' of documentation required to process each element of payload and all the components of every Shuttle mission. Other such documentation included guidelines for operating the Shuttle in the environment of low Earth orbit, safety and integrity information, and the safety of payloads carried on an Orbiter, in particular the possibility of an abort or emergency landing with a full payload bay of hardware and consumables.[3]

SPACE STATION PROCESSING FACILITY

Having been authorized to create a space station, in 1984 NASA selected the Kennedy Space Center to be the site for both processing and launching all station elements. The following year, proposals were sought to determine exactly how this operation should be executed. This review process continued for the next two years, and in 1987 it was decided to construct a purpose-built facility in the KSC Industrial Area, to the east of the Operations & Checkout Building. Construction of the facility for what by then was Space Station Freedom, officially started in April 1991 and was to cost approximately $72 million. Official dedication of the facility occurred on June 23, 1994, and by then the project had become the International Space Station. In 1994, with construction of the High Bay still underway, the Test Control and Monitoring System software team, the first occupants of the building, moved into their second floor offices. Three years after the formal opening of the facility, the south wall of the High Bay was equipped with a visitor's window.

Constructed as a 457,000 sq ft (42,456.68 sq m) three-story building, the SSPF contained two processing bays, an airlock, operational control rooms, a number of laboratories, logistics areas, offices, and a cafeteria. Non-hazardous hardware and Shuttle payloads were readied in the processing areas, the airlock, and laboratories designed to 100,000 class (measurement of air cleanliness) clean room standard.

The Space Station Processing Facility under construction in 1992.

As with many facilities at the Cape, the numbers for the SSPF were impressive. The Intermediate Bay (I Bay) measured 338 ft (103.02 m) long by 50 ft (15.24 m) wide with a ceiling height of 30 ft (9.14 m), had a pair of 5 ton bridge cranes, and was certified as a 100,000 class clean room. The High Bay (H Bay) was 362 ft (110.33 m) long by 105 ft (32.00 m) wide with a ceiling height of 61 ft (18.59 m) and a pair of 30 ton bridge cranes. This was designated as a Class 1 Division 2 area and could be separated into as many as eight different processing areas to give a range of configuration options to suit processing requirements. There were facilities for ammonia servicing, gaseous nitrogen (6,000 psi or 414 bars) and gaseous helium (also 6,000 psi), as well as compressed air at 125 psi and potable water. The electrical services provided 480 volts 3-phase power at 60 Hz and an uninterruptible power supply at 450 kVA. The structure hosted 140,000 sq ft (13,020 sq m) of office space, nine independently operated control rooms, fifteen offline laboratories, two chemical labs and two dark rooms, as well as a certified offline laboratory for "planetary protection processing" that was classified as a 100 class clean work area.[4]

Planning The Process

Prior to processing any hardware, a raft of systems requirements were defined, both for the ISS elements and for the Shuttle. For station hardware, the planned use of technical apparatus needed to be defined in order to allow the necessary operational support to be specified, based on the configuration of the hardware and how it fitted into the existing schedule. On this basis the KSC Utilization Integrated Product Team (UIPT) worked in conjunction with both the developers and contractors of the payload and a multitude of program organizations to approve a set of schedules for processing each element. Then the UIPT would generate the plans to satisfy all the safety and schedule requirements at the Cape. The result was a Work Authorization Document (WAD).

This was an ongoing and fluid process that ran throughout the planning, processing, and mission cycle. A plan originated at the design phase of the payload could continue through the early stages of mission planning and processing, and in some cases require activities subsequent to the flight, such as routine Operations and Maintenance (O&M) and ongoing tasks running across several missions.

Organizing The Process

As prime contractor, Boeing's facility at Canoga Park in California was responsible not only for the US segment hardware but also developing the electrical system for the US segment. Over 200 Remote Power Control Modules were fabricated for this task. They were delivered to Marshall Space Flight Center in Huntsville, Alabama, for installation in the ISS modules. Marshall was also responsible for the principal manufacturing and quali-fication testing of US segment elements. All the manufactured elements of the US segment would arrive at MSFC from the various subcontractors and, after undergoing rigorous testing, would be integrated into the structure of that module.

Processing The Hardware

Hardware for the ISS usually arrived at the Cape on cargo planes touching down at the Shuttle Landing Facility. After having its identification confirmed, and being checked for transit damage, each item would be added to the processing inventory and taken to the Space Station Processing Facility or the Operations & Checkout Building. Most ISS hardware items arrived at the SSPF hardware inspection areas, while those requiring an airlock were received either in the airlock facility itself or in the High Bay if the airlock was currently in use. Acceptance was followed by months of testing and integration, as well as compatibility tests with assorted other experiments and items of hardware.

One of the challenges of assembling the ISS was that few of the components would actually meet on Earth in order to participate in such compatibility testing prior to their installation on-orbit. A digital pre-assembly process was therefore completed as part of the processing cycle, to reveal any potential physical incompatibilities that would pose serious problems on-orbit. Computer models of the physical interfaces were created in order to allow compatibility and mating tests and simulations to detect and address any conflicts or inconsistencies while the hardware was still on the ground. These multiple element tests were conducted in the facilities of the SSPF, sometimes with astronauts present to gain hands-on experience of the hardware that they were to handle on-orbit.

Early on in the planning of the space station there had been a significant difference between the designers of the hardware which was to be sent into space and the people who would process the hardware for launch. This difference centered on the shipping of flight-ready hardware to the Cape and launching it without testing its compatibility with other modules; the so-called 'ship and shoot' philosophy that was favored by the designers as opposed to the extensive program of pre-launch integrated tests preferred by payload processers. In 1995, designers at JSC began (quite rightly) to consider the idea of integrated testing at the Cape to ensure each piece of hardware was compatible with the next, and thus the whole construction. In the spring of 1996 additional testing and evaluation was added to the Interface Verification Tests for mission STS-88 (2A), including expanded Shuttle Avionics Integration Laboratory (SAIL) and cargo testing. Beyond 1997, with the flight hardware all now being processed at the Cape, integrated testing was formally incorporated into the payload processing activity for ISS elements. This was agreed by program management, primarily because all of the systems would have to function on-orbit over a long period of time in a manner that ensured the safety of the station crew. Using this Multi-Element Integration Testing (MEIT) process, each element of hardware would take about three years to progress from planning through to certifying the payload for flight.

A further process called the Mission Sequence Test (MST) which became known as 'end to end' integrated testing involved running a full-up configuration of the Tracking and Data Relay Satellite Systems (TDRSS) communications using equipment that was nominally ready to be sent up to the ISS while it was still on the ground. This provided flight simulation testing of critical communication links before apparatus was launched into space.

Node 1 Unity is moved by overhead crane to the payload canister for transfer to the pad as work continues on other hardware in the background.

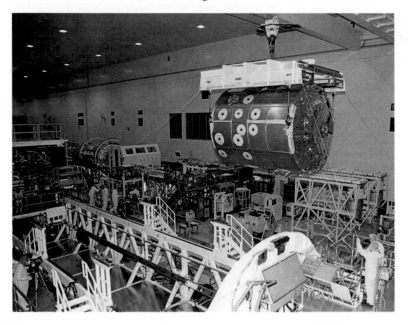

MPLM Leonardo is moved to the weight and balance scales in the SSPF in preparation for insertion into the payload bay of STS-102.

The S5 segment of the Integrated Truss Structure is relocated by overhead crane in the SSPF during preparations for STS-118.

Flight Hardware Processing

On July 7, 1995, the Russian-built Docking Module that a Shuttle was to install on Mir arrived by Russian cargo aircraft at KSC and was taken to the SSPF, thus becoming the first piece of flight hardware to be processed there. It was carried to Mir aboard Atlantis in November of that year. In June 1997 the first US-built element for the ISS to arrive at the SSPF was Node 1, named Unity. After due processing, it was launched in December 1998 in the payload bay of Endeavour on Assembly Mission 2A.

Thereafter, with the exception of payloads delivered by Russian spacecraft, *all* major hardware components for the ISS, including some international partner payloads, were processed through the SSPF. In addition, each of the Multi-Purpose Logistics Modules (MPLM) was loaded with cargo in the SSPF prior to being placed into the payload bay of the designated Orbiter.

Details of major payload delivery dates for ISS hardware are shown in Table 4.3 and a full list of the major payloads delivered by Shuttles to space stations is given in Table 4.4.

The Multi-Element Integrated Testing (MEIT) for the ISS was undertaken in three phases from December 1998 through September 2003. These tests were performed on flight hardware which had been soft-mated on the High Bay floor in order to replicate the configuration intended for orbit. During these tests, all the cabling and fluid lines were connected between the elements to form an integrated system. This was the first time that many of the international components had been tested in this manner. The tests verified the mechanical connections and the flow of electrical power and fluids between modules, and also exercised the relevant flight software.

Table 4.3 Space Station Major Payload Delivery Dates

On Dock KSC	Payload (STS & assembly designations in brackets)
INTERNATIONAL SPACE STATION	
NODES	
1997 Jun 23	Node 1 *Unity* (STS-88 2A)
2003 Jun 1	Node 2 *Harmony* (STS-120 10A)
2009 May 20	Node 3 *Tranquility* (STS-130 20A)
IN STORAGE	*Node 4 Docking Hub System*
PRESSURIZED MATING ADAPTERS (PMA)	
1997 Jul 25	PMA-1 (STS-88 2A)
1997 Oct	PMA-2 (STS-88 2A)
1998 Feb 20	PMA-3 (STS-92 3A)
CREW LIVING QUARTERS	
Cancelled & Recycled	*Habitation Module*
LABORATORY MODULES	
1998 Nov 16	US *Destiny* Lab (STS-98 5A)
2006 May 30	Columbus Module (STS-122 1E)
2003 May 30	Japanese Kibo PM (STS-124 1J)
2007 Mar 12	Japanese Kibo ELM-PS (STS-123 1J/A)
2008 Sep 24	Japanese Kibo EF (STS-127 2J/A)
Cancelled 2005	*Centrifuge Accommodation Module*
DOCKING MODULES	
2009 Dec 17	Russian MIM-1 *Rassvet* (STS-132 ULF-4)
Cancelled 2002	*CRV Berthing and Docking Modules x 2*
STOWAGE MODULES	
1998 Jul 31	MPLM-1 *Leonardo*
1999 Aug 4	MPLM-2 *Raffaello*
2001 Feb 1	MPLM-3 *Donatello*
2010 Apr 20	Permanent Multipurpose Module (formerly MPLM *Leonardo*, STS-133 ULF-5)
AIRLOCK	
20001 Sep 13	Joint Airlock *Quest* (STS-104 7A)
OBSERVATION FACILITY	
2004 Oct 7	Cupola (STS-130 20A)
INTEGRATED TRUSS ASSEMBLY (ITA)	
1998 Feb 17	Z1 (STS-92 3A)
2000 Jul 18	P6 (STS-97 4A)
1999 Jun 12	S0 (STS-110 8A)
1999 Oct 6	S1 (STS-112 9A)
2000 Jul 26	P1 (STS-113 11A)
2000 Jul 30	P3/P4 (STS-115 12A)
2001 Jul 19	P5 (STS-116 12A.1)
2000 Dec 17	S3/S4 (STS-117 13A)
2002 Mar 13	S5 (STS-118 13A.1)
2001 Dec 11	S6 (STS-119 15A)

(continued)

Table 4.3 (continued)

On Dock KSC	Payload (STS & assembly designations in brackets)
MOBILE SERVICING SYSTEM (MBS)	
2000 Aug	MBS (including CETA Carts A and B)
REMOTE MANIPULATOR SYSTEMS	
1999 May 16	SSRMS *Canadarm2*
2007 Jun	SPDM *Dextre*
2006 Dec	Kibo RMS
MAJOR EXTERNAL SCIENTIFIC EXPERIMENTS	
2010 Aug 26	Alpha Magnetic Spectrometer-02 (AMS-02) (STS-134 ULF-6)
CREW RESCUE VEHICLE	
Cancelled 2002	*CRV 1 through 4 were originally proposed*

NOTES: Unflown ISS Hardware

Node 4: This fourth connecting node started out as the Structural Test Article (STA) and was intended to become Node 1. However, structural flaws were discovered during construction and as a result Node 2 became Node 1 and the STA was placed in storage at KSC. In 2011 it was considered for possible launch by an Expendable Launch Vehicle, but at the time of writing (March 2017) remains in long-term storage.

Habitation Module: As originally envisaged this module was to be the ISS main living quarters for the crew with a galley, toilet, shower, sleep stations, and medical facilities. Following the Columbia accident in 2003 it was definitively cancelled. On February 14, 2006 NASA decided to recycle the Habitation Module for ground based life support research in support of future programs

Centrifuge Accommodations Module: A Japanese designed module, but owned by NASA and cancelled in 2005. As well as the centrifuge the module would also have included a Life Sciences Glove Box and would have operated as a U.S. facility once attached to the ISS.

Crew Rescue Vehicle: Up to four CRVs were to be delivered by the Space Shuttle each with a three-year orbital life time

Berthing and Docking Modules: Two BDM's were to be delivered by the Space Shuttle to support the CRVs

Prior to the December 1998 MEIT, an ISS Flight Emulator (an accurate mock-up) was developed to stand in for the hardware which was already on-orbit, including the Unity Node, PMA-1 and PMA-2. Three MEIT and one Integrated Systems Test (IST) were then carried out for these elements. The IST for Node 2 consisted of Node 2 and emulators for the Unity Node and the US Destiny laboratory, both of which were then on-orbit, as was part of the ISS Flight Emulator.[*]

The three-phase MEIT program included:

- MEIT-1: Planning began in 1997 and testing in January 1999. The hardware used included the US Destiny laboratory, Z-1 truss, P6 truss, and the Node 1 emulator. Once Destiny had been launched, an emulator was built to stand in for the flight laboratory. This was required because the laboratory controlled several other modules.

[*]There were several ISS test configurations within each MEIT, and after MEIT-3 the configuration morphed into the ISS Flight Emulator as the station grew.

Table 4.4 Major International Space Station Payloads Carried By Shuttle Orbiters 1998–2011

STS	Orbiter OV	Year	Station	Station Flight	Payloads	Notes Mass figures are approximate & rounded up	Transferred to station	Transferred to orbiter	Returned to Earth
88	105	1998	ISS	2A	Unity Node 1/PMA1 & 2	25,479 lb/11,555 kg	1998 Dec 6	No	No
96	103	1999	ISS	2A.1	Spacehab 13/DM 7	Remained in payload bay	N/A	N/A	Yes
101	104	2000	ISS	2A.2a	Spacehab 14/DM 8	Remained in payload bay	N/A	N/A	Yes
106	104	2000	ISS	2A.2b	Spacehab 15/DM 9	Remained in payload bay	N/A	N/A	Yes
92	103	2000	ISS	3A	Zenith Truss (Z1)	18,427 lb/8,357 kg	2000 Oct 13	No	No
					PMA-3	2,549 lb/1,156 kg	2000 Oct 16	No	No
97	105	2000	ISS	4A	P-6 Solar Arrays	34,817 lb/15,790 kg	2000 Dec 2	No	No
98	104	2001	ISS	5A	U.S. Destiny Lab	33,481 lb/15,184 kg	2001 Feb 10	No	No
102	103	2001	ISS	5A.1	Lab Cradle Assembly	300 lb/136 kg	2001 Mar 11	No	No
					MPLM-1/Leonardo 1	25,007 lb/11,341 kg	2001 Mar 11	2001 Mar 18	Yes
100	105	2001	ISS	6A	SSRMS Canadarm2	6745 lb/3,059 kg	2001 Apr 22	No	No
					MPLM-2/Raffaello 1	20,226 lb/9,173 kg	2001 Apr 23	2001 Apr 27	Yes
104	104	2001	ISS	7A	Quest Joint Airlock	13,301 lb/6,032 kg	2001 Jul 14	No	No
105	103	2001	ISS	7A.1	MPLM-3/Leonardo 2	20,226 lb/9,450 kg	2001 Aug 13	2001 Aug 19	Yes
					MISSE 1 & 2	Approx. 216 lb/98 kg each	2001 Aug 16	2005 Jul 30	STS-114
108	105	2001	ISS	UF-1	MPLM-4/Raffaello 2	20,226 lb/9,173 kg	2001 Dec 8	2001 Dec 14	Yes
110	104	2002	ISS	8A	S0 Truss	26,782 lb/12,146 kg	2002 Apr 11	No	No
					Mobile Transporter	1826 lb/828 kg	2002 Apr 11	No	No
					Airlock Spur (no mass given)	Estimated @ 842 lb/382 kg	2002 Apr 14	No	No
111	105	2002	ISS	UF-2	Mobile Base System	3182 lb/1,443 kg	2002 Jun 11	No	No
					MPLM-5/Leonardo 3	2,300 lb/1,043 kg	2002 Jun 8	2002 Jun 14	Yes
112	104	2002	ISS	9A	S1 Truss	28,107 lb/12,747 kg	2002 Oct 10	No	No
					CETA Cart A	534 lb/242 kg	2002 Oct 10	No	No
113	105	2002	ISS	11A	P1 Truss	27,511 lb/12,477 kg	2002 Nov 26	No	No
					CETA Cart B	534 lb/242 kg	2002 Nov 28	No	No
114	103	2005	ISS	LF-1	MPLM-6/Raffaello 3	18,068 lb/8,194 kg	2005 Jul 29	2005 Aug 7	Yes
					MISSE 5	Approx. 216 lb/98 kg	2005 Aug 3	2006 Sep 15	STS-115

121	103	2006	ISS	ULF1.1	MPLM-7/Leonardo 4	9,507 kg	2006 Jul 6	2006 Jul 14	Yes
					MISSE 3 & 4 (MISSE-4 was deployed during ISS-13)	Approx. 216 lb /98 kg	2006 Aug 3	2007 Aug 18	STS-118
115	104	2006	ISS	12A	P3/P4 Truss	34,892 lb/15,824 kg	2006 Sep 11	No	No
116	103	2006	ISS	12A.1	Spacehab 16/SM 8	Remained in payload bay	N/A	N/A	Yes
					P5 Truss	4,110 lb/1,864 kg	2006 Dec 12	No	No
117	104	2007	ISS	13A	S3/S4 Truss	35,686 lb/16,184 kg	2007 Jun 11	No	No
118	105	2007	ISS	13A.1	Spacehab 17/SM 9	Remained in payload bay	N/A	N/A	Yes
					S5 Truss	4,000 lb/1814 kg	2007 Aug 11	No	No
					External Stowage Platform 3	5,678 lb/2,575 kg	2007 Aug 14	No	No
120	103	2007	ISS	10A	Harmony Node 2	31,580 lb/14,322 kg	2007 Oct 26	No	No
122	104	2008	ISS	1E	ESA Columbus Lab	26,632 lb/12,078 kg	2008 Feb 11	No	No
					EuTEF	770 lb/350 kg	2008 Feb 15	2009 Sep 1	STS-128
123	105	2008	ISS	1J/A	Kibo JEM ELM-PS	18,496 lb/8,388 kg	2008 Mar 14	No	No
					Dextre	3,31 lb/1,556 kg	2008 Mar 13	No	No
					OBSS	842 lb/382 kg	2008 Mar 22	2008 Jun 3	STS-124
					MISSE-6A & 6B	Approx. 216 lb/ 98 kg each	2008 Mar 22	2009 Sep 1	STS-128
124	103	2008	ISS	1J	Kibo JEM-PM	32,572 lb/14,772 kg	2008 Jun 2	No	No
					Kibo RMS-Main & small fine arms	Approx. 2,139 lb/970 kg	2008 Jun 5	No	No
126	105	2008	ISS	ULF-2	MPLM-8/Leonardo 5	28,015 lb/12,705 kg	2008 Nov 17	2008 Nov 26	Yes
119	103	2009	ISS	15A	S6 Truss	31,119 lb/14,113 kg	2009 Mar 18	No	No
127	105	2009	ISS	2J/A	JEM EF	8,330 lb/3,778 kg	2009 Jul 18	No	No
					ICC-VLD	9,041 lb/4,100 kg	2009 Jul 19	2009 Jul 24	Yes
					ELM-ES	2,646 lb/1,200 kg	2009 Jul 21	2009 Jul 26	Yes
128	103	2009	ISS	17A	MPLM-9/Leonardo 6	27278 lb/ 12,371 kg	2009 Aug 31	2009 Sep 7	Yes
					Lightweight MPESS Carrier	Remained in payload bay	N/A	N/A	Yes
					Ammonia Tank Assembly	1,702 lb/772 kg	2009 Sep 3 (new)	2009 Sep 1 (old)	Yes old unit

(continued)

Table 4.4 (continued)

STS	Orbiter OV	Year	Station	Station Flight	Payloads	Notes Mass figures are approximate & rounded up	Transferred to station	Transferred to orbiter	Returned to Earth
129	104	2009	ISS	ULF-3	Express Logistics Carrier 1	14,103 lb/6,396 kg	2009 Nov 18	No	No
					Express Logistics Carrier 2	13,530 lb/6,136 kg	2009 Nov 21	No	No
					MISSE 7A & 7B	Approx. 216 lb/ 98 kg each	2009 Nov 23	2011 May 20	Yes
130	105	2010	ISS	20A	Tranquility Node 3	40,000 lb/18,144 kg	2010 Feb 12	No	No
					Cupola	4,145 lb/1,880 kg	2010 Feb 12	No	No
131	103	2010	ISS	19A	MPLM-10/Leonardo 7	27,273 lb/12,371 kg	2010 Apr 8	2010 Apr 16	Yes
					Lightweight MPSS Carrier	Remained in payload bay	N/A	N/A	Yes
132	104	2010	ISS	ULF-4	Rassvet MRM-1	17,640 lb/8,000 kg	2010 May 18	No	No
					ICC-VLD	8,330 lb/3,778 kg	2010 May	No	No
133	103	2011	ISS	ULF-5	ExPRESS Logistics Carrier 4	7,676 lb/3,481 kg	2011 Feb 24	No	No
					PMM Leonardo	28,359 lb/12,861 kg	2011 Mar 1	No	No
134	105	2011	ISS	ULF-6	AMS-02	15,303 lb/6,940 kg	2011 May 19	No	No
					ExPRESS Logistics Carrier 3	13,475 lb/6,111 kg	2011 May 18	No	No
					MISSE 8	Approx. 216 lb/98 kg each	2011 May 20	*2013 Jul 9*	*Yes SpaceX-3*
					OBSS	842 lb/382 kg	2011 May 27	No	No
135	104	2011	ISS	ULF-7	MPLM-11/Raffaello 4	20,947 lb/9,500 kg	2011 Jul 11	2011 Jul 18	Yes

- MEIT-2: This was said to be the largest payload test in history, with in excess of 3,000 cables. It involved the S0 truss, Mobile Transporter and Mobile Base System, S1 truss, P1 truss, P3 truss, P4 truss and the US laboratory emulator.
- MEIT-3: Finished in 2007, this test involved Node 2, the Japanese Experiment Module and the US laboratory emulator.

The three phases revealed significant problems that could only have been found by such an integrated, multi-element flight configuration. Influencing both hardware and software, these problems were isolated and corrected prior to any item leaving for the pad. Had this testing not been accomplished, serious malfunctions and other problems could have hampered the on-orbit assembly, thus delaying or extending the assembly timescales. At worst, such issues could have posed significant threats to the hardware, to crew safety, and perhaps to the integrity of the whole program.

In all, over 5,000 verifications were performed during these test phases.[5] Some of the more serious problems identified pre-flight were:

- Breakers tripped while testing the P6 element, resulting in power losses. This required a redesign of the Assembly Power Converter Unit (APCU). Had this had not been discovered prior to launch, activation of P6 on-orbit by STS-97 would not have been possible, necessitating a re-flight of that unit.
- During MEIT-2, video lines were discovered to have been swapped between the Trailing Umbilical Systems of the US orbital segment and the MBS that was supplied by Canada. If this fault had not been remedied prior to launch it would have impaired the routing of video signals between the SSRMS and the Robotic Work Station in the US Destiny laboratory, and an extra EVA would have been needed in order to swap over the two harnesses.
- Activation of critical elements of Destiny was mandated to take less than 2 hr, but during the MEIT it actually took 36 hr. Had this occurred on-orbit, thermal loading limitations would have been exceeded, perhaps resulting in the loss of the module before it could be fully activated.
- A mismatch between the mass property units of the Robotic Work Station and the Command and Control of Destiny could have caused attitude errors and an excessive use of propellant.
- Control Moment Gyroscopes powered down unexpectedly during a test of the Command and Control switchover for Destiny. If this had occurred on-orbit it would have required an unplanned use of propellant to regain attitude control.

Over the three phases, there were significant reductions to the number of issues that were raised. For MEIT-1, 884 Interim Problem Reports (IPR) were recorded. This fell to 334 for MEIT-2 and to 64 for MEIT-3. Some of this reduction was due to improved factory level testing and qualification.

For the ISS the SSPF was used to process:

STS-88	Node 1 Unity with PMA-1 and PMA-2
STS-92	S0 truss with PMA-3
STS-97	P6 solar array truss
STS-98	US Lab Destiny
STS-102	MPLM Leonardo

STS-100	MPLM Raffaello and Canadarm2
STS-104	Joint Airlock Quest
STS-105	MPLM Leonardo
STS-108	MPLM Raffaello
STS-110	S0 Central Truss Structure
STS-111	Mobile Base System
STS-112	S1 Starboard Side Thermal Radiator truss
STS-113	P1 Port Side Thermal Radiator truss
STS-115	P3/P4 solar array trusses
STS-116	P5 truss
STS-117	S3/S4 solar array trusses
STS-118	S5 truss
STS-120	Node 2 Harmony
STS-122	European Science Module Columbus
STS-123	Japanese Experiment Logistics Module, Pressurized Section
STS-124	Japanese JEM laboratory Kibo
STS-119	S6 solar array truss
STS-127	Japanese Exposed Facility
STS-129	Express Logistics Carrier with 30,000 lb (13,608 kg) of parts
STS-130	Cupola
STS-131	MPLM Leonardo
STS-132	Russian-built Rassvet
STS-133	Permanent Multipurpose Module Leonardo
STS-134	Alpha Magnetic Spectrometer-02
STS-135	MPLM Raffaello

Side-by-side processing for Node 2 (Harmony) and the Japanese Kibo laboratory in the SSPF.

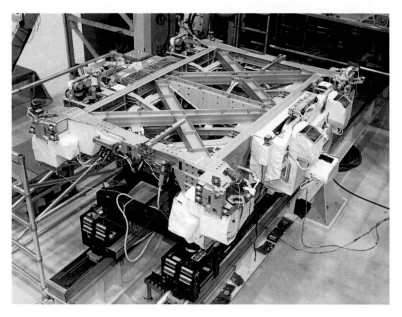

The Mobile Transporter is prepared in the SSPF as part of the STS-110 payload.

A New Tenant

By 2013, two years after the retirement of the Shuttle, things were changing at KSC, and not just with the switch to the Constellation program. In April the small satellite software and electronic systems company Micro Aerospace Systems, based in nearby Melbourne, Florida, began leasing office and laboratory facilities in the almost empty SSPF.[6] With most of the hardware of the ISS now on-orbit and the remaining Shuttle Orbiters in museums NASA's immediate need for the huge facility had passed. Rather than sitting idle, the vast space available has drawn other customers to use it, until its resources can be utilized once more for a future checkout and processing program of hardware destined for space. Nevertheless, the SSPF still has a role in supporting the ISS because it prepares the cargo for commercial resupply spacecraft.

Postponements, Delays And Scrubs: ISS, 1997–2002

- *STS-88 (2A)*: Planned for December 4, 1997, but postponed first to July 9 then December 3 1998. Launch scrubbed for 24 hr on December 3 due to a problem with hydraulic pump no. 1. It launched successfully on December 4.
- *STS-96 (2A.1)*: Due to launch December 9, 1998, it was postponed to May 13, 1999 owing to changes to multiple ISS flights. It was then moved farther back to May 20, slipped to May 24, then advanced to May 20 again. These changes followed the decision to roll back to the VAB on May 16 to repair hail damage on ET foam (648 divots, 459 requiring repair) but the return to the pad did not occur until May 21. It was finally launched on May 27.

Supplied by ESA from the Alenia Spazio factory in Turin, Italy, the multi-windowed Cupola is prepared for delivery by STS-130.

- *STS-101 (2A.2a)*: Scheduled for August 5, 1999, it was slipped to October 14 and then to December 2, 1999, both delays resulting from ongoing Discovery wire inspections and repairs. In April, the launch was postponed again due to Commander Jim Halsell injuring his ankle during training. This was followed by the replacement of the Rudder Speed Brake PDU on Atlantis with one from Columbia. The launch was scrubbed three further times by weather concerns and delays in launching the GOES-Atlas III. It did not finally leave the ground until May 19, 2000.

The Alpha Magnetic Spectrometer is seen here in 2010, suspended by a crane during preparations for STS-134. In the background is MPLM Leonardo, which would be modified for permanent attachment to the ISS.

- *STS-106 (2A.2b)*: This launch was originally due on August 19, 2000 but was postponed to September 8, on which date it was launched successfully.
- *STS-92 (3A)*: Originally planned for July 23, 1998, the launch was slipped six times. Initially it was slipped to January 14, 1999, then to June 17, and finally to December 2 due to delays in the ISS flight manifest. Then it was slipped to June 14, 2000, to September 21, and finally to October 5, 2000, due to delays in launching the Russian Zvezda Service Module. The first launch attempt on October 5 was slipped to October 9 after analysis of STS-106 launch imagery indicated that an ET-to-Orbiter attachment bolt had failed to retract correctly. The second launch attempt was scrubbed by high winds, and the October 10 attempt was canceled after a piece of ground support equipment fell onto a strut in the stack. It was launched on October 11, 2000.
- *STS-97 (4A)*: The six postponements to the original launch date of April 8, 1999 for this mission were all caused by slippage in launching the Russian Zvezda Service Module. It did not finally launch until December 1, 2000.
- *STS-98 (5A)*: This was also postponed six times by the delays to Zvezda. Its original May 20, 1999 launch date was slipped to January 18, 2001. That in turn was postponed by the decision to return the stack to the VAB to X-ray inspect cables in the SRBs; these were replaced. It was finally launched on February 7, 2001.
- *STS-102 (5A.1)*: The five postponements from the original launch date of March 16, 2000 were due to the Zvezda delays, to the replacement of nine damaged RCS thrusters, to STS-98 launch slippage, and to inspections and replacement of damaged cables in the SRBs. It was launched on March 8, 2001.

- *STS-100 (6A)*: Targeted for launch on December 2, 1999, this mission also suffered six postponements because of the delays to previous missions. The launch occurred on April 19, 2001.
- *STS-104 (7A)*: This mission endured three launch postponements from the original date of August 2000, initially to February 2001, then to May, and finally to July of that year. It launched on July 12, 2001.
- *STS-105 (7A.1)*: The intended June 21, 2001 launch date was postponed three times and the attempt on August 9 was scrubbed due to weather violations. It was successfully launched the next day.
- *STS-108 (UF1)*: Set for October 4, 2001, this launch was initially postponed twice. The attempt on November 29 was then scrubbed because of a problem docking a Progress freighter at the ISS. The work to remedy that issue led first to a 24 hr delay, then to a 48 hr delay, and finally to a launch scrub in order to accommodate an unplanned EVA (this was called an ISS Technical Scrub, the first use of that category). The attempt to launch on December 4 was scrubbed due to weather concerns. It launched the next day.
- *STS-110 (8A)*: Set for January 17, 2002, a trio of postponements culminated on January 10 with a delay during ground processing to remove an OMS pod. The launch attempt on April 4 was scrubbed while loading the ET because of a leak that developed in a hydrogen vent line on the MLP. It was finally launched on April 8, 2002.
- *STS-111 (UF2)*: Launch was set for May 2, 2002, but postponed to May 31 to allow the EVA crew extra time to train to replace a failed SSRMS Wrist Roll Joint. After being advanced to May 30, the launch was scrubbed on May 30 and 31 by weather issues. On May 30, pressure differences were detected in the left OMS pod of the Orbiter; this was replaced on pad and the launch set for June 4. This was delayed 24 hr to allow additional work in the OMS pod area. It finally got off the ground on June 5, 2002.
- *STS-112 (9A)*: The intended launch in June 2002 was postponed by the need to make post-STS-110 inspections on SSME liquid hydrogen flow liners on all of the Orbiters and the Main Propulsion Test Article (MPTA). In a separate issue, some cracks were discovered in the MLP jacking cylinder bearings of Crawler Transporter #2. These postponements meant STS-112 and STS-113 would fly before STS-107 because Columbia had more cracks to repair than Atlantis or Endeavour. The October 2 launch was scrubbed because Hurricane Lili in the Gulf of Mexico was threatening MCC-Houston. Control of the US segment of the ISS was transferred to the back-up facility in Moscow until the storm threat passed on October 3. Launch occurred on October 7, 2002.
- *STS-113 (11A)*: Postponed from July 2002 by inspections of liquid hydrogen flow liners, STS-113 was nevertheless moved ahead of STS-107 because that Orbiter had more cracks needing repair than Endeavour. The launch was reset for November 6,

At Pad 39B the Payload Canister carrying Spacehab and the P5 truss for STS-116 is hoisted to the Payload Changeout Room on the Rotating Servicing Structure.

then slipped 24 hr. That attempt was scrubbed by an oxygen leak in the PCS-2 and the launch was next set to November 18. It was delayed again to November 22 after an access platform struck the RMS, damaging the Kevlar honeycomb and causing minor damage to the boom. The November 22 launch was also scrubbed, this time by weather infringements. It was launched the next day.

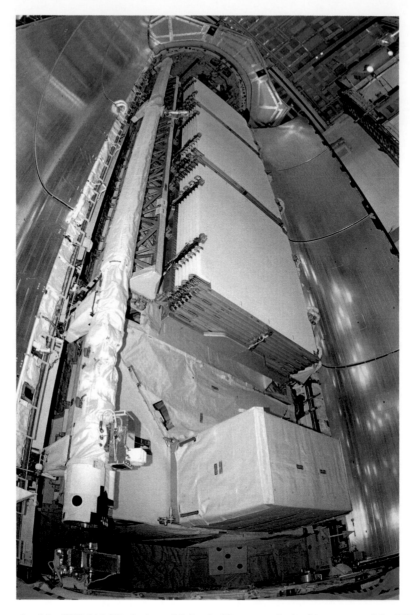

The payload for STS-112 fills the bay of Atlantis. Uppermost is the S1 segment of the ITS that would be attached to the S0 segment already on the Destiny laboratory. At the rear of the payload bay (bottom of frame) inside a protective carrier is the Crew and Equipment Translation Cart (CETA) A, the first of two human-powered carts to ride along the rails of the ITS to serve as mobile work platforms for EVA astronauts. At left, next to the port longeron (note yellow EVA handrails) is the stowed SSRMS showing the end effector and wrist TV camera of the arm at the bottom.

Postponement, Delays And Scrubs: Post-Columbia Return-To-Flight

- *STS-114 (LF1)*: On the schedule planned in June 12, 2001, Atlantis was to fly as ULF1 on an ISS crew rotation flight set for launch on January 16, 2003. In September 2002, it was delayed to March 1, 2003 owing to fleet-wide issues with cracks in engine flow liners. After the loss of Columbia the mission was postponed six times and changed to an ISS logistics mission LF1. The crew rotation was canceled and the Orbiter switched to Discovery. Another four launch date postponements followed. On May 26, 2005, the stack was rolled back to the VAB to be swapped with the ET/SRBs meant for STS-121 because of the late requirement to install a heater on upper bellows of the ET liquid oxygen feed line to prevent icing during ascent. STS-114 was returned to the pad on June 15 for launch on July 13. That launch was scrubbed by a failed ET fuel sensor and reset for July 26, when it lifted off successfully.
- *STS-121 (ULF1.1)*: The launch manifested for this mission on November 15, 2004 using Discovery slipped twice because of the loss of Columbia. Another five postponements were caused by the need for better lighting conditions for photo-documentation of the ascent (part of the revised launch planning post-Columbia). Attempts to launch on July 1 and 2, 2006, were thwarted by bad weather. It finally left the ground on July 4.
- *STS-115 (12A)*: Planned for April 10, 2003, it was postponed initially by the need to repair cracks in engine flow liners and then a further twelve times by the loss of Columbia and the delays in returning to flight with STS-114 and STS-121. The launch attempt on August 27, 2006 was scrubbed by a risk of lightning strikes, and the attempt on August 29 was scrubbed at L-37 hr by a weather infringement from Tropical Storm Ernesto. The stack was midway to the VAB when the decision was taken to roll it out to the pad again as soon as Ernesto had died down. The launched occurred on September 9.
- *STS-116 (12A.1)*: Set for launch on May 6, 2003, two launch postponements arose from repairing engine flow liner cracks and a further five by the loss of Columbia. The STS-116 mission by Atlantis was deleted from the manifest on March 22, 2004 and reinstated on December 9 with Discovery for a February 9, 2006 launch date. Five further postponements were due to "latest planning decisions." The launch was advanced from December 14 to December 7, but weather infringements postponed it for 48 hr. It finally got off the ground on December 10.

Attempting To Resume Assembly: 2008–2009

- *STS-117 (13A)*: This mission was set to launch on September 5, 2003. It was delayed first by the SSME flow liner crack repairs, then the loss of Columbia caused four additional postponements. The mission was deleted in its original form on October 3, 2003. On March 17, 2005 it was returned to the manifest with a launch scheduled no earlier than May 18, 2006. Subsequent changes to the manifest caused three further delays, and in November 2006 the intended launch date of March 16, 2007 was again postponed, this time due to delivery and processing issues relating to the ET. On February 26, 2007, after damage by a hail storm on the pad, the launch was put "under review" while the stack was rolled back to the VAB for repairs. It finally launched on June 8.

- *STS-118 (13A.1)*: Scheduled to launch on October 9, 2003, this mission was postponed to November 13, 2003 by the engine flow liner repairs. The loss of Columbia resulted in two postponements, then the flight was deleted from the manifest on May 28, 2003. It was restored on July 14, 2004 with a launch date of September 14, 2006. On November 10, 2005, manifest constraints imposed a "to be determined" status. Three more postponements were caused by other manifest constraints, ET delivery and processing issues, and as a consequence of the STS-117 roll back. Adjustments to the planned launch date advanced it from August 9 to August 7, 2007 to preclude a range conflict, but it was then delayed again on June 28, 2007 for "cabin leak checks" and other processing work. It finally launched on August 8.

- *STS-120 (10A)*: Originally scheduled for launch on February 19, 2004, this was another mission which was delayed and then deleted from the manifest (post-Columbia) in May 2003. When it was restored on June 1, 2006 it was targeted for an August 9, 2007 launch. This slipped to September 7 by ET delivery and processing schedules, then advanced to August 26 to prevent docking conflicts with Soyuz and ATV operations at the ISS. It was postponed again on August 7 to October 20 due to the STS-117 roll back and then slipped by three days to October 23 to maintain the standard minimum interval between the undocking of Soyuz TMA-10 (having had its landing opportunities revised) and the next spacecraft docking. It launched on October 23 as planned.

- *STS-122 (1E)*: Originally manifested for launch on October 17, 2007, this mission was postponed on April 16, 2007 owing to the rollback of STS-117. This was followed by two "technical scrubs" in December prompted by false readings in the engine cutoff sensor systems during the fueling of the ET. It was reset for February 7, 2008, on which date it was successfully launched.

- *STS-123 (1J/A)*: The launch date of December 8, 2007 slipped to February 14, 2008 because of the STS-117 roll back and was postponed again to March 11, 2008 by the slippage of STS-122. It was launched on that date.

- *STS-124 (1J)*: This mission's scheduled launch date of February 28, 2008 also slipped as a result of the STS-117 roll back, and then again because of a delay in delivering the ET. On April 3 it was postponed because of adverse weather which again affected the delivery of ET-128. It was successfully launched on 31 May.

- *STS-126 (ULF2)*: Set for September 18, 2008, this launch slipped to October 16 due to delays with STS-122 and then to November 10 because of delays in delivering ET-127 and ET-129. On September 8 the launch was postponed to November 12 due to Hurricane Faye impacting on the HST payload readiness for STS-125. On September 24, Hurricane Ike caused a further delay to STS-125, which in turn postponed STS-126 to November 16. On October 19, with the crew released from Launch-on-Need responsibilities for STS-125 because that mission had been pushed down the manifest, the launch was advanced to November 14 (local time), when it was launched successfully.

- *STS-119 (15A)*: The effort to get this flight off the ground was protracted. In January 2003 the mission was planned for launch on January 15, 2004 but it was postponed twice after the loss of Columbia and finally deleted from the manifest on May 28, 2003. On being restored on November 8, 2006, it had a launch date of October 4, 2007. This was followed by seven postponements over the next two years caused by delays to STS-122, ET delivery schedules, additional testing and analysis of engine flow control valves and the lack of a consensus by the MMT to agree to its launch. A technical scrub also occurred on March 11, 2009 because of a gaseous hydrogen vent line leak. The mission was finally launched on March 15.

The Florida weather was always an issue for an on-time Shuttle launch, as this image from STS-134 preparations clearly demonstrates.

The Final Stretch: 2009–2010

- *STS-127 (2J/A)*: The original launch date was April 23, 2009, but it slipped to May 15, 2009 due to delays in delivering ETs to KSC. It was then postponed to June 13 while the manifests of other flights were reprioritized. It served as the STS-400 Launch-on-Need rescue mission for STS-125, but wasn't called upon to fly in that role. It endured two technical scrubs due to a hydrogen leak in the Ground Umbilical Carrier Plate and two further scrubs for the weather before finally being launched on July 15.
- *STS-128 (17A)*: The original launch date of July 30, 2009 was postponed four times to accommodate changes to the launch of STS-125 (which was the final HST service mission), slippage to STS-127 and assessments of the wider KSC processing program. The first attempt to launch on August 25 was scrubbed by the weather. The attempt the next day was canceled when a valve in the Main Propulsion System failed to operate as expected. The third attempt, scheduled for August 27, was

delayed 24 hr to gain additional time to work on the MPS valve. It launched on August 28 (local time).

- *STS-129 (ULF3)*: Originally planned for October 15, 2009, this was another mission to fall victim to re-planning of the STS-125 Hubble mission. It was first postponed on December 4, 2008 until November 12, 2009, and again on October 29, 2009 to November 16 due to additional revisions to the planning schedule, on which date it was launched successfully.

- *STS-130 (20A)*: When manifested in November 2008 it was set for launch on December 10, 2009, but was postponed twice when the launch schedules and operational resources were re-evaluated at KSC as the program began to wind down. The launch on February 7, 2010 was scrubbed by weather violations. It lifted off the next day.

- *STS-131 (19A)*: Planned for a March 18, 2010 launch, this was postponed on March 9 until April 5 due to cold weather conditions that delayed the transfer of the Orbiter from the OPF to the VAB, leaving insufficient time to meet the March 18 date. It got off the ground on April 5 without difficulty.

- *STS-132 (ULF4)*: The planned launch of May 13, 2010 was put back by a day on May 4 to rule out having a Soyuz undocking from the ISS on the same day that a Shuttle docked. The launch on May 14 was successful.

The Last Stacks Standing

- *STS-133 (ULF5)*: The original launch set for July 29, 2010 was postponed to September 16 due to adjustments required to "flight product planning." It was slipped again on July 1 by delays in supplying critical spares for the ISS that were to be carried in the PMM. There were then four launch scrubs. The first on October 29 was due to helium and nitrogen leaks in one of the OMS pods and the second was called on November 2 by the L1 management meeting in response to issues with an SSME controller. The third, on November 4, was due to weather. The fourth, on November 5, was called when a leak occurred in the Ground Umbilical Carrier Plate during fueling. This pushed the launch into 2011. Later, a 24 hr delay was inserted in order to allow ATV-2 to dock, in accordance with the flight rule that required an interval of 72 hr before the Shuttle arrived at the station. It was finally launched on February 24, 2011.

- *STS-134 (ULF6)*: In June 2009 this mission was planned with a July 29, 2010 launch date. It was postponed on July 1, 2010 to November 26 to gain time to replace a magnet in the AMS-02 payload. Shortly thereafter it was delayed to no sooner than February 26 by conflicts with other launches intended to visit the ISS, temperature issues associated with the station's orbit in January, and Air Force range conflicts with a number of unmanned launches. Three more postponements slipped the launch into April 2011, due to STS-133 ET issues and the launch turnaround time required at the single remaining active Shuttle launch pad (39A) following the launch of STS-133. The third delay, to April 29, was to avoid a clash with a Russian Progress ship arriving at the ISS. That launch was also scrubbed due to a failed APU fuel line heater. The launch was then rescheduled a further three times by ongoing repairs and electrical testing issues before it finally got off the ground on May 16.

- *STS-135 (ULF7)*: On January 31, 2011, the former STS-335 rescue mission (for STS-134) was revised to STS-135 with launch set for June 28, 2011. On May 24 it was postponed to July 8 by delays in launching STS-134, on which date it lifted off. This was the final Shuttle launch processing sequence.

LAUNCH-ON-NEED

After the loss of Columbia, the Launch-on-Need (LON) mission category was created to support a disabled or damaged Orbiter that was unable to return safely to Earth and would require a crew rescue. From STS-114 in 2005, the STS-300 series designation was assigned to the LON missions. STS-300 would have been the designation for the first such mission flown, with number designation increasing with each such mission.

Crews for LON flights were initially made up of the flight deck crew (CDR, PLT, MS1 and MS2) from the most recently flown mission, on the basis that they would be ahead of those still in training. All potential rescue missions would have been outfitted to return with ten or eleven astronauts, depending on how many were launched on the mission that required rescuing. LON missions were expected to last up to eleven days. Luckily, none were ever needed.

When a problem became apparent with STS-115, it became necessary to identify missions assigned to Orbiters for LON planning manifests. Atlantis became STS-300 and it would support both STS-114 and STS-121. Discovery was designated STS-301 to support STS-115. Beginning with STS-116, the rescue designations were based on the regular mission for which the Orbiter was being processed. Thus the LON mission on standby to rescue STS-116 was STS-317 because the next manifested flight for the LON Orbiter was STS-117. The crews for the missions would have included a CDR, PLT and two experienced MS (preferably with RMS and EVA experience) from the flight deck crew. After STS-124, the LON crew was made up of members of the next crew to fly because the station was sufficiently stocked with supplies to support a full Shuttle crew under safe-haven conditions until it could be rescued. For STS-135, the final Shuttle mission, there was no LON available, so the four-person crew received training and equipment to enable them to remain on the ISS until they could return to Earth in a series of Soyuz vehicles (see Chapter 10).

Crewing for the LON missions is given in Table 4.5.

INTO THE WIDE BLUE YONDER

With the payload bay loaded, the Shuttle fueled on the pad, and everything checked out and ready to go, the launch sequence would progress to the point where the launch was cleared for the final countdown. For the crew, this was where it all came together. This was *Launch Day*.

Table 4.5 STS Launch-On-Need Crew Assignments 2005–2011

STS Mission	STS Orbiter	LON Mission	LON Orbiter	LON Crew				
				Commander	Pilot	MS	MS	
STS-114	Discovery	STS-300	Atlantis	Lindsey	Kelly M.	Fossum	Sellers	
STS-121	Discovery	STS-300	Atlantis	Jett	Furguson	Tanner	Burbank	
STS-115	Atlantis	STS-301	Discovery	Polansky	Oefelein.	Curbeam	Fuglesang (ESA)	
STS-116	Discovery	STS-317	Atlantis	Sturckow	Archambault	Swanson	Reilly	
STS-117	Atlantis	STS-318	Endeavour	Kelly S.	Hobaugh	Mastracchio	Williams D. (CSA)	
STS-118	Endeavour	STS-322	Discovery	Melroy	Zamka	Parazynski	Wilson	
STS-120	Discovery	STS-320	Atlantis	Frick	Poindexter	Walheim	Schlegel (ESA)	
STS-122	Atlantis	STS-323	Discovery	Gorie	Johnson G. C.	Foreman	Linnehan	
STS-123	Endeavour	STS-324	Discovery	Kelly M.	Ham	Garan	Fossum	
STS-124	Discovery	STS-326	Endeavour	Furguson	Boe	Bowen	Stefanyshyn-Piper	
STS-125[1]	*Atlantis*	*STS-400*	*Endeavour*	*Furguson*	*Boe*	*Bowen*	*Kimbrough*	
STS-126	Endeavour	STS-326	Discovery	Archambault	Antonelli	Acaba	Swanson	
STS-119	Discovery	STS-319	Endeavour	Polansky	Hurley	Wolf	Cassidy	
STS-127	Endeavour	STS-317	Discovery	Sturckow	Ford	Forrester	Hernandez	
STS-128	Discovery	STS-318	Atlantis	Hobaugh	Wilmore	Melvin	Bresnik	
STS-129	Atlantis	STS-329	Endeavour	Zamka	Virts	Hire	Robinson	
STS-130	Endeavour	STS-330	Discovery	Poindexter	Dutton	Mastracchio	Metcalf-Lindenburger	
STS-131	Discovery	STS-331	Atlantis	Ham	Antonelli	Reisman	Good	
STS-132	Atlantis	STS-332	Discovery	Lindsey	Boe	Drew	Bowen	
STS-133	Discovery	STS-333	Endeavour	Kelly M.	Johnson G. H.	Fincke	Vittori (ESA)	
STS-134	Endeavour	STS-334	Atlantis	Furguson	Hurley	Magnus	Walheim	
STS-135	Atlantis	None	None	Several (2+) Soyuz TMA would have been used if required				

[1] STS-125 was the final servicing mission (SM-4) to the Hubble Space Telescope (HST) and not an ISS mission.
The STS-400 crew was drawn from the planned STS-126 mission.

The sight of a fully fueled Shuttle, up close and personal, was eloquently described by Clayton Anderson in his 2015 book *The Ordinary Spaceman*.[7] Viewing Atlantis on the pad on June 8, 2007, ready to carry him on his first flight into orbit as a member of the STS-117 crew, he described the vehicle as "standing tall against the background of a near cloudless sky. I swallowed and said a silent prayer." Before riding the elevator to the Orbiter Access Arm, he was awed by the scale of the 200 ft (70 m) vehicle in front of him and was vividly reminded of its raw power. As the cryogenic liquids "audibly seethed" through the Orbiter's propulsion system, it appeared the vehicle was a living being. "I'd heard it many times and now I understood. She really did seem to breathe, as if alive." As Atlantis, restrained on the pad, groaned and creaked under the strain of its loaded mass, Anderson likened the scene to that of "a rodeo bull awaiting its break into the arena."

With the crew safely strapped into their seats, the point was reached at which the months, or even years of planning, preparation and dedication of the many teams of workers, engineers, designers, experimenters, scientists, testers, trainers, controllers, managers, and flight crew now culminated in the ignition of the engines to begin the latest Shuttle mission bound for a space station.

Notes

1. AIS interview with Janet Kavandi October 11, 2015, plus E-mail dated November 13, 2015
2. E-mail to AIS from Steve Hawley February 2017
3. Details of some of these documents were explained by the author in *The Hubble Space Telescope From Concept to Success*, "Protection against contamination," pp. 124–127, Springer-Praxis, 2016
4. Historical Survey and Evaluation of the Space Station Processing Facility, John F. Kennedy Space Center, Brevard County, Florida, prepared for NASA KSC Environmental Management Branch by Archeological Consultants Ltd, Sarasota, Florida, September 2010
5. Integrated Testing of Space Flight Systems, Stephanie Sowards and Timothy Honeycutt, Integrated Test & Verification, Ground Operations Project, NASA KSC, KSC-2008-014R, April 8, 2008
6. New Tenant moves into KSC Space Station Processing Facility, Irene Klotz, Space News, April 29, 2013 Spacenews.com last accessed May 3, 2016
7. *The Ordinary Spaceman, From Boyhood Dreams to Astronaut,* by Clayton C. Anderson, Nebraska University Press, 2015, pp. 197

5

Getting There

On behalf of the launch team,
and all the thousands of people here at KSC,
we're very, very proud that we finished strong
from the launch perspective.

Mike Leinbach, Shuttle Launch Director,
following the launch of STS-135, July 8, 2011

The loss of Columbia in 2003, while not part of the ISS assembly program, seriously delayed the completion of the station and signaled the end of the Shuttle program as soon as the assembly could be finished. Despite this tragic accident, the sequence of Shuttle launches for the assembly and resupply of the station proceeded remarkably smoothly. The process was not entirely trouble free though. Various issues involving the Orbiter's electrical wiring, main engines, and thermal protection system added to processing delays. Another constraint was the availability of Orbiters (particularly at times when one was having its systems upgraded by a scheduled Maintenance Down Period), the delivery of hardware, and starting in 2005 the need for the *next* Shuttle to be ready to mount a rescue mission if damage were to prevent the current flight from returning to Earth. All of this added to the three decades of experience in preparing a Shuttle stack for on-time launch while also dealing with the weather and humidity in Florida. However, the ascent profiles of the thirty-seven ISS assembly missions were remarkable trouble free, with only a few recording any significant issues.

FROM PAD TO ORBIT

During the launch of STS-112 in 2002 the back-up pyrotechnic separation system was used to free one of the SRBs from the launch platform because the primary charge had failed to sever the holding bolts and ground connections. The STS-114 launch in 2005, the first post-Columbia Return-To-Flight mission, became the most documented launch in

© Springer International Publishing Switzerland 2017
D.J. Shayler, *Assembling and Supplying the ISS*, Springer Praxis Books,
DOI 10.1007/978-3-319-40443-1_5

history. An array of cameras on the ground and on board high-altitude aircraft joined ground-based radar systems, sensors, and laser tracking to study the ascending Shuttle from every conceivable angle. In addition, a television system on the ET transmitted a spectacular view of the ascent, including the jettisoning, 2 min into the flight, of the spent SRBs and of the Orbiter after ET separation about 6 min later. The analysis of this imagery established that the ET was still shedding bits of foam insulation. This caused sufficient concern for mission managers to ask for further investigation of the situation once the Shuttle was on-orbit. Fortunately, the vehicle was able to return to Earth safely. Further foam debris impacts were seen on the second Return-To-Flight mission, STS-121, a year later, but none that were deemed to pose a serious risk.

The first night launch in over four years (due to the post-Columbia requirement to obtain detailed imagery of the ascent) was STS-116 in late 2006. It carried additional safety monitoring of the Space Shuttle Main Engines, known as the Advanced Health Management System (AHMS). On this mission, the performance data from the three engines was recorded only to evaluate the system, but for future missions this system would be able to shut down an ailing engine automatically if it detected an imminent failure. Fortunately, this intervention was never triggered on the remaining missions.

Shortly after STS-117 achieved orbit in early 2007, an initial inspection of the left-hand OMS pod indicated that a 25.4×64.77 in (10×25.5 cm) section of the thermal protection system had been pulled away from adjacent tiles. A detailed inspection the next day with a camera on the OBSS extension of the RMS confirmed the damage. It was subsequently repaired by spacewalker Danny Olivas.

Following STS-124's launch in June 2008, a routine post-launch walk around of the Pad 39A area revealed severe damage to the forty-year-old east wall of the north flame trench. The heat resistant brickwork across a 75×20 ft (22.86×6.10 m) section of the trench had been 'blown out' by the exhaust of the Shuttle engines. Then in November that year, it was noticed that a small amount of thermal blanket had come loose on the aft of Endeavour during the STS-126 ascent. Later analysis determined that it was still safe for re-entry though.

Mission Control Houston, Shuttle-ISS

Mission Control Houston (MCC) traditionally occupied the second floor of the block officially known as Building 30 on the NASA JSC campus near Clear Lake, 30 miles south of central Houston. This handled all fights from 1965 to the mid-1990s, when a new set of flight control rooms were commissioned in an adjacent office block called Building 30-South (30S) to handle Shuttle and ISS operations. The new building also housed payload and mission support rooms, plus some contractor and specialist team rooms that supported the flights. Building 30S officially became operational in 1998 after a period of sharing flight control roles with the old FCR. In April 2011, just prior to the retirement of the Shuttle, the original Building 30 was renamed the Christopher C. Kraft Jr. Mission Control Center but it kept its old radio call sign. Kraft was a now-retired NASA engineer and manager who had been instrumental in establishing flight control operations for Project Mercury and the ethos of the Mission Operations Room. Indeed, Kraft established the tradition of 'Mission Control.' The new control rooms in 30S were labeled Red, White and Blue. From 1996 the new White Room was used for Shuttle missions through to 2011. The Blue Room is used for ISS operations. The Red Room has traditionally been used for training flight controllers.

"Tower cleared. Houston now controlling," is the call on July 8, 2011 as Atlantis lifts off and responsibility for the final Shuttle mission, STS-135, passes to the controllers in Mission Control-Houston.

Table 5.1 Shuttle Flight Directors For ISS Assembly Missions 1998–2011

STS	Mission	MCC	Lead Flight (#)	Ascent/Entry	Orbit 1	Orbit 2	Planning (Orbit 3)	Orbit 4	MOD
88	2A	FCR White	Antares (29)	Shannon	CASTLE (L)	Engelauf	Algate	Shannon	Bantle
96	2A.1	FCR White	Turquoise (28)	Ham	HALE (L)	Dye	Reeves	-	Bantle
101	2A.2a	FCR White	Regulus (31)	Shannon	ENGELAUF (L)	Beck, Cain	Shaw, Cain	-	Bantle
						Beck, Shaw, Cain switched shifts during flight			
106	2A.2b	FCR White	Regulus (31)	Hale	ENGELAUF (L)	Dye	Beck	-	Reeves
92	3A	FCR White	Altair (24)	Hale (Ascent) Cain (Entry)	Castle	Shannon	SHAW (L)	Austin	Heflin
97	4A	FCR White	Alpha (23)	Hale (Ascent) Cain (Entry)	REEVES (L)	Engelauf	Beck	-	Bantle
98	5A	FCR White	Antares (29)	Cain	CASTLE (L)	Beck	Austin	-	Bantle
102	5A.1	FCR White	Midnight (38)	Hale	SHANNON (L)	Hill	Dye	-	Bantle
100	6A	FCR White	Regulus (31)	Cain	ENGELAUF (L)	Beck	Austin	-	Heflin
104	7A	FCR White	Atlas (40)	Hale	HILL(L)	Hale	Shannon	-	Castle
105	7A.1	FCR White	Iron (36)	Shannon	DYE (L)	Beck	Austin	-	Hale
108	UF1	FCR White	Turquoise (28)	Cain	HALE (L)	Hill	Koerner	-	Heflin
110	8A	FCR White	Ares (41)	Cain	HANLEY (L)	Dye	Stich	-	Heflin
111	UF2	FCR White	Atlas (40)	Shannon	HILL (L)	Ceccacci	Beck	-	Castle
112	9A	FCR White	Regulus (31)	Shannon	ENGELAUF (L)	Koerner	Curry	-	Castle
113	11A	FCR White	Regulus (31)	Shannon	ENGELAUF (L)	Koerner	Curry	-	Castle
114	LF1	FCR White	Atlas (40)	Cain	HILL (L)	Ceccacci	Koerner	Beck	Engelauf
121	ULF1.1	FCR White	Intrepid (57)	Stich	CECCACCI (L)	Knight	Dye	-	Engelauf
115	12A	FCR White	Iron (36)	Stich	DYE (L)	Koerner	Lumney, B	-	?
116	12A.1	FCR White	Intrepid (57)	Stich (Ascent only) Knight (Entry only)	CECCACCI (L)	Abbott	LaBrode	Jones, R	Engelauf

(continued)

Table 5.1 (continued)

STS	Mission	MCC	Lead Flight (#)	Ascent/Entry	Orbit 1	Orbit 2	Planning (Orbit 3)	Orbit 4	MOD
117	13A	FCR White	Topaz (56)	Knight	KOERNER (L)	Lunney, B.	Jones, R	Sarafin	Engelauf
118	13A.1	FCR White	Aquarius (55)	Stich	ABBOTT (L)	Jones R. (FD1-6) Sarafin (FD7-EOM)	Moses (FD1-undock) Dye (Prelaunch-post-landing)	LaBrode	Engelauf
120	10A	FCR White	Pegasus (46)	Knight, Lunney, B (Entry)	LABRODE (L)	Moses (FD2-13) Abbott (FD1, 14, & wave off)	Sarafin (FD-13) Ceccacci (Prelaunch, FD1, 14, & waveoff)	Dye	Engelauf
122	1E	FCR White	Kodiak (66)	Knight (Ascent) Lunney, B (Entry)	SARAFIN (L)	Ceccacci	Dye	Abbott	Engelauf
123	1J/A	FCR White	Apex (67)	Lunney, B (Ascent) Jones, R (Entry)	MOSES (L)	LaBrode	Abbott	Jones, R Ceccacci	Engelauf
124	1J	FCR White	Aquarius (55)	Knight (Ascent) Jones, R (Entry)	ABBOTT (L)	Sarafin	Dye Ceccacci	LaBrode	Mccullough
126	ULF2	FCR White	Kodiak (66)	Lunney	SARAFIN (L)	Ceccacci (FD1-12) Dye (FD 13-EOM)	Dye (FD 1-13) Alibaruho (FD14-EOM)	Jones, R	Mccullough
119	15A	FCR White	Iron (36)	Jones, R (Ascent) Lunney (Entry)	DYE (L)	Sarafin (FD 1-12) Ceccacci (FD 13-EOM)	Labrode (Prelaunch-FD1) Knight (FD2-8) Lunney (FD9-EOM)	Ceccacci	Mccullough
127	2J/A	FCR White	Iron (36)	Lunney, B	DYE (L)	Alibaruho	Horlacher Sarafin	Jones, R	Mccullough

128	17a	FCR White	Intrepid (57)	Jones, R	CECCACCI (L)	Alibaruho	Horlacher	Sarafin	Mccullough
129	ULF3	FCR White	Kodiak (66)	Lunney	SARAFIN (L)	Horlacher	Dye	Alibaruho	Mccullough
130	20A	FCR White	Defiant (59)	Knight	ALIBARUHO (L)	Horlacher	Edelen	Dye	Mccullough
131	19A	FCR White	Sigma (65)	Lunney	JONES, R (L)	Sarafin	Kerrick	Horlacher	Mccullough
132	ULF4	FCR White	Kodiak (66)	Jones, R (Ascent) Ceccacci (Entry)	SARAFIN (L)	Edelen	Kerrick	Dye	Mccullough
133	ULF5	FCR White	Onyx (54)	Jones, R (Ascent) Ceccacci (Entry)	LUNNEY, B (L)	Kerrick	LaBrode	Dye (plus Prelaunch)	Mccullough
134	ULF6	FCR White	Viper (76)	Jones, R (Ascent) Ceccacci (Entry)	HORLACHER (L)	Dye	Alibaruho	Jones, R	Mccullough
135	ULF7	FCR White	Defiant (59)	Jones, R (Ascent) Ceccacci (Entry)	ALIBARUHO (L)	LaBode	Dye	N/A	Mccullough

[REF: Data courtesy: Space Shuttle Mission Summary, Robert D. ('Bob') Legler and Floyd V. Bennett, Mission Operations, DA8, NASA JSC, Houston, Texas, NASA TM-2011-216142, September 2011]

Each mission required the crew to keep up with the housekeeping, which was one of the more mundane roles of a Shuttle astronaut.

A view of the additional lap top computers and other devices that adorned the aft flight deck and supplemented the general purpose computers of the Orbiter.

A view of the Mission Control Center in Houston during STS-135, the final Shuttle mission. On duty at the Capcom console is veteran astronaut Shannon Lucid, who first took that role for STS-51I in August 1985. Occupying the Flight Director console is Planning Shift FD Paul Dye.

Director Of Operations In Russia

Continuity from the Shuttle-Mir program to the assembly of the ISS was provided by the NASA JSC Office at the Cosmonaut Training Center named for Yuri A. Gagarin (TsPK) in Star City, Moscow. A succession of senior astronauts served as the Director of Operations in Russia (DOR) with a small team supporting the astronauts in training for long-duration expeditions and visiting Shuttle crews. The astronauts who served in this capacity during the years of ISS assembly provided a direct link to senior Russian space officials (see Table 5.2).

Shuttle At ISS

In June 1993, an $18 million contract was awarded to the Russian space engineering company RKK Energiya to supply a number of APAS-89 docking systems to enable US Shuttles to dock at Mir. Rockwell International, NASA's prime contractor for the Shuttle, took receipt of the first flight unit on September 11, 1994.[1] It was integrated into the Orbiter Docking System meant for use with Space Station Freedom, which at that time was then being fabricated at the company's plant in Downey, California. In December 1994 the ODS was delivered to KSC to be mounted in the payload bay of Atlantis during preparations for STS-71, the first Shuttle-Mir docking mission. In all, Shuttles made nine dockings with Mir without encountering any difficulties.

Table 5.2 Director Of Operations In TsPK, Russia 1998–2012

Name	Date Start Tour	Date End Tour
Terrence W. Wilcutt	1998 July	1999 February
Joe F. Edwards	1999 February	1999 July
Donald A. Thomas	1999 July	2000 May
Scott J. Kelly	2000 May	2001 January
William S. McArthur	2001 January	2001 August
Chris A. Hadfield	2001 August	2003 January
Kenneth D. Cockrell	2003 January	2004 January
Kevin A. Ford	2004 January	2005 January
Douglas H. Wheelock	2005 January	2006 January
Eric A. Boe	2006 January	2006 September
Douglas G. Hurley	2006 September	2007 October
John McBrine ( )*	*2007 October*	*2008 December*
B. Alvin Drew	2008 December	2009 December
Mark L. Polansky	2009 December	2012 May

NOTE: (*) John McBrine was not an astronaut
[Information courtesy of Michael Cassutt and Bert Vis]

The APAS-95 system for the ISS was essentially a modified form of the APAS-89 where the active ring on the Shuttle projected forward to capture the passive ring of a Pressurized Mating Adapter of the station. Once the system was aligned and stable, it would retract, drawing the spacecraft together so that twelve pairs of structural hooks (one active and one passive per pair) could latch to establish an airtight seal.

Orbital Docking System (ODS): With a mass of 3,500 lb (1,575 kg), this comprised an airlock, a supporting truss structure, a docking base, and the Russian APAS topping out the 13.5 ft (4.1 m) tall structure. The ODS was installed in the payload bay behind the tunnel adapter. The external airlock formed an airtight internal tunnel between the Shuttle and the station after the docking. The truss provided a strong base upon which to mount all elements of the docking system. Attached to the payload bay, it included rendezvous and docking aids such as trajectory control systems, cameras, and docking lights.

Aft Flight Deck Panels: Two control panels designated A (Aft) 7 and A8 on the aft flight deck of the Orbiter were linked up to nine avionics boxes in the sub-floor of the external airlock in order to provide power and logic control of the various mechanical components. The flight deck panels included circuit breaker protection, the switches (A7) to apply logic and power to the Russian-supplied APAS (A8) control panel, the avionic boxes, docking lights and valves.

SHUTTLE DOCKING PROFILE

As described in the companion volume *Linking the Space Shuttle and Space Station: Early Docking Technologies from Concept to Implementation*, several Shuttle flights contributed to the profiles for the ISS assembly missions, most notably the servicing missions to the

Hubble Space Telescope. These were then trialed by the Shuttle-Mir docking missions, with Orbiters being routinely operated in conjunction with a space station without major incident. But then in the wake of the Columbia accident it was decided to create the opportunity to visually inspect the thermal protection system on the 'belly' of the Orbiter, either by using the RMS fitted with an extension boom that carried cameras or directly by the crew of the space station.

It will be recalled that there were three main methods available to the Shuttle for docking with the ISS:

- R-Bar. This is employed when the active spacecraft (Shuttle) approaches a passive target (the ISS) along an imaginary line (bar) aligned with Earth's radius (R, the radial vector). An approach from 'above' with the target in between the active vehicle and Earth is called the 'negative R-Bar' and an approach from 'below' where the active vehicle is positioned between the target and Earth is known as the 'positive R-Bar.'
- V-Bar. This describes a 'horizontal' approach along the passive vehicle's velocity vector (V) from 'behind,' parallel to the velocity of the target (in front).
- Z-Bar. In this 'horizontal' case the active spacecraft approaches from the side of the orbital plane of the target.

As the Orbiter completed its final approach the Commander initiated the Rendezvous Pitch Maneuver, flipping the Orbiter over in order to allow the station crew to inspect, image, and report on the condition of the underside thermal protection system.

Rendezvous Pitch Maneuver

After the loss of Columbia, NASA instigated a requirement to inspect the Orbiter's surface, in particular the thermal protection system, for signs of damage during the launch and ascent phase. Using the information gathered by the network of support cameras, on

ISS-27 FE Cady Coleman inspects Endeavour during STS-134, demonstrating one vantage point from which to watch an Orbiter approaching the ISS.

board cameras, and visual inspections by the crew using the RMS and OBSS system, managers could determine whether to proceed as planned, attempt to repair TPS damage, or declare the Orbiter unsafe to attempt re-entry and instruct its crew to use the ISS as a safe haven until a rescue flight could be arranged.

Part of this analysis came from the station crew performing a visual inspection and obtaining imagery while the Orbiter executed the Rendezvous Pitch Maneuver, a slow back-flip maneuver flown by the Commander to display the belly of the spacecraft. In the docked phase of the mission the ground teams would draw up any action plan well in advance of the nominal end of the mission.

The back-flip was officially the R-Bar Pitch Maneuver (RPM), but it was widely known as the Rendezvous Pitch Maneuver. The Shuttle approached along the R-Bar, the imaginary line from the center of Earth to the ISS. When 600 ft (180 m) away the Shuttle executed the 360° back-flip, which took about 10 min. Because this operation was conducted very close to the station it required highly skilled piloting because the Commander would lose sight of the station when exposing the belly of the Orbiter for inspection. Care had to be taken to prevent a collision. This maneuver always looked spectacular, especially when the movie was speeded up. This done, the Orbiter would prepare to dock. This would usually involve moving from directly beneath the station into a position in front of it, then completing the approach.

The RPM was developed by Stephen R. Walker, an aerospace technologist of the Flight Design and Dynamics Division (DM34) and Mark B. Schrock and Jessica A. LoPresti, two members of the United Space Alliance engineering staff in DM34, all of whom were engineers at JSC. The challenge for the Orbiter crew was being unable to see the ISS for a portion of the maneuver. "The unique thing is that we knew from the beginning we could do this maneuver," observed Walker, who also had experience on the Rendezvous Console in Mission Control working shifts during Shuttle rendezvous activities in conjunction with free-flyers and some of the Hubble servicing missions. It did not require any new hardware, it was simply a matter of developing a procedure to execute the flip in a confined envelope.[2]

The APAS docking system on PMA-2 photographed by the STS-108 crew.

The ranging scale on the screen can be seen as the Orbiter approaches the ISS for docking.

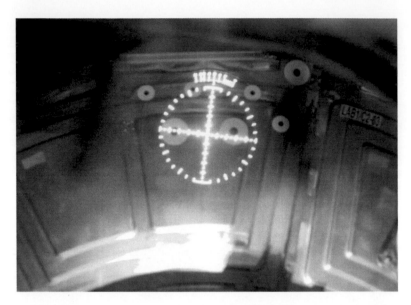

The orientation angle is displayed on a screen as the Commander aligns the docking target.

On the aft flight deck the crew congratulate themselves after successfully docking with the ISS.

THE DOCKED PHASE

Once the docking was achieved, the crew could prepare to enter the space station and set about their assigned tasks, either independently or jointly with the residents. In the early stages of ISS assembly, with the first resident expedition aboard, some activities such EVAs were carried out by the Orbiter crew prior to opening the hatches to access the station and be greeted by their hosts.

It's All In The Moves

With the hatches finally open, the emphasis shifted to the transfer of hardware and supplies between the two vehicles. Three areas came under this joint activity. After being outlined here, these activities will be dealt with individually in the remaining chapters.

Shuttle-Based EVAs: With the delivery of the Quest airlock by STS-104 in 2001, spacewalks by Shuttle crew could be conducted from the station rather than from the Orbiter. From 2002, all EVAs by Shuttle crew at the ISS (with the exception of STS-114 in 2006) were made using the Quest airlock.

RMS Activities: The Remote Manipulator System (RMS) of the Orbiter was used to transfer payloads and large hardware elements from the payload bay across to the ISS. Initially, these transfers were directly onto the nascent station but as the structure grew it became necessary to install the station's own RMS system, Canadarm2, to operate in areas beyond the reach of the Shuttle RMS. And Canadarm2 continued activities when no Shuttle was present. The two robotic arms were instrumental in supporting the EVA program.

Logistics Transfer: While activities were underway outside, there was an extensive program of work inside the Orbiter and the ISS. The primary internal objective was to transfer masses of material from the Orbiter's internal compartments, Spacehab, or the Multi-Purpose Logistics Modules into the station. This task occupied most of the time the crew were actually inside the station, lasting several days. The transfer of logistics had to be carefully planned and orchestrated, as did the return of science samples and miscellaneous unwanted items to the Orbiter. These controlled movements involved a strict timescale and a checklist that had to have all its boxes ticked before the hatches could be closed for the final time. This program of moves involving several astronauts, and sometimes ISS residents, was choreographed by the Shuttle crewmember assigned as the loadmaster. Their responsibility was to ensure that items departed the Orbiter in the proper sequence and were stowed in the correct position on board the station, and then that items to be returned to Earth were stowed on the spacecraft in a manner that would preserve its center of mass during re-entry.

Extra Power

The time that any Shuttle mission could spend on-orbit was limited by the amount of consumables it could carry. During the first decade of Shuttle operations the average duration of a mission was between seven and ten days, but from 1992 the addition of the Extended

Docking achieved. Endeavour is seen linked to the PMA-2 on the Harmony Node 2. In the foreground at left is the Japanese Kibo laboratory, and to the far right, partially out of frame, is the European Columbus laboratory.

Duration Orbiter (EDO) pallet in the rear of the payload bay containing 368 lb (167 kg) of liquid hydrogen and 3,124 lb (1,417 kg) of liquid oxygen allowed flights of up to sixteen days. This facility was used on fourteen missions that focused upon research in microgravity and life sciences as well as terrestrial and astronomical observations. The EDO flights were between June 1992 (STS-50) and January 2003 (STS-107).

Other studies and innovations throughout the program helped to further extend the time an Orbiter could remain on-orbit. These included a laundry for crew apparel, an improved waste collection system, and the addition of solar panels.[3]

There was consideration given to adding a second EDO pallet to Endeavour as a generic upgrade to permit the Orbiter to remain in space for up to 28 days. However, with the upcoming ISS program it was decided to remove the EDO capacity to save weight and thereby enable more logistics to be carried to the station. Once the Station-to-Shuttle Transfer System (SSPTS, pronounced 'Spits') was introduced in 2006, and with the retirement of the Shuttle fleet pending in 2011, the EDO pallet was no longer required.

The SSPTS system enabled the docked Orbiter to draw solar power from the ISS to reduce the drain on its own fuel cells. It replaced the Assembly Power Converter Unit (APCU) with the Power Transfer Unit (PTU). In fact this system was installed only in Discovery and Endeavour, enabling them to remain docked at the station for four days longer than Atlantis.

During the early stages of ISS assembly, the APCU converted the Shuttle's 28 volt direct current main bus power to 124 volt direct current to make it compatible with the 120 volt direct current of the US segment of the station, thus supplementing the power provided by the Russian modules. The PTU could improve this process by converting the 120 volt direct current supplied by the station to the Orbiter's 28 volt direct current main bus power and transferring an additional 8 kW of power to the Orbiter. This two-way system allowed each vehicle to use the other's power systems when required, but by the time the SSPTS was up and running the station did not need to draw any power from the Orbiter. To accommodate the new system, PMA-2 at the front of Destiny was rewired during the STS-116 visit. The first mission to use the system operationally was STS-118.

Re-Boost

One advantage of having the Shuttle docked to the ISS was that, providing there were sufficient reserves, the OMS engines could perform a series of small burns to alter the orbit of the complex either to avoid orbital debris or to stave off orbital decay. Shuttle-ISS missions generally had about 511.6 lb (232 kg) of propellant available for re-boost maneuvers, whilst ensuring there was enough remaining for the Orbiter to complete its planned undocking, fly-around and separation sequence, any subsequent maneuvers in the run up to the de-orbit burn…and *still* have some left for contingencies.

A Growing Station

Some lucky Shuttle crewmembers made return visits to the ISS during its assembly, in some cases after a gap of several years, and everyone found the changes to the station remarkable.

Ellen Ochoa first visited it in May 1999, when there was just the Zarya module and the Unity Node. When she returned three years later on STS-110 she was struck by the difference.[4] "It was just incredible to see how the station had grown between STS-96 and STS-110. I got a chance to go back and see all the pressurized modules and it just seemed absolutely huge on-orbit.* When you floated from one end to the other, it just seemed to go on forever. It was about 150 feet [45.7 m] of pressurized area. I wouldn't say it was disorienting…because we ensured that 'up and down' remained in the same orientation as you go from module to module. That wasn't the case on Mir. We tried to keep that orientation aboard the ISS to make it less disorientating. You don't have the sensation of going from one module to another and having to flip to whichever way is up. That remains pretty constant. But if you do make a ninety degree turn, to go down into the Russian docking compartment or something, you do have to pause to get your bearings, to make sure you know which way you're heading."

Asked if she still recognized the Unity Node and Zarya after what she had seen on STS-96 Ochoa replied, "The node had changed quite a bit, mainly because by then we were stowing a lot of equipment in the node. And it was no longer the first module we came to. We entered the big US laboratory [Destiny] first. Zarya seemed pretty much the same but the new modules are so much bigger and that's where the focus of living and working was – the Russian service module and the US lab – and I had never seen those on-orbit before. You know what to expect from pictures and descriptions, but it was wonderful to see a person actually working in the lab and spend a lot of time with the robotics workstation."

After the hatch opening and welcoming ceremony, the real work for a visiting Shuttle crew begins.

---

*This was *prior to* the installation of two additional Nodes and the European and Japanese laboratories which made the ISS considerably larger.

Habitability On Station

In the early years, life aboard the embryonic International Space Station was very basic, with most of the visiting crews spending their non-working hours inside the Orbiter. For some, that early exposure to the ISS would pay off years later, when they returned for a second visit or to serve as a member of a resident crew.

Ed Lu felt that his preparation with Yuri Malenchenko to fly STS-106 in 2000 was very helpful in developing the bond for subsequently training to undertake a long flight together in 2003. Also, because their first visit was to outfit the newly docked Zvezda, their training was focused more on the subsystems of the station than any other Shuttle crew before or since.[5]

On the early ISS assembly flights the Shuttle crews did not always enter the station immediately after docking, they worked on their own program, which more often than not involved spacewalks, prior to joining the resident crew inside the station for a few hours on the final days of the visit.

Some crewmembers of the STS-127 assembly mission and the ISS-20 station residents enjoy a meal on the ISS: [clockwise from left] Mike Barratt, Roman Romanenko, unidentified, Koichi Wakata (floating above), Robert Thirsk, Julie Payette, Frank De Winne and Chris Cassidy. Not clearly visible or out of frame are Mark Polansky, Doug Hurley, Dave Wolf, Tim Kopra, Tom Marshburn and Gennady Padalka.

On STS-98 Tom Jones experienced something similar, as he recalled in 2006, "We had our docking day, and we had a long afternoon with the hatches open, but then we closed the hatches for the spacewalk the next day, reopening them again at the end of that day. Then we did the whole of the next day working together on the activation of the Destiny lab. Then we had the hatches closed for several days to do the second and third spacewalks. We opened them up after the third spacewalk and did a full day and the next day of transfer and installation work prior to undocking the next morning. So we had probably a total of two and half days with the hatches open.

"Each crew had a detailed flight plan that they followed. The station guys had their own flight plan. They didn't really show it to us. We didn't really care about their flight plan except when there was an overlap and we did things in conjunction. Similarly they had no interest at all in the Shuttle flight plan. They just said, 'Okay, we understand that you guys will be working over here. Let us know if we can help. We'll stay out of your way.' And vice versa, of course. We'd consult about every big job that we were going to be doing, for info purposes, but we were ruled by separate control centers: Moscow for the resident crew and MCC-Houston for us. It proved very easy to get around each other. I would see Shep [Bill Shepherd, ISS-1 Commander] over there hooking up his various computers while I was linking up some communication line or removing bolts from safety equipment or something. I would say here is what I am doing and I will be over here for the next half an hour. It was very easy to de-conflict those things."

For Jones one of the few disappointments was that the schedule was so packed full of work, there was little time to have a movie night or get together for a space station meal where they could enjoy a couple of hours together, exchanging stories. That just never happened. As Jones reflected, "We would cook up a meal and float over to the station and gather in the quarters to eat some of the things that we'd brought and share some of our stuff with them. You might find two of the station guys there still working. And our crew would never all be there at one time because we were all cast in different directions. But one by one we would socialize there. We all slept on the Orbiter except for the last night, when I think three of us slept over on the station because we wanted to.

"We found it a struggle to even find time for a late night coffee break together. I was so beat that I just had to go to bed because I was to make a spacewalk the next day. So I went back to my side of the ship and went to sleep. On the day before undocking, I still said I'm just completely exhausted and I gotta go to sleep and I'll say goodnight to you guys now. I saw them the next morning before we closed the hatches. But it was tough to find that social time, and that's one of my regrets."[6]

DOWN TO BUSINESS

With the Shuttle crew safely aboard the station for a period ranging from six to twelve days depending on the flight objectives, the activities were divided into the transfer of large elements of hardware using the Shuttle RMS, or as on later missions, the Shuttle arm working in conjunction with the arm that was installed on the ISS. In addition to the extensive robotics activities there was normally a series of demanding spacewalks by members of the Orbiter crew, sometimes supplemented by members of the resident crew. Finally, there was the internal transfer between the vehicles of tons of hardware and supplies, or the return of unwanted equipment, waste and experiment samples, an operation described in the next chapter.

Notes

1. NASA News 94-151, September 12, 1994
2. Steve Walker: Flight Controller Readies Shuttle for First Flip, Return to Flight Feature, NASA website September 20, 2004, http://www.nasa.gov/vision/space/preparingtravel/steve_walker_prt.htm
3. Stretching the Shuttle, David J. Shayler, *Orbiter* 83, AIS Publications, March 1995 pp. 45–69
4. AIS interview with Ellen Ochoa, March 2, 2004
5. AIS interview with Ed Lu, June 2005
6. AIS interview with Tom Jones, August, 3, 2006

6

Crew Transfers and Loadmasters

> *The satellite space station, regardless of the form it may
> eventually have…would be put together in piecemeal fashion.
> Materials and equipment would be brought up by a fleet of
> cargo-carrying rocket space-ships. Even after the station
> was completed, a number of commuting rocket ships
> would be necessary to bring needed supplies as well
> as replacements for the crew.*
>
> From *Space Ships & Space Travel* by Frank Ross Jr, 1956

Written four decades before the Shuttle flew to Mir, these words accurately forecast the need to develop a regular resupply system and logistics infrastructure to keep a station in operation. In 1978 the Soviets introduced the first unmanned variant of their Soyuz, named Progress, which routinely supplied cargo, consumables, and propellants to the Salyut 6, 7, and Mir stations, and is today servicing the ISS. During the nearly twenty years the ISS has been in space, a variety of unmanned cargo craft have supplemented the capabilities of the Space Shuttle to stock it with supplies and new hardware and to remove unwanted items. But none of these spacecraft has been able to match the great volume of the Orbiter's payload bay, nor its capacity to transport an array of different logistics owing to its particular portfolio of resources.

The vast payload bay of the Shuttle, the robotic manipulator, and a crew's ability to perform multiple EVAs were crucial to the assembly of the ISS. But assembly was not the only reason for sending the Shuttle there: that same payload bay gave the Orbiter a tremendous advantage over other resupply vehicles.

And because the operation of both Mir and the ISS required the exchange of crews, the Shuttle could ferry three residents to or from a station just by reducing the Orbiter's crew to the CDR, PLT and two Mission Specialists.[*]

---

[*]The largest number of people ever carried on a Shuttle was eight. On STS-61A in 1985 Challenger launched and landed with eight astronauts to operate Spacelab-D1. On STS-71 in 1995, the first Shuttle-Mir docking, Atlantis launched with seven people, two of whom remained aboard Mir while the Shuttle retrieved three and therefore landed with eight people.

© Springer International Publishing Switzerland 2017
D.J. Shayler, *Assembling and Supplying the ISS*, Springer Praxis Books,
DOI 10.1007/978-3-319-40443-1_6

Table 6.1 lists the loadmasters for the Shuttles which docked at Mir or the ISS. The duties of this person would commence shortly after docking, and over the next several days they would ensure everything was correctly transferred. This job would end only when the hatches were closed in preparation for undocking.

Table 6.1 Loadmaster Assignments And Logistics Allocations
[Where known – Table does not include major external payloads]

STS-88	Loadmaster	Krikalev (Zarya specialist)		Assistants	
ISS	Cargo mass to station	lb		Cargo mass from	lb
		kg		station	kg
NOTES					
STS-96	Loadmaster	Ochoa		Assistants	
ISS	Cargo mass to station	lb	4,228	Cargo mass from	lb 197
		kg	1,917	station	kg 89
NOTES	Transfers to ISS: 661 lb by EVA; 2,881 lb by IVA and water transfers of 686 lb (7 CWC)				
STS-101	Loadmaster	Weber		Assistants	
ISS	Cargo mass to station	lb	3,371	Cargo mass from	lb 1,391
		kg	1,530	station	kg 631
NOTES	2,657 dry cargo (by IVA) plus 4 CWC with H_2O 387 lbs, and 327 lb external (by EVA)				
STS-106	Loadmaster	Burbank		Assistants	
ISS	Cargo mass to station	lb	5,399	Cargo mass from	lb 948
		kg	2,448	station	kg 430
NOTES	Transfer IVA to station included 10 CWC (H_2O 780 lb)				
STS-92	Loadmaster	Lopez-Alegria		Assistants	
ISS	Cargo mass to station	lb	1,098	Cargo mass from	lb
		kg	498	station	kg
NOTES					
STS-97	Loadmaster	Garneau		Assistants	
ISS	Cargo mass to station	lb	1,457	Cargo mass from	lb 227
		kg	660	station	kg 102
NOTES	Transfers to station included 773 lb of hardware plus 7 CWC with 684 lb of H_2O				
STS-98	Loadmaster	Ivins		Assistants	
ISS	Cargo mass to station	lb	3,036	Cargo mass from	lb 872
		kg	1,377	station	kg 395
NOTES	Transfers of dry cargo by IVA included H_2O 10 CWC 993 lb; plus external by EVA 368 lb				
STS-102	Loadmaster	Thomas, A.		Assistants	Voss. J.S.
ISS	Cargo mass to station	lb	10,629	Cargo mass from	lb 1,649
		kg	4,821	station	kg 747
NOTES	9,649 dry cargo plus H_2O 980 lb on 10 CWC				
STS-100	Loadmaster	Guidoni (ESA)		Assistants	Rominger
ISS	Cargo mass to station	lb	6,346	Cargo mass from	lb 1,608
		kg	2,878	station	kg 729
NOTES	Transferred 1,380 lb of H_2O in 14 CWC				
STS-104	Loadmaster	Kavandi		Assistants	

(continued)

Table 6.1 (continued)

ISS	Cargo mass to station	lb	7,380	Cargo mass from	lb 626
		kg	3,348	station	kg 284
NOTES	Dry cargo 6,483 lb plus 897 lb of H_2O in 9 CWC				
STS-105	Loadmaster	Barry		Assistants	Culbertson
ISS	Cargo mass to station	lb	9,102	Cargo mass from	lb 3,802
		kg	4,128	station	kg 1,725
NOTES	Dry cargo to ISS included MPLM 6,314lb, middeck 1794 lb and 10 CWC with 994 lb of water; cargo transfer back to OV included MPLM 2,564 lb, and middeck 1,238 lb				
STS-108	Loadmaster	Godwin		Assistants	Gorie, Onufriyenko
ISS	Cargo mass to station	lb	6,244	Cargo mass from	lb 4,156
		kg	2,832	station	kg 1,885
NOTES	Cargo to ISS included MPLM 5,249 lb, Middeck 995 lb; Cargo to Shuttle included MPLM 3,007 lb, middeck 1,149 lb				
STS-110	Loadmaster	Bloomfield?		Assistants	
ISS	Cargo mass to station	lb	2,228	Cargo mass from	lb 2,607
		kg	1,010	station	kg 1,182
NOTES	Middeck cargo transfers included: Consumables 1,46 lb O_2, 45 lb N_2, and 1,465 lb of water (1,367 lb in 14 CWC and 68 lb in three PWR)				
STS-111	Loadmaster	Lockhart?		Assistants	Cockrell?
ISS	Cargo mass to station	lb	9,512	Cargo mass from	lb 6,343
		kg	4,314	station	kg 2,877
NOTES	Cargo to ISS included: MPLM 8,062 lb, Middeck 1,450 lb; Consumables transferred included: H_2O 884.9 lb (8 CWC with 788.9 lb and 4 PWR with 86.0 lb) Consumables transferred included O_2 34 lb for Joint Air Lock (JAL), N_2 18.9 lb tank transfer; return cargo included MPLM 4,668 lb and middeck 1,675				
STS-112	Loadmaster	Magnus		Assistants	Yurchikhin
ISS	Cargo mass to station	lb	1,444	Cargo mass from	lb 1,351
		kg	655	station	kg 613
NOTES	Dry cargo plus consumable transfers: H_2O in 16 CWC with 1,603.7 lb and 3 PWR with 54.4 lb; N_2 68.2 lb, O_2 60 lb				
STS-113	Loadmaster	Lopez-Alegria		Assistants	
ISS	Cargo mass to station	lb	2,160	Cargo mass from	lb 2,250
		kg	979	station	kg 1,020
NOTES	Consumables transferred to ISS: H_2O 690 lb (672 lb in 7 CWC plus 18 lb in one PWR), O_2 32 lb, 6 LiOH cans				
STS-114	Loadmaster	Lawrence		Assistants	Camarda
ISS	Cargo mass to station	lb	3,695	Cargo mass from	lb 6,600
		kg	1,676	station	kg 2,993
NOTES	Consumables transferred to ISS: H_2O 1,855.2 lb in 18 CWC and 5 PWR, N_2 36.7 lb, O_2 60.85 lb				
STS-121	Loadmaster	Wilson		Assistants	Reiter (ESA)
ISS	Cargo mass to station	lb	9,287	Cargo mass from	lb 6,051
		kg	4,212	station	kg 2,744
NOTES	Consumables transferred to ISS: H_2O 1,545.8 lb (1,454.9 in 15 CWC and 90.0 lb in 4 PWR) N_2 74.2 lb				

(continued)

Table 6.1 (continued)

STS-115	Loadmaster		Stefanyshyn-Piper		Assistants		
ISS	Cargo mass to station	lb	1,126		Cargo mass from	lb	993
		kg	510		station	kg	450
NOTES	Consumables transferred to ISS: H_2O 1,110.5 lb (1,043.8 lb in 11 CWC and 66.1 lb in 4 PWR), O_2 103 lb						
STS-116	Loadmaster		Higginbotham		Assistants		Polansky, Patrick
ISS	Cargo mass to station	lb	4,877		Cargo mass from	lb	4,911
		kg	2,212		station	kg	2,228
NOTES	Consumables transferred to ISS: O_2 69 lb, N_2 47.2 lb, H_2O 261.6 lb (201.9 lb. in two CWC and 59.7 lb in three PWR)						
STS-117	Loadmaster		Archambault?		Assistants		Sturckow?
ISS	Cargo mass to station	lb	1,277		Cargo mass from	lb	1,528
		kg	579		station	kg	693
NOTES	Consumables transferred to ISS: H_2O 1,656 lb, O_2 89 lb, N_2 33.2 lb; LiOH three used to ISS and three new to STS						
STS-118	Loadmaster		Morgan		Assistants		Drew
ISS	Cargo mass to station	lb	5,062		Cargo mass from	lb	3,297
		kg	2,296		station	kg	1,495
NOTES	Consumables transferred to ISS: H_2O 918.6 lb, O_2 77 lb, N_2 33.8 lb, plus 30 new LiOH cans from STS to ISS, and 12 cans (9 old and 3 used) from ISS to STS						
STS-120	Loadmaster		Wheelock		Assistants		Wilson, Tani
ISS	Cargo mass to station	lb	2,254		Cargo mass from	lb	2,020
		kg	1,023		station	kg	917
NOTES	Consumables transferred to ISS: H_2O 939.1 lb, O_2 30 lb, N_2 31.6 lb						
STS-122	Loadmaster				Assistants		
ISS	Cargo mass to station	lb	3,777		Cargo mass from	lb	3,585
		kg	1,713		station	kg	1,626
NOTES	Consumables transferred to ISS: H_2O 1,386 lb, O_2 95 lb, N_2 27 lb						
STS-123	Loadmaster		Doi (JAXA for Kibo)		Assistants		Linnehan
ISS	Cargo mass to station	lb	1,432		Cargo mass from	lb	1,565
		kg	649		station	kg	709
NOTES	Consumables transferred to ISS: H_2O 608 lb, O_2 608 lb						
STS-124	Loadmaster		Hoshide (JAXA for Kibo)		Assistants		
ISS	Cargo mass to station	lb	1,787		Cargo mass from	lb	1,807
		kg	810		station	kg	819
NOTES	Consumables transferred to ISS: H_2O 569 lb, O_2 121 lb, N_2 15 lb						
STS-126	Loadmaster		Pettit		Assistants		
ISS	Cargo mass to station	lb	16,390		Cargo mass from	lb	3,642
		kg	7,434		station	kg	1,651
NOTES	Consumables transferred to ISS: O_2 25 lb						
STS-119	Loadmaster		Phillips & Arnold		Assistants		
ISS	Cargo mass to station	lb	1,843		Cargo mass from	lb	1,963
		kg	836		station	kg	890
NOTES	Consumables transferred to ISS: H_2O 1,142 lb						
STS-127	Loadmaster		Marshburn		Assistants		Wolf
ISS	Cargo mass to station	lb	6,796		Cargo mass from	lb	5,853
		kg	3,082		station	kg	2,654

(continued)

Table 6.1   (continued)

NOTES	Consumables transferred to ISS: H_2O 1,225 lb, O_2 45 lb, N_2 12 lb			
STS-128	Loadmaster	Fuglesang (ESA)	Assistants	Hernandez
ISS	Cargo mass to station lb 18,548		Cargo mass from	lb 4,789
	kg 8,413		station	kg 2,172
NOTES	Consumables transferred to ISS: H_2O 1,243 lbs			
STS-129	Loadmaster		Assistants	
ISS	Cargo mass to station lb 31,789 (IVA & EVA)		Cargo mass from	lb 2,110
	kg 14,419		station	kg 957
NOTES	Consumables transferred to ISS: O_2 40 lb, N_2 11 lb			
STS-130	Loadmaster	Virts	Assistants	Hire
ISS	Cargo mass to station lb 1,991		Cargo mass from	lb 1,803
	kg 903		station	kg 817
NOTES	Consumables transferred to ISS: O_2 24 lb, H2O 1,095 lb			
STS-131	Loadmaster	Yamazaki	Assistants	Wilson
ISS	Cargo mass to station lb 12,060		Cargo mass from	lb 6,639
	kg 5,470		station	kg 3,011
NOTES	Consumables transferred to ISS: O_2 94.5 lb, H_2O 975 lb			
STS-132	Loadmaster	Sellers	Assistants	Ham, Antonelli
ISS	Cargo mass to station lb 9,765		Cargo mass from	lb 8,229
	kg 4,429		station	kg 3,732
NOTES	Consumables transferred to ISS: O_2 72 lb, N_2 10.5 lb, H_2O 1325 lb			
STS-133	Loadmaster	Barratt	Assistants	Stott, Boe, Drew, Lindsey
ISS	Cargo mass to station lb 6,607		Cargo mass from	lb 2,599
	kg 2,996		station	kg 1,178
NOTES	Consumables transferred to ISS: O_2 182 lb, N_2 26 lb, H_2O 931 lb			
STS-134	Loadmaster	Vittori	Assistants	Kelly
ISS	Cargo mass to station lb 14,067		Cargo mass from	lb 2,235
	kg 6,380		station	kg 1,013
NOTES	Consumables transferred to ISS: O_2 295 lb, N_2 18 lb			
STS-135	Loadmaster	All 4 crew members	Assistants	ISS resident crew
ISS	Cargo mass to station lb 25,758		Cargo mass from	lb 4,943
	kg 11,683		station	kg 2,242
NOTES	Consumable transfers to ISS: O_2 65 lb, N_2 111 lb, H_2O 1,652, cargo included 2,677 lb (1,214.3 kg) of food, enough for the nominal resident crew for 1 year			

[REF: Data courtesy: Space Shuttle Mission Summary, Robert D. ('Bob') Legler and Floyd V. Bennett, Mission Operations, DA8, NASA JSC, Houston, Texas, NASA TM-2011-216142 September 2011]

THE CARRIERS

The final use of the Spacelab long module was for the Neurolab mission of STS-90 in 1998. It was flown two months prior to the final Shuttle-Mir mission and eight months ahead of the first ISS assembly mission.

The priority of the early ISS flights was to achieve a point at which the station was able to support a crew of three, with regular resupply by unmanned cargo vessels from Russia and deliveries by visiting Shuttles. Since the large science modules were not to be installed until later, the initial focus was on construction and resupply. The science program would come into its own when the crew could be increased to six residents.

Although the Spacelab module was well configured for science missions, and was used by STS-71 on the first docking with Mir to investigate the state of adaptation of the retiring crew to the space environment, the module was unsuitable for logistic and utilizations flights. The later flights to Mir and the early resupply missions to the ISS used the Spacehab mid-deck augmentation module as a pressurized carrier. However, its payload had to be maneuvered through narrow hatches and tunnels into the station. The introduction of the Italian-built Multi-Purpose Logistics Module (MPLM) for the ISS was a major step forward because it could be hoisted out of the Orbiter and mated with the station and accessed through a large hatch. Furthermore, it could transport the same racks as were used aboard the station, making transfers much simpler. Then once unloaded, racks could be filled with unwanted items for the return flight to Earth. This provided a return-to-Earth capability much greater than the limited and unrecoverable Progress and the even more limited but recoverable Soyuz return payload.

Spacehab: This mid-deck augmentation module was developed in the 1980s in order to satisfy the need for additional locker space on Orbiters for small science experiments and additional storage space. When the space station was finally authorized as the ISS, it was realized that Spacehab would be an excellent facility for transporting additional supplies to the station and returning materials to Earth. Seven of the nine Shuttle-Mir docking missions used a Spacehab module to carry supplies to the Russian station. Of these, two had the single module configuration and the others flew the double module. Five double modules configurations were flown to the ISS, three (STS-96, -101, and -106) within the first two years of flight operations as the core configuration expanded, and the others (STS-116 and -118) flew in 2006 during the recovery from the loss of Columbia. By then, however, the far greater capacity of the MPLM was carrying most of the racks and supplies to the station.

Spacehab Specifications

Description	Single Module (SM)	Double Module (DM)
Volume	308 cu ft	1,100 cu ft
Height	11.2 ft	11.2 ft
Width	13.5 ft	13.5 ft
Length	Approx. 10 ft	Approx. 20 ft
The height and width of both the single and double module were constrained by the dimensions of the Orbiter payload bay whilst still offering visibility out of the aft flight deck windows.		
Launch Mass (un-laden)	10,000 lb	20,000 lb
Cargo Capacity (max)	3,000 lb	10,000 lb
Lockers Only	61 max	61 max
Racks with Lockers	1 rack with 51 lockers or, 2 racks with 4 lockers	61 lockers with 4 rooftop locations or, 4 racks (2 powered), plus floor stowage for large, unique items
Windows/Viewports	None	1
Construction	Aluminum structure with multi-layered insulation	
Sub-systems	Lighting, ventilation, limited power supply, command and data, fire detection and suppression, vacuum venting	
Connection with Orbiter	Utilizes Spacelab tunnel and adapters allowing connections to the Orbiter mid-deck and Shuttle docking and airlock facilities	

Multi-Purpose Logistics Module: Work on constructing three MPLMs began in April 1996 at the Alenia Aerospace facility in Turin, Italy, as part of that nation's contribution to the ISS. NASA was assigned ownership in exchange for Italian research time aboard the station.

Sometimes described as 'moving vans' for the Shuttle these modules were meant to transport large quantities of apparatus, experiments, supplies and other materials to the ISS, and to return miscellaneous unwanted items, outdated hardware and general trash. The Shuttle could carry only one module at time and its interior was inaccessible while in the payload bay. Once the Shuttle was docked at the ISS a robotic arm would lift the module from the bay and attach it to a vacant berth on one of the nodes. Over the next several days, members of the two crews would first empty the module of its cargo and then refill it with unwanted items. Then the arm would transfer the module back to the bay. The ability to berth the MPLM with the ISS made the cargo transfer much easier, not least because there were no tunnels to navigate and the hatch was the same size as others on the US segment of the station. An MPLM berthed at a node module offered ready access to the Destiny, Columbus and Kibo laboratories.

To enable the MPLM to function outside of the payload bay of the Orbiter, it had provisions for life support, fire detection and suppression, electrical distribution and computer operations. It could also be refrigerated to carry fresh food or life sciences experiments to the ISS. Like Spacehab, the environmental utilities were sufficient to permit two astronauts to work in the compartment simultaneously, together with any flora or fauna that was part of the science payload.

Each cylindrical module was 12.5 ft (3.81 m) in diameter and 21 ft (6.40 m) long, with an empty mass of 4.5 tons. It could carry sixteen laboratory racks loaded up with apparatus, supplies and/or logistics. The large hatch meant that racks could readily be moved in or out of the station, enabling them to be returned to Earth for modification, servicing, refitting and reuse. Each rack was 73 in (182.88 cm) high and 42 in (106.68 cm) wide, and had a volume of 53 cu ft (1.50 cu m) that could typically carry 1,540 lb (698.5 kg) of experiment hardware. They were outfitted on the ground and transferred into the station as complete payloads. This eliminated the need for the Shuttle crew to spend valuable time unloading the racks; that could be left to the residents who would already have a set of racks ready for the Shuttle to return to Earth.

The first MPLM (named Leonardo) was flown from Italy to the United States on a Beluga transport aircraft, arriving at KSC in August 1998. The second (Raffaello) was delivered in August 1999. The third and final module (Donatello) arrived in February 2001. The names were chosen by the Italian Space Agency from three great talents in Italian history. Leonardo da Vinci was a remarkable multi-talented inventor, scientist, civil engineer, architect, artist, military planner and weapons designer. A depiction of human grandeur by artist Raffaello Sanzio da Urbino is famous for its clarity of form and ease of composition. Donato di Niccolò di Betto Bardi founded Renaissance style sculpture and is recognized as one of the greatest talents of all time in his art.[*]

---

[*]Despite urban myth, the MPLMs were not named in honor of the fictional Teenage Mutant Ninja Turtles!

Of the three modules, only Leonardo and Raffaello flew missions to the ISS:

- Leonardo flew seven two-way missions on STS-102, -105, -111, -121, -126, -128 and -131, and was then refitted as the PMM and permanently attached to the ISS by STS-133.
- Raffaello flew four two-way missions on STS-100, -108, -114 and -135, and was then retired.
- Donatello was a more advanced design which could supply continuous power to payloads, but it was never required. Some of its parts were cannibalized to modify Leonardo to function as the PMM.

Permanent Multipurpose Module: With the retirement of the Shuttle fleet imminent and the MPLMs soon to become redundant, NASA decided to modify Leonardo to be permanently attached to the ISS as an additional 2,472 cu ft (69.96 cu m) of storage at little extra cost to the program.

Thales Alenia Space, the prime contractor for the MPLMs, modified the module to withstand a decade of continual exposure to the space environment rather than just the few days that it had endured on each of its seven previous missions. This was done by fitting micrometeoroid shields to its exterior; rerouting electrical harnesses to simplify crew access and thus permit on-orbit maintenance; adding user friendly restraints and payload retention devices; updating the software; and adding life extension certification for all the apparatus and subsystems.

The launch of PMM Leonardo by STS-133 was also exploited to deliver to the ISS an additional 28,353 lb (12,861 kg) of supplies.

Unpressurized Carriers

The Shuttle used two types of unpressurized payload carrier during its operations with Mir and the ISS.

Spacelab Pallet: Complementing the pressurized Spacelab modules were the 'U' shaped pallets to transport a range of instruments and equipment, such as telescopes, sensors and antennas that could be exposed directly to the space environment. Pallets could be used either in conjunction with a pressurized module or in pallet-only form, when they were supported by a subsystems enclosure nicknamed an igloo. There was also the option of carrying an Instrument Pointing System when a payload required to be pointed more accurately than the Orbiter could achieve.

Spacelab pallets were aluminum framed structures 13.1 ft (4 m) across and 10 ft (3 m) long. Apparatus was bolted on to so-called hard points. They could accommodate several configurations of support structures, and hence offer bespoke attachments for particular instruments, experiments, hardware, or other items. Pallets could be linked together to produce trains. As many as five pallets could fit into the bay, but this was never done.

Integrated Cargo Carrier (ICC): Six variants of ICC were available for use on Shuttle missions to the ISS but only four variants were ever flown. They offered a flexible, modular capability to match any unpressurized payload or hardware. Each could accommodate payload on its top (front) or bottom (back), with power or data services being supplied by either the Orbiter or the Spacehab module as convenient.

Each of the six ICC types included EVA handrails and foot restraint attachment sockets, and was capable of carrying either active or passive payloads. The ICC-G, ICC-GD, and ICC-L types were constructed from a flat aluminum box-beam pallet structure and a keel yoke assembly for transportation of unpressurized cargo in the payload bay region above the Spacehab tunnel. This used the payload area that was normally lost when the module was flown. Each of the ICC units could support the facility called the Spacehab and Oceaneering-Space Systems (SHOSS) Box.

The units flown were ICC-G (seven times), CC-GD (twice), ICC-L (once), and ICCVLD (twice).

Table 6.2 Integrated Cargo Carrier Versions

ICC Type	ICC Code	Mass (lb/kg)	Width (inches/cm)	Length (inches/cm)	Capacity (lb/kg)	Surface area (square inches/cm per side)
Generic	ICC-G	1,850	165	90	6,000	14,850
Generic Deployable	ICC-GD	1,930	165	90	6,000	14,850
Light (Lite)	ICC-L	1,203	165	45	4,000	7,425
Bay 13 Carrier	BTC	1,000	45	165	2,500	7,425
Vertical	ICC-V	2,435	165	165	7,000	20,025
Vertical Light (Lite)	ICC-VL	1,820	165	105	7,000	13,725

ICC Flight Assignments			
STS-96	2A.1	ICC-G	Strela (Up); ORU Transfer Device (up); SHOSS Box (Up & Down)
STS-101	2A.2a	ICC-G	Strela (Up); SHOSS Box (Up and Down) SOAR (Up and Down)
STS-106	2A.2b	ICC-G	SOAR (Up & Down), SHOSS Box (Up & Down)
STS-102	5A.1	ICC-G	ESP-1, with PFCS, LCA, EAS, RU remained on orbit
STS-105	7A.1	ICC-G	EAS up MISSE 1 & 2 up
STS-114	LF1	ICC-GD	ESP-2, with VSSA, MBSU, RC, UTA remained on orbit
STS-121	ULF1.1	ICC-G	PM (up) TUS- RA (new up and old down FGB
STS-116	12A.1	ICC-G	SMDP 2/3/4; SMDP Adapter up and STP-H2 launch canister up/down
STS-118	13A.1	ICC-GD	ESP-3 with P/R-J; CMG; NTA; BCDU; ATS FSE all up
STS-122	1E	ICC-L	EuTEF & SOLAR (both up) and NTA up and down
STS-127	2J/A	ICC-VLD	SGANT, LDU, PM all three up & 6 new P6 batteries up and 6 old P6 batteries down
STS-132	ULF4	ICC-VLD	EOTP, SGANT, & Boom (up) 6 P6 Batteries new (up) and 6 P6 old batteries down

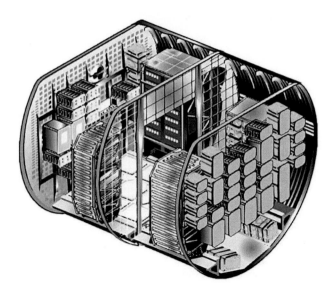

Typical locker detail inside Spacehab, in this STS-57. (Courtesy Spacehab Inc.)

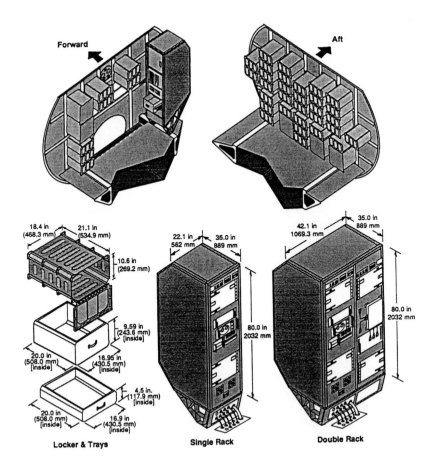

Orbiter mid-deck locker, tray, and Spacehab rack dimensions. (Courtesy Spacehab Inc.)

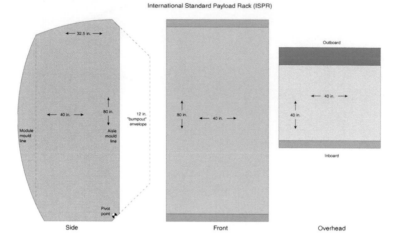

Detail of the International Standard Payload Rack (ISPR) dimensions.

THE STOWAGE FACILITIES

A variety of facilities and apparatus to store, carry, and transfer cargo to and from the Orbiter and the ISS in a secure and safe manner were trialed by Shuttle-Mir missions. Those that were carried forward to the assembly of the ISS are described below.

Lockers: The standard mid-deck lockers for the Shuttle and Spacehab were 10.6 in (26.92 cm) tall, 18.4 in (46.73 cm) wide, and 21.1 in (53.59 cm) deep. There were two types of tray available for insertion into the lockers. The large type was 9.59 in (24.35 cm) tall, 16.95 in (43.045 cm) wide, and 20.0 in (50.8 cm) deep. The small unit had the same width and depth but was only 4.5 in (11.43 cm) tall. The trays could readily slide in and out of the lockers, with two of the smaller ones fitting into a single locker space. A hinged door flap and twist catches secured each locker.

Mounting Plates: As an alternative to lockers, experiments could also be affixed to mounting plates. These flat plates had two optimum hole configurations for mounting on the mid-deck or a Spacehab. A single plate had a rated capacity of 60 lb (27.2 kg), and a double plate was twice that. Their inclusion depended upon whether the number of required locker spaces was available. In combination with experiment apparatus in lockers, the mounting plates offered standard interfaces and a little more flexibility in the design of carriers for experiments.

Payload Racks: The payload racks of the Spacehab module were supplied as single or double configurations. Measuring 80.0 in (203.2 cm) tall, 22.1 in (56.41 cm) wide, and 35.0 in (88.9 cm) deep, a single rack could carry 655 lb (297 kg) of payload. The double rack was the same as a single, except for being 42.1 in (106.9 cm) wide with a capacity of 1,210 lb (548.8 kg). Irrespective of its size, a rack could provide 1 kW of power, water cooling and vacuum venting. The actual configurations were typically a compromise, since each single rack occupied the volume of ten lockers in the module. The design of the rack also offered standard 19 inch (48.26 cm) hole patterns to allow payloads to be mounted on the front panel. The rack design was compatible with both the European Spacelab module and the facilities developed for the ISS.

Soft Stowage Bags (SSB): These could either be mounted on the front of a payload rack or the floor of the Spacehab module. They were used for additional soft transfer items, often the final cargo to be stowed prior to launch, such as clothing, linen, or items that were too large for the crew transfer bags.

Crew Transfer Bags (CTB): These came in four sizes, referred to as half, single, double, and triple; the size being related to the mid-deck locker equivalent (MLE) dimensions, namely:

CTB	Maximum capacity	Dimensions in inches	Dimensions in cm
Half	30 lb/13.6 kg	16.75 × 9.75 × 9.25	42.54 × 24.76 × 23.49
Single	60 lb/27.2 kg	16.75 × 19.75 × 9.25	42.54 × 50.16 × 23.49
Double	120 lb/54.4 kg	18.75 × 19.75 × 18.75	47.62 × 50.16 × 47.62
Triple	180 lb/81.6 kg	18.75 × 19.75 × 28.0	47.62 × 50.16 × 71.12

A bag had an end opening facility, a clear label with a general description of what was enclosed (clothing, tools, and so on), and Lexan windows to display the contents without having to open the bag. A content label printed in English was duplicated and placed in pockets on the sides of the bag. A detailed list itemized the contents and the toxicity of the most toxic items. Each bag was assigned a bar code identification label, an operations nomenclature, and toxicity. To aid the crewmembers in rapidly storing a bag in the proper place on a station during transfer (and taking into account the limited time available), each bag was color coded for fast temporary storage, pending sorting later. On the ISS this would include salmon for Unity, tan for Zarya, etc. White labels were reserved for bags that were packed on-orbit and marked 'Go Home' for return to Earth. Because these were normally manually filled by the crews on-orbit, plain white cards were used to update the contents as each bag was filled.

Contingency Water Containers (CWC): Water generated as a waste product by the Orbiter's fuel cells was stored in containers and transferred to Mir to supplement the limited supply there. Transfer bags proved very convenient in this role. This was also done in the early years of ISS assembly, prior to setting up the water reclamation unit aboard the station. Even after that, bagged water served as a back-up. However, after the retirement of the Shuttle a urine recycling system was available and extra water could be delivered as required aboard unmanned cargo vehicles. A CWC bladder resembled a duffle bag and could hold approximately 90 lb (44 liters) of fluid. Payload Water Reservoirs (PWR) were similar, but rather smaller. The Russian containers that are still being delivered by Progress resupply ships hold about 48.5 lb (22 liters).

A MOVING EXPERIENCE

The various types of carrier and stowage facilities developed to enable the Shuttle to transport cargo to and from a space station were a new experience for the Americans because, apart from Skylab, the crews of Mercury, Gemini, Apollo, and standalone Shuttle missions knew that everything they would need was somewhere aboard their vehicle. Each crew devoted a significant amount of training time to identifying what was stored where and for what purpose. Even in Skylab, the Apollo CSM delivered only a limited range of fresh supplies; nothing on the scale the Shuttle was capable of transporting.

The dictionary says that the word 'logistics' involves the moving and organizing of supplies and equipment, by implementing a plan or its operation. For the Shuttle, this had to be planned in detail to ensure that each element of cargo ended up in the correct location, and that nothing was inadvertently left in the Shuttle or the station. It was not simply a matter of grabbing the next item that came in reach and driving it through the tunnel to be dumped into some likely looking place on the other craft. The process had to be planned and care had to be taken with perishable items, scientific instruments, or results of science experiments. At all times the mass of the Shuttle had to be calculated to control the changing center of mass for re-entry. There was also the limited time the Shuttle could remain docked to the station.

On each flight to a station, a quantity of soft cargo, hardware, or consumables were transferred from the habitation modules and storage locations of the Orbiter to the host. And there were objects to be returned to Earth, if only to be discarded after the flight to alleviate the constant accumulation of trash and no longer wanted equipment.

One thing the Shuttle possessed was volume, and when a Spacehab or MPLM was carried this greatly increased the capability for cargo going into and out of the station. Indeed, several missions were designated utilization flights, rather than cargo flights, this word simply meaning "to make practical use of." Everything on a logistics flight was packed for a specific purpose.

As with many aspects of the joint US-Russian program, the Shuttle's support of the Mir space station was a fast learning curve in preparation for the tons of logistics that would be needed by the ISS.

STS 134 MS Mike Fincke displays two of the Crew Transfer Bags used to relocate small items of cargo to and fro between the Orbiter and the ISS.

The first MPLM to fly, Leonardo, is shown in the payload bay of Discovery during STS-102 in March 2001. Note the External Stowage Platform at bottom containing several spares for transfer by EVA to the ISS and the RMS with elbow camera at the bottom of the image.

Plan Against Reality

In 2002, Mission Specialist Steve Smith visited the ISS as a member of the STS-110 crew. A veteran of three previous solo Shuttle missions, he found entering the station for the first time a whole new experience in terms of working volume. Accustomed to the confines of the Orbiter flight deck and mid-deck, and in several cases the transfer tunnel to a Spacelab or Spacehab module, each time he entered the ISS he took a few seconds to orientate himself.[1] A Shuttle docked at the ISS has been said to resemble two adjoining houses set at odd angles, so that things change as you transfer from one to the other.

STS-100 MS Umberto Guidoni refers to crew documentation inside the second MPLM Raffaello. The compactness of the module is evident. Note the crew instruction on the transfer bag at right: 'Cinch straps tightly prior to re-entry.'

STS-96 in 1999 was only the second Shuttle to visit the ISS. Ellen Ochoa served on that crew as the first official loadmaster. "My primary job on that flight was to transfer about 4,000 pounds [1,815 kg] of supplies from the Shuttle to the station," she recalled. "Most of it was inside Spacehab and we moved that internally, but there was also some equipment outside [on carriers in the bay] that was moved by spacewalking."[2] Ochoa also pointed out that on this early mission the sequence of storage in the Spacehab was not as organized as it should have been. This was surprising because with this being the thirteenth Spacehab flight and the seventh of the Double Module, it could be presumed that NASA would know the most efficient way to pack the module. "The supplies were packed in a series of racks in Spacehab, and it somehow got stowed just as equipment became available. It didn't get stowed in any order that was particularly helpful to us up in space. Most of it was destined for the Russian Zarya Control Module. Each item had to be placed into a particular location. The locations had been analyzed ahead of time to agree with the docking loads which were going to be required when the Zvezda Service Module automatically docked later, so the Russians had very strict requirements about what could be put where and exactly how it must be tied down. As loadmaster, I had to make sure that all of this equipment got transferred across and to ensure that each item was in the correct place and tied down in the manner that was specified by the Russian experts."

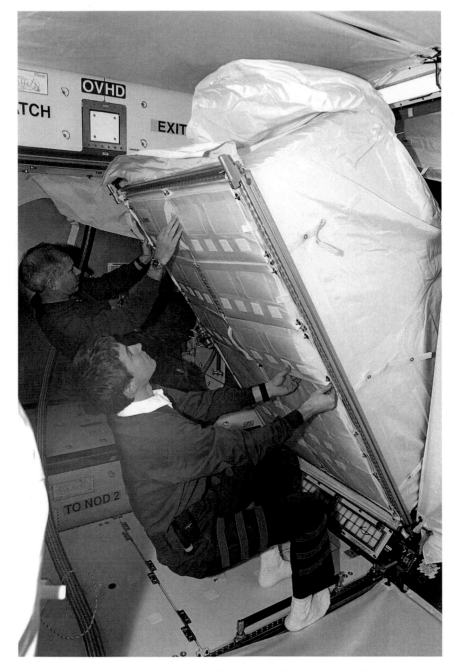

The transfer of large science racks to the ISS was simplified by the fact that the hatch of an MPLM provided direct access to the station.

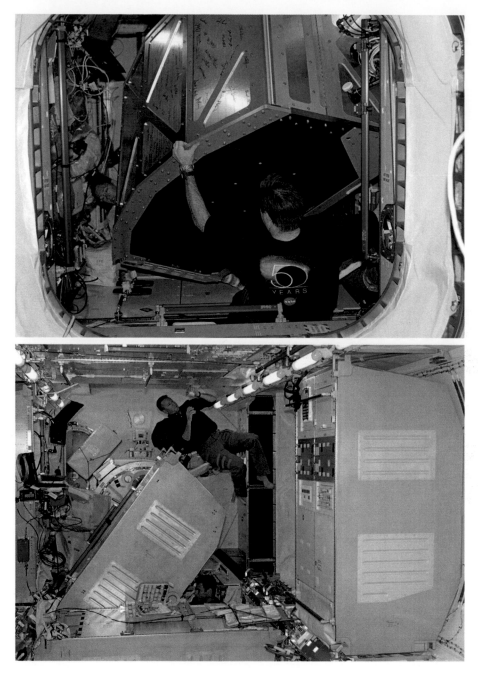

Care had to be taken in transferring a rack not to damage the station, the rack, its contents, and the fingers of crewmembers (top). Slotting the racks into place was usually a two-person task, as here during the STS-124 mission (bottom).

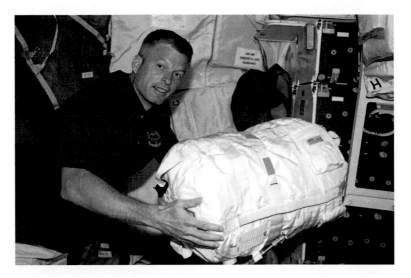

STS-117 MS Steve Swanson moves one of the smaller stowage bags from the mid-deck of Atlantis.

The mid-deck of Endeavour during STS-126 with stowage bags surrounding the hatch to the ISS. MS Steve Bowen refers to his checklist. MS Sandra Magnus is visible in the background.

With the STS-114 cargo transfers complete, MS Soichi Noguchi celebrates the retrieval of the MPLM by the RMS.

The 'Bad Air' Incident On STS-96

Of course not everything progressed smoothly during the Shuttle docking missions. There were minor failures and setbacks but they were mostly worked around by the crew or controllers on the ground. One of the earliest incidents was reported during STS-96 in 1999. There were reports that when the crew were working in the Zarya module they had encountered difficulty with some of the open panels restricting or disrupting the nominal flow of air around the module. This was said to have caused headaches and nausea for some astronauts. As a result the flow of air supplied by the environmental subsystem of the Orbiter, already supporting the Spacehab, had to be increased to cleanse the air in the Russian module.[3]

Recalling this incident, Mission Specialist Ellen Ochoa said, "We knew that there were going to be issues with airflow, because we'd worked with the Russian experts before flying. There were a lot of NASA people on the ground working the problem, not just the crew. We were deciding which panels could be opened on Zarya and for how long, because it really requires the panels to be shut for the airflow to operate in the manner it was designed. There were a lot of constraints and rules about how long you could open panels in Zarya, and things like that, so we were aware there would be an issue with airflow. There were some crewmembers that experienced headaches but it's not always clear exactly what that is from, and you don't always correlate it with a particular thing at the time. I think it was a little more in retrospect that we recognized that it was affecting people who spent a lot of time in Zarya. We could never really say for sure. We did talk about it after we got back, and people did try to sort of analyze it further to understand whether that would have been the cause. It wasn't something that we could determine absolutely afterward, but we decided to try to understand whether there was a way to increase the airflow in order to potentially mitigate that problem for future flights."[4]

During the work inside Zarya the astronauts had installed 'mufflers' over the fans to reduce the noise levels, but the mufflers caused some of the air-ducting to collapse. An example of this was recorded on video by CDR Kent Rominger. Six years later, when Rominger was asked about the situation, he did not believe the air quality in Zarya was 'bad.' He claimed it was more probably a result of open panels blocking the natural air flow in some regions of that module. Perhaps the interrupted airflow produced regions of 'stale' air that caused people to suffer headaches and malaise because if they spent time in a location that had poor circulation, their exhaled carbon dioxide would build up.[5]

For the next mission, STS-101, several 'quick fix' solutions were implemented. In addition, air sampling kits were available to document the problem. Ironically STS-88, the first mission to the ISS, reported no such incidents. However, that mission hadn't a Spacehab in the payload bay to impose an extra load upon the Shuttle's environmental control system.

By the time the first resident crew was aboard the ISS, some eighteen months after the STS-96 incident, the Zvezda Service Module was in place with the environmental systems for the Russian segment of the station. Nevertheless, after that every Shuttle mission and visiting Soyuz craft collected samples for post-flight analysis in order to monitor the condition of the internal atmosphere of the station as it was expanded and as the numbers of resident and visiting crewmembers varied.

EXPERIMENTS AND INVESTIGATIONS

It would be far beyond the scope and confines of this present volume to discuss *all* the experiments, research projects, and tests that were performed aboard the Shuttle during missions to Mir and the ISS. The reader is strongly urged to consult the press kits and mission reports for each of the forty-seven missions to obtain detailed accounts of their operational activities, and to NASA's various websites for post-flight results. A search of the internet will usually also yield conference and technical papers on such topics.[6]

One example of these investigations was SDTO 13004-U Russian Vehicle Docking-Undocking Loads on the ISS, which was carried out by the STS-119 mission in March 2009. The structure of the ISS had at least a fifteen-year lifespan requirement to assure the safety of the hardware and the lives of resident crews and visitors. The first Russian element, Zarya, was launched in November 1998 and the second, Zvezda, in July 2000. The station expanded over the next eleven years with the fifteen-year lifespan intended to see it through to 2015. The experiments performed for this SDTO helped to keep the understanding of the station's structure up to date, allowing its nominal life expectancy to be extended through to at least 2024.

There were also studies to determine the structural lifespan based on the use of the ISS in 2009, and to determine the capacity to extend its operational life and to expand the physical structure. The structural life estimates were based on worse-case loading conditions, using finite element models of structures and forcing-function estimations.

This data was then used to analyze actual measurements of the forces during docking and undocking, thruster firings, and other activities, in order to better understand the loads imparted upon the structures. Videos of crew activities, such as the transfer of logistics or installation of components of the station were also included in this study.

- *Education-Outreach-Education Astronaut Project*: Many demonstrations and experiments were supplied by educational establishments across the US, and indeed worldwide. They were a feature of several standalone Shuttle missions and continued with the station docking and assembly missions. They could be carried out within the Hitchhiker, GAS or Student Experiment programs, or a given educational project such as the Education Astronaut Project (EAP) that was part of the NAS-Elementary and Secondary Education Program to inspire or motivate studies to attain higher levels of Science, Technology, Engineering and Mathematics (STEM) or later Science, Technology, Engineering, Arts and Mathematics (STEAM). Following on from the Teacher-in-Space program the EAP included selecting teachers having experience from kindergarten through twelfth grade to train as fully fledged NASA astronauts. By sending educators into space, NASA believed that the skills gained as a classroom teacher could be applied to offer new avenues for the imagination and ingenuity of teachers and students, and hopefully inspire students to follow careers in key subjects. Barbara Morgan, who served as back-up to Teacher-in-Space finalist Christa McAuliffe, lost aboard Challenger in 1986, later became a Mission Specialist and flew with the STS-118 crew as the first Educator Astronaut in Space. She was followed in this role by Joseph ('Joe') M. Acaba and Richard R. ('Ricky') Arnold, both on STS-119, and by Dorothy ('Dottie') M. Metcalf-Lindenburger on STS-131.
- *Small Satellites*: A number of non-recoverable satellites were spring-ejected from the payload bay of an Orbiter on an ISS docking mission, usually after undocking.
- *Outreach and Commemoration*: Each Shuttle crewmember prepared a small Personal Preference Kit (PPK) as part of their own equipment. This included up to twenty personal, non-commercial items that had to fit within a bag that was only about 3 in (75 cm) square. There were also up to ten items that had less stringent regulations in the Official Souvenir Kit (OSK). This contained souvenirs from institutions, official organizations, or Very Important People. Many of these items were offered as gifts during post-flight tours and visits. There were also items associated with "promoting the space program." One was the IMAX Cargo Bay Camera, a 2.535 in (65 mm) color motion picture camera that was carried on STS-88 to record the first stage of assembling the ISS on-orbit.

SUMMARY

For each ISS assembly mission, the relocation of hardware by the use of robotic arms and spacewalks were the headliners. As important as these were to the overall success of the program, we must not neglect the internal transfer of tons of supplies, logistics and other materials both into and out of the various pressurized modules of the station and Orbiter compartments.

This was a major logistics challenge, because nothing like it had been attempted in space before, even though for over twenty years the Russians had used small Progress craft to ensure the longevity of their Salyut and Mir stations. The ISS was an entirely different proposition, being far larger, more complex, and an international venture. It required not only to be built in space but also to be supplied, and teams of crews from several countries had to learn to live and work together to pursue common goals. The ISS is the largest international construction project in space history. It also became an excellent example of multi-cultural teamwork, spread over many years and across the globe. What some might have thought impossible on Earth, and seemed at times to be unattainable, would eventually prove to be very successful in space.

Notes

1. Smith 2006
2. AIS interview with Ellen Ochoa, March 2, 2004
3. *Disasters and Accidents in Manned Spaceflight*, David J. Shayler, Springer-Praxis, 2000, pp. 223–227
4. Ochoa 2004
5. AIS interview with Kent Rominger, May 23, 2006
6. For an extensive listing of small experiments, DTOs/DSOs and other research investigations flown on Shuttle missions in general, see *Space Shuttle Mission Summary* by Robert D. ('Bob') Legler and Floyd V. Bennett, Mission Operations, DA8, NASA JSC, Houston, Texas, NASA TM-2011-216142, September 2011, and *The Space Shuttle Almanac*, by Joel W. Powell and Lee Robert Brandon-Cremer, 2011.

7

Building a Space Station

*To operate the remote arm, I had to look out the aft
cockpit windows. I wasn't tall enough to see out the
window so I had to bungee myself to the switch
panel. That worked well in weightlessness and
replaced the footstool I had used on Earth.*

Rhea Seddon, MS STS-51D, from *Go for Orbit*

Robotics played a key role in the creation of the International Space Station during the assembly between 1998 and 2011. It continues to this day, thanks to the Space Station Robotic Manipulator System (SSRMS) that was delivered by STS-100 in April 2001. The apparent ease with which this was achieved was honed during several early Shuttle missions and by pioneering robotic arm operators who had to devise a different way of working several feet or meters away from where the workplace was situated, as pointed out above by Rhea Seddon. The development of the RMS and its operation first on the Shuttle and then aboard the ISS, was one of the unsung success stories of the series.[1]

A MANIPULATOR FOR SHUTTLE

The concept of a robotic manipulator was introduced by science fiction, and was later considered by studies of possible advanced space programs. In the late 1960s, such a device was even considered as an addition to the AAP's Orbital Workshop, later called Skylab, in order to support extensive EVA operations. But the idea was not pursued for the final design of that station. For the Shuttle, the idea of manipulators assisting in the deployment and retrieval of payloads and other hardware, supporting spacewalkers and assisting in the assembly of large space structures was proposed almost from the word go.

In August 1972, eight months after NASA was given formal approval to develop the Shuttle concept, the Manned Spacecraft Center (which was soon to become the Johnson Space Center) in Houston, Texas, awarded a contract to Martin Marietta to assess the potential for cargo handling using robotic manipulators on board the Shuttle. In 1975 the

© Springer International Publishing Switzerland 2017
D.J. Shayler, *Assembling and Supplying the ISS*, Springer Praxis Books,
DOI 10.1007/978-3-319-40443-1_7

agency signed a Memorandum of Understanding with Canada. The Canadians had done some preliminary work in this area and expressed interest in participating in the Shuttle program by supplying robotic arms. Made by Spar Aerospace of Toronto, this became known as the Canadarm.

The development of the Shuttle RMS to assist with orbital operations provided the developers with an opportunity to gather data and learn lessons that would be useful in designing robotic systems for space stations and other structures meant for use in deep space exploration.

A Helping Hand

The Shuttle RMS was originally seen as a satellite deployment and retrieval tool, but it was soon realized the system was capable of supporting a very wide range of activities. Of the eventual 135 Shuttle missions the RMS was carried on 91 (67.4%) although not all employed it. In addition to the inspection activities and supporting of spacewalking astronauts envisaged by the early concepts, the arm was used for transferring payloads into and out of the bay. Early artistic impressions of the Shuttle depicted a manipulator installed on each side of the bay, but in reality the RMS was mounted only on the port longeron. When the OBSS was developed later, this was carried on the starboard side.

The final two-boom jointed structure carried aboard the Shuttle measured a total of 50 ft (15.24 m) in length and was capable of moving with six degrees of freedom. The design operated like a human arm and shoulder joint, with a 21 ft (6.4 m) upper boom and elbow joint, and a 23 ft (7.0 m) lower boom that was attached to a wrist joint with an end effector that acted as the hand. The controls were located on the right of the aft flight deck, where specially trained RMS operators used a combination of direct vision out of the aft flight deck or the overhead windows, as well as television cameras in the payload bay and further cameras on the arm's elbow, wrist and end effector in order to maneuver the end effector over a grapple pin on the object that was to be grasped. Then a system of three snare wires were withdrawn into a protective retractable sheath inside the end effector, closing around the pin to establish a firm grip. The operator used hand controllers to indicate the desired arm actions, and software interpreted these signals to determine how best that could be achieved in terms of a sequence of joint motions. The software monitored the condition of the arm every 80 milliseconds and in the event of a failure it would apply brakes to all joints and inform the operator of the problem.

The RMS had a number of operating modes: Automatic, Manual Augmented, Single Drive, Direct Drive, and Back-Up Drive. The first three included servo joint rate control and were supported by software on a general purpose computer of the Orbiter. As part of the preparation for a mission, the maximum allowable rates of joint motion for each payload (which could vary considerably) were loaded into the software. The Direct and Back-Up modes operated the arm with a reduced capability that applied a fixed voltage. These were for contingency situations, and permitted only single joint control for cases where computer support of motions had been lost.

The Automatic modes included both pre-programmed and automatic sequences that were commanded by the astronaut 'flying' the RMS. Each pre-programmed sequence

Table 7.1 Shuttle Remote Manipulator System Assignments On ISS Assembly Missions

Flight	Mission	Orbiter	RMS Serial #	OBSS Serial #
STS-88	2A	Endeavour	202	n/a
STS-96	2A.1	Discovery	303	n/a
STS-101	2A.2a	Atlantis	202	n/a
STS-106	2A.2b	Atlantis	202	n/a
STS-92	3A	Discovery	301	n/a
STS-97	4A	Endeavour	303	n/a
STS-98	5A	Atlantis	202	n/a
STS-102	5A.1	Discovery	301	n/a
STS-100	6A	Endeavour	303	n/a
STS-104	7A	Atlantis	202	n/a
STS-105	7A.1	Discovery	301	n/a
STS-108	UF1	Endeavour	303	n/a
STS-110	8A	Atlantis	202	n/a
STS-111	UF2	Endeavour	303	n/a
STS-112	9A	Atlantis	202	n/a
STS-113	11A	Endeavour	201	n/a
STS-114	LF1	Discovery	301	201
STS-121	ULF1.1	Discovery	303	202
STS-115	12A	Atlantis	301	201
STS-116	12A.1	Discovery	303	202
STS-117	13A	Atlantis	301	201
STS-118	13A.1	Endeavour	201	202
STS-120	10A	Discovery	202	203
STS-122	1E	Atlantis	301	201
STS-123	1J/A	Endeavour	201	203 (U)
STS-124	1J	Discovery	202	203 (D)
STS-126	ULF2	Endeavour	201	202
STS-119	15A	Discovery	202	203
STS-127	2J/A	Endeavour	201	202
STS-128	17A	Discovery	202	203
STS-129	ULF3	Atlantis	301	201
STS-130	20A	Endeavour	201	202
STS-131	19A	Discovery	202	203
STS-132	ULF4	Atlantis	301	201
STS-133	ULF5	Discovery	202	203
STS-134	ULF6	Endeavour	201	202
STS-135	ULF7	Atlantis	301	201

RMS #	Delivered to NASA	ISS Missions
201	1981 April [15 pre-ISS missions]	7 missions
301	1983 January [21 pre-ISS missions]	10 missions
302	1983 December [5 pre-ISS missions]	Lost in the explosion of Challenger
303	1985 March [9 pre-ISS missions]	7 missions
202	1993 August [4 pre-ISS missions]	13 missions

(continued)

Table 7.1 (continued)

OBSS #	Delivered to NASA	ISS Missions
201	2004 December 22	7 missions
203	2006	7 missions

Information courtesy of MDA Corporation, Brampton, Ontario, Canada.
With special thanks to Lynn Vanin, Manager, Public Affairs.
[Ref: Email from L. Vanin to AIS December 4, 2015]

was evaluated, developed, and tested during ground preparations, then loaded into the Orbiter's computer. In the case of the operator-commanded auto sequence modes, the data specifying the final position and orientation needed to be input prior to initiating the arm motion. Auto sequence modes were normally used for surveys of the Shuttle's surface or payload, and for more complicated and lengthy maneuvers. By contrast the Manual Augmented modes required real-time operator inputs, using the Translational Hand Controller (THC) and Rotational Hand Controller (RHC). Grappling a payload, berthing or unberthing a payload, and tracking and capturing a free-flyer could all be done this way. The point of resolution and the coordinate axis could be varied to suit the task and the target. This mode was also used for coarse positioning of the Shuttle's RMS. For refined, individual control of each arm joint the operator chose Single Drive mode and employed this to uncradle the arm from its mount and then position it with a high degree of accuracy.

Over the thirty-year history of the Shuttle there were only three RMS failures, with only one of these preventing the completion of the planned task:

- STS-41B (February 1984): A wrist yaw communicator failed resulting in loss of all primary operator modes. The scheduled deployment of a SPAS payload had to be abandoned.
- STS-51I (August 1985): The elbow joint electronics shorted out to ground. A work-around by the crew enabled them to use the back-up drive for the elbow and single drive for the other joints.
- STS-57 (June 1993): An incorrectly installed cable connector in the Special Purpose End Effector (SPEE) precluded the transfer of power and data to the EURECA free-flyer. The satellite was berthed safely but the data transfer had to be achieved by using a payload bay connector.

There were less significant failures and problems, but all were worked around by the combined efforts of ground engineers and the flight crew to enable the assigned tasks to be achieved without losing the functionality of the arm.

The success of the RMS during ISS assembly was a tribute to its simple design, the precise manufacturing and testing, extensive crew and ground controller training, and the inclusion of back-up procedures and modes to support a wide variety of operations.

Arming The Shuttle

In all, Spar Aerospace supplied five RMS flight units (201, 202, 301, 302 and 303) to NASA between April 1981 and August 1993.* Unit 302 was lost aboard Challenger in January 1986 and was not replaced. Following the loss of Columbia in February 2003, all subsequent missions also carried the Orbiter Boom Sensor System (OBSS), a 50 ft (15.24 m) long structure that was used in conjunction with the Shuttle RMS to inspect the thermal protection system for any breach of the heat shield prior to re-entry. When the Shuttle was retired, the OBSS was transferred to the ISS to supplement the robotic arm there.

The RMS proved itself many times prior to the Shuttle starting space station work, especially in supporting EVA operations and the capture and maneuvering of massive items such as the Hubble Space Telescope, Solar Max, LDEF, and some commercial and free-flying satellites.

Deployment And Retrieval Operations

The RMS was first carried on board STS-2 in November 1981. It was further tested by STS-3 and STS-4 in March and June 1982 respectively, and by STS-8 in 1983. Its first real use was in June 1983, when STS-7 released and recaptured the SPAS-1 free-flyer. The system performed similar release-and-capture tasks on a number of other missions.

The satellites deployed using the RMS include the Earth Radiation Budget Satellite (ERBS) by STS-41G in 1984, the Gamma Ray Observatory (GRO) by STS-37 in 1991, and the Upper Atmosphere Research Satellite (UARS) by STS-48 in 1991.

Another notable flight was STS-41C in 1984, on which the RMS first deployed the Long Duration Exposure Facility (LDEF) and then picked the ailing Solar Maximum Mission satellite out of orbit, placed it in a cradle in the payload bay, and then helped spacewalkers to repair the satellite for redeployment. Six years later, STS-32 used the RMS to retrieve the LDEF for return to Earth.

The RMS would support most Shuttle-based EVAs. A variety of pre-planned and contingency operations gave NASA great confidence that the RMS would be a vital component in the assembly and maintenance of a space station. Indeed, it was readily apparent that the station would require its own robotic arm. The RMS also developed into a useful inspection tool.

The RMS helped spacewalkers many times throughout the Shuttle program. Most often, the arm grasped a foot restraint on which one of the EVA crew stood. The arm would then maneuver the astronaut into position to work. On occasion a spacewalker would manually grasp an item and the arm would swing them into position. The arm was a firm platform that countered the forces imparted upon the astronaut while they were undertaking a task. Over the years, a variety of tool holders and stanchion arms were developed to aid an astronaut riding on the arm.

The RMS was also an essential tool in the deployment and servicing of the Hubble Space Telescope. This payload was deployed during the STS-31 mission in 1990, then

---

* Spar Aerospace was acquired by McDonald Dettweiller Associates (MDA) in 2005.

retrieved and redeployed five times for servicing across the next nineteen years. It was during the challenging EVAs of such missions that a method of direct communication between the RMS operator and the crewmember riding the RMS was devised to better coordinate the movement of the arm to the desired location.

By the time the Shuttle was ready to start assembling the ISS there was a vast store of knowledge and experience about operating the RMS from the aft flight deck of the Orbiter.

RMS ON ISS ASSEMBLY MISSIONS

An RMS unit was carried by every Shuttle-ISS assembly flight from 1998 to 2011. The OBSS was carried by every post-Columbia flight, starting with the STS-114 Return-To-Flight mission in 2005. The work carried out by these units was impressive: some forty-six ISS elements, totaling 1,185,297 lb (537,656 kg) in mass, were maneuvered by both the Canadarms of the Orbiters and the improved Canadarm2 that was installed on board the ISS in 2001.[2] As shown in Table 7.3, by the retirement of the Shuttle the RMS had handled thirty ISS elements (some 537,288 lb or 243,668 kg mass) including seventeen hand-offs of hardware to Canadarm2, which itself had handled thirty-four ISS elements (648,115 lb or 293,988 kg mass) during assembly.

When ISS assembly began in 1998, the Shuttle RMS was the primary system for the construction work. It mated the Unity Node to the Zarya Control Module, added the US Destiny laboratory, and installed the first segments of the Integrated Truss Structure. In 2001 the first elements of Canadarm2 were delivered to enhance the robotics capability at the station. Upgraded servo electronics on the RMS improved its controllability in manipulating heavier payloads. The orchestration between the two arms was a sign of the maturity of robotics in human space flight, and essential for adding components to the ISS at positions that were beyond the reach of the arm on the docked Shuttle.

RMS Operators

All Shuttle missions that carried the RMS included at least one crewmember who was trained to operate the RMS, preferably, but not essentially, with past experience of the system. There was also usually at least one, and often two back-up crewmembers who were qualified with the RMS. The issue of gaining that experience, which surfaced on the HST service missions, was that an astronaut had to handle the RMS, be it laden or unladen, in order to get a 'feel' for how the real thing differed from the simulator. This on-orbit training was initially frowned upon by NASA, lest something untoward occur. But it was ultimately accepted that the best way to train to use the RMS was to use it in space. Ground training and simulations helped but, as with most aspects of space flight, there was nothing better than on-orbit experience.

The minimal RMS operations at Mir left a hurdle to be overcome in order to rely on using the Shuttle's arm to initiate ISS construction. As a result, most, but not all, of the early station assembly missions had experienced RMS operators assigned as prime for robotic activities.

Table 7.2 Shuttle RMS Operators ISS Assembly Missions 1998–2011

STS	Mission	Orbiter	Prime RMS	Backup/Alternate	SSRMS
ISS ASSEMBLY AND RESUPPLY MISSIONS					
88	2A	Endeavour	Currie	Cabana	
96	2A.1	Discovery	Ochoa	Jernigan	
101	2A.2a	Atlantis	Weber	Halsell	
106	2A.2b	Atlantis	Mastracchio	Altman	
92	3A	Discovery	Wakata (JAXA)	McArthur	
97	4A	Endeavour	Garneau (CSA)	Bloomfield	
98	5A	Atlantis	Ivins	Cockrell	
102	5A.1	Discovery	Thomas A.	Kelly J.	
100	6A	Endeavour	Hadfield (CSA)/ Parazynski	Ashby, Parazynski	Hadfield (SSRMS specialist)
104	7A	Atlantis	Kavandi	Lindsey	Helms, Voss (ISS-2 crew)
105	7A.1	Discovery	Forrester	Horowitz (EVA)	None?
108	UF1	Endeavour	Kelly M. (EVA)	Godwin (MPLM); Onufriyenko (RSA - BUp EVA)	None?
110	8A	Atlantis	Ochoa	Bloomfield/Frick	Ochoa
111	UF2	Endeavour	Cockrell	Perrin (CNES)	Walz, Whitson
112	9A	Atlantis	Magnus	Ashby	Magnus
113	11A	Endeavour	Wetherbee	?	Whitson, Bowersox, Pettit,
114	LF1	Discovery	Thomas A.	Camarda, Lawrence, Kelly	Kelly, Lawrence
121	ULF1.1	Discovery	Nowak	Wilson	Nowak, Wilson
115	12A	Atlantis	Burbank	Ferguson/MacLean	Stefanyshyn Piper, MacLean
116	12A.1	Discovery	Patrick	Polansky	Higginbotham
117	13A	Atlantis	Forrester	Swanson, Olivas	Archambault
118	13A.1	Endeavour	Caldwell	Morgan	Hobaugh. Morgan
120	10A	Discovery	Zamka	Parazynski, Wilson, Nespoli	Wilson, Tani
122	1E	Atlantis	Melvin	Love?	Melvin, Love, Eyharts, Tani
123	1J/A	Endeavour	Johnson G.H.	Gorie, Doi	Johnson G.H., Behnken, Riesman, Eyharts
124	1J	Discovery	Ham	Nyberg, Garan, Fossum	Nyberg (also worked with new Japanese RMS), Hoshide (also worked Japanese RMS); Chamitoff
126	ULF2	Endeavour	Boe	Pettit, Kimbrough	Pettit, Boe
119	15A	Discovery	Antonelli	Phillips, Acaba,	Phillips, Magnus, Wakata

(continued)

Table 7.2 (continued)

STS	Mission	Orbiter	Prime RMS	Backup/Alternate	SSRMS
127	2J/A	Endeavour	Payette (CSA)	Kopra, Hurley, Polansky	Hurley, Wakata (also Japanese RMS), Kopra (also Japanese RMS)
128	17A	Discovery	Hernandez	Ford, Forrester	Ford
129	ULF3	Atlantis	Wilmore	Melvin	Wilmore, Melvin
130	20A	Endeavour	Hire	Patrick, Zamka, Robinson Virts	Hire, Virts, Patrick
131	19A	Discovery	Dutton	Metcalf-Lindenburger	Wilson, Dutton, Yamazaki
132	ULF4	Atlantis	Antonelli,	Sellers, Reisman, Ham	Riesman, Sellers
133	ULF5	Discovery	Lindsey	Boe, Drew	Barratt, Stott
134	ULF6	Endeavour	Fincke	Feustel, Johnson G.H., Vittori	Chamitoff, Johnson, G.H.
135	ULF7	Atlantis	Hurley	Magnus, Ferguson	Hurley, Magnus

Balancing The Boom

With the Return-To-Flight after the loss of Columbia in 2003 a further task for the RMS operator to master was the use of the Orbiter Boom Sensor System (OBSS). This was a 50 ft (15.24 m) single-arm boom extension to the RMS intended to enable it to scan the thermal protection of the Orbiter using a variety of cameras and lasers. It had electrical connections to enable the end effector to pass power and data to and from this suite of instruments.[*] During the flight to the station, the OBSS was used to survey the surface of the Orbiter, particularly the leading edges of the wings (damaged on Columbia) and the underbelly. Evaluations by STS-114 and STS-121 not only qualified the system for this inspection task, it was also shown that an EVA astronaut could 'ride' the OBSS to conduct a repair on the Orbiter. Furthermore, the Shuttle RMS could if required grasp the ISS and maneuver the vehicle so that Canadarm2 could inspect the Shuttle's surface. Astronaut training had to be adjusted to simulate the various uses of the OBSS.

The total mass of the manifested payload that made up the Japanese Kibo laboratory was too large to be launched by a single Shuttle, so it had to be distributed across three flights. In fact, some of this hardware was so voluminous that on the first Kibo mission, STS-123, the requirement to fly the OBSS on each Shuttle flight introduced a problem. The OBSS was temporarily transferred to the station by STS-123 to ensure there would be room for the large Kibo laboratory in the payload bay of STS-124. Having installed the module on the ISS, that Orbiter retrieved the OBSS from the station to perform the appropriate inspection and then stowed the boom on the payload bay wall for return to Earth.

---

[*] Each of the three OBSS units flew seven times. Unit 201 was delivered to NASA on December 22, 2004 and carried by STS-114, -115, -117, -122, -129, -132, and -135. Unit 202 arrived in 2005 and was carried by STS-121, -116, -118, -126, -127, -130, and -134. It was this unit that was permanently transferred to the ISS by STS-134. Unit 203 was delivered in 2006 and was carried by STS-120, -123, -124, -119, -128, -131, and -133.

Post-Columbia, crews on the way to the ISS used the RMS and OBSS to survey the thermal protection system of the Orbiter. At the station the combination of the RMS and OBSS, together with the SSRMS, enhanced the robotics program far beyond anything that had been attempted in space before.

Robotics At A Stretch

From STS-88 through to STS-104 the single Shuttle RMS was sufficient to support all the robotics and EVA activities at the ISS. But by the summer of 2001 the station was beginning to extend beyond the reach of the arm of docked Shuttle. The installation of Canadarm2, the Dextre system, and airlocks on both the US and Russian segments of the ISS permitted robotic operations and spacewalks to be undertaken well beyond the original operating envelope of the Shuttle. In a sense the RMS of the Shuttle served as the 'proof of concept' for designing the far more sophisticated SSRMS.

Shuttle crewmembers would be assisted by station residents in working the SSRMS as well as the Shuttle arm to hand-off payloads for installation on the ISS or stowage in the payload bay. This greatly expanded the training, as shown in the example from the crew assignment list that was posted shortly prior to the STS-116 mission in December 2016:

Robotics	Activity	Prime	Back-up
SRMS	TPS inspection	Patrick	Polansky
SRMS	P5 unberthing	Patrick	Polansky
SSRMS	P5 hand-off	Williams S.	Higginbotham
SSRMS	P5 install	Williams S.	Higginbotham
SRMS	EVAs 1, 2 & 3	Patrick	Polansky
SSRMS	EVAs 1 & 2	Williams S.	Higginbotham
SSRMS	EVA 3	Lopez-Algeria	Higginbotham

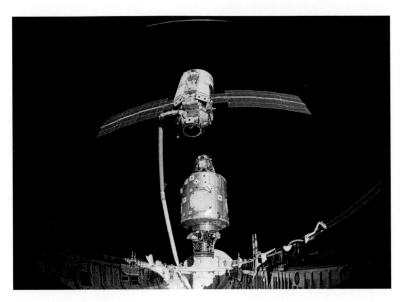

The first two elements of the ISS are assembled on-orbit as STS-88 berths the Unity Node with the Zarya Control Module previously launched by the Russians.

STA-88 RMS operator Nancy J. Currie.

STS-88: The First ISS Robotics Maneuvers

The first use of the RMS during ISS assembly was by STS-88, which was launched on December 4, 1998. As Endeavour rendezvoused with the Zarya Control Module, Jerry Ross and Jim Newman set up the Space Vision System that would provide precise data to the RMS operator about the position and orientation of hardware in the payload bay. Then Nancy Currie checked out of the arm itself.

Two days after launch, Currie used the RMS to hoist the 12.8 ton Unity Node from the aft payload bay, rotate it and place it just over the extended docking ring on the top of the docking system at the front of the payload bay. In a similar maneuver to that for the Docking Module that was transported to Mir, Currie put the arm into 'limp' mode and CDR Robert Cabana used the RCS jets to drive the spacecraft up towards Unity's Pressurized Mating Adapter 2. This soft docking was followed by a hard docking that created a rigid connection. After the docking system had been pressurized the hatches were opened and Cabana and Ross placed caps over the vent valves in preparation for the formal entry into Unity later in the mission.

Once Endeavour was station keeping with its payload bay 10 ft (3 m) from Zarya, Currie extended the RMS to grapple the Russian module. At 21 tons, Zarya was the heaviest object yet handled by the Shuttle's arm, exceeding the record of the Gamma Ray Observatory by approximately 7,000 lb (3,175 kg). She positioned Zarya directly above the docking unit of PMA-1 at the far end of Unity with a separation of several inches, then placed the arm into limp mode again. At that point, Cabana repeated the thruster firing operation to join the two ISS modules together.[3]

Updating The RMS

Prior to attempting to use the RMS to assemble a station, studies and evaluations of its performance and capabilities needed to be analyzed as part of preliminary planning for ISS assembly. During these studies it was found that, in its present design, the RMS on the Shuttle would not be capable of controlling the massive payloads envisaged for the station. It would require updates to the electronics on the arm – in particular the Servo Power Amplifiers in each joint – and improvements to the self-checking software that protected the arm against hardware failures. Without these upgrades, there would have been an elevated likelihood of failures on-orbit when handling massive payloads, with the potential of serious damage to hardware or harm to the astronauts. Further planned enhancements included several visual and cueing aids, most notably the Space Vision System and Centerline Berthing Cue System to give clearance and load information to the RMS operator while manipulating large items in confined spaces.[4]

USING THE ARMS AT THE ISS

In addition to the US Destiny laboratory, the Shuttle RMS was used to transfer the parts of the SSRMS across to the ISS. Once Canadarm2 was operational, it assisted with the assembly of the ITS, some segments of which incorporated the folded solar arrays that were to power the station's systems.

Moving Logistics

The Shuttle RMS also handled the first five transfers of MPLMs from the payload bay to the station and back again (STS-102, -100, -105, -108, and -111), totaling a mass of

108,080 lb (49,016 kg) in the assembly sequence. The remaining MPLM movements (STS-114, -121, -126, -128, and -131) were executed by the SSRMS, as was installing the PMM module when that was delivered by STS-133. The Shuttle arm was also used in the movement of several unpressurized equipment carriers and facilities, both on its own and in conjunction with Canadarm2.

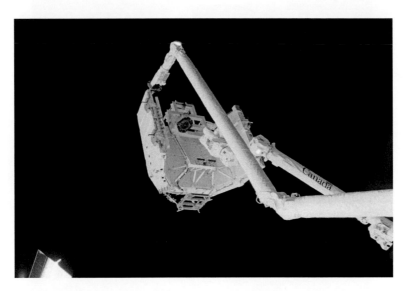

The first part of Canadarm2 is transferred to the station during the STS-100 mission.

Delivering SSRMS

The first part of the 58 ft (17.6 m) Canadarm2 was delivered to the ISS by STS-100 in 2001 and installed on a grapple fixture on the Destiny laboratory to undergo a detailed checkout. With an end effector at each end, the SSRMS could 'walk' across the station from one grapple fixture to another. When the 3,182 lb (1,443 kg) Mobile Base System (MBS) was delivered by STS-111 in 2002, this was mounted on a set of rails on the S0 segment of the ITS attached to Destiny. Then Canadarm2 stepped off Destiny and onto the MBS. In 2008 STS-123 delivered the 3,431 lb (1,556 kg) Dextre 'hand' unit whose dexterity would enable a variety of tasks to be performed outside that would otherwise have required an astronaut to make a spacewalk. Dextre can be stowed on a convenient grapple fixture until picked up by the arm for a specific task.

Robotic Hand-Offs

Table 7.3 specifies the seventeen hand-offs between the Shuttle RMS and the SSRMS that occurred between 2002 (STS-113) and 2011 (STS-134). The hardware transferred included elements of the P1, P3/4, P5, and S3/S4, S5, and S6 truss segments. The Z1 truss

The Shuttle RMS supported spacewalks during the early ISS assembly flights. Here STS-96 MS Tammy Jernigan rides the foot restraint on the end of the arm.

Table 7.3 Canadarm And ISS Assembly Operations 1998–2011

STS CANADARM

STS	ISS Flight	RMS Unit	ISS Element	Mass (lb)	Mass (kg)
88	ISS 1A	202	FGB	42,637	19,340
88	ISS 1A	202	Unity Node 1	25,474	11,555
92	ISS 3A	301	Z1 Truss	18,424	8,357
92	ISS 3A	301	PMA 3	2,548	1,156
97	ISS 4A	303	P6 Truss	34,810	15,790
98	ISS 5A	202	Destiny US Lab	33,474	15,184
102	ISS 5A.1	301	MPLM 1 Leonardo	25,002	11,341
100	ISS 6A	303	SSRMS/SLP	6,743	3,059
100	ISS 6A	303	MPLM 2 Raffaello	18,759	8,509
105	ISS 7A.1	301	EAS	1,410	640
105	ISS 7A.1	301	MPLM 1 Leonardo	20,833	9,450
108	ISS UF-1	303	MPLM 2 Raffaello	20,223	9,173
111	ISS UF-2	303	MPLM 3 Leonardo	23,243	10,543
113	ISS 11A	201	P1	27,507	12,477
115	ISS 12A	301	P3/P4 Truss	34,885	15,824
116	ISS 12A.1	303	P5 Truss	4,109	1,864
117	ISS 13A	301	S3/S4 Truss	35,679	16,184
118	ISS 13A.1	201	S5	3,999	1,814
118	ISS 13A.1	201	ESP-3	5,676	2,575
123	ISS 1J/A	201	JEM ELM-PS	18,492	8,388
119	ISS 15A	202	S6	31,113	14,113
127	ISS 2J/A	202	ELM-ES	2,645	1,200
127	ISS 2J/A	201	JEM-EF	8,329	3,778
127	ISS 2J/A	201	ICC-VLD	9,038	4,100
129	ISS ULF3	301	ELC1	14,034	6,366
129	ISS ULF3	301	ELC2	14,014	6,357
132	ISS ULF4	301	MRM1 Rassvet	17,636	8,000
133	ISS ULF5	202	ELC4	7,674	3,481
134	ISS ULF6	201	ELC3	13,472	6,111
134	ISS ULF6	201	AMS	13,300	6,940
			TOTAL	537,182	243,668
30 ISS ELEMENTS					

RMS – SSRMS HANDOFFS

ISS CANADARM 2

ISS Element	Mass (lb)	Mass (kg)	STS	ISS Flight
Quest Airlock	13,298	6,032	104	ISS 7A
S0	26,777	12,146	110	ISS 8A
MBS	3,181	1,443	111	ISS UF2
S1	28,102	12,747	112	ISS 9A
P1	27,507	12,477	113	ISS 11A
MPLM	18,064	8,194	114	ISS LF1
MPLM	20,959	9,507	121	ISS ULF1.1
P3/P4 Truss	34,885	15,824	115	ISS 12A
P5 Truss	4,109	1,864	116	ISS 12A.1
S3/S4 Truss	35,679	16,184	117	ISS 13A
S5	3,999	1,814	118	ISS 13A.1
ESP-3	5,676	2,575	118	ISS 13A.1
Node 2 Harmony	31,574	14,322	120	ISS 10A
P6 Relocate	34,810	15,790	120	ISS 10A
Columbus ESA Lab	26,627	12,078	122	ISS 1E
SPDM	3,430	1,556	123	ISS 1J/A
JEM PM – Kibo Lab	32,566	14,772	124	ISS 1J
MPLM 1 Leonardo	28,009	12,705	126	ISS ULF2
S6	31,113	14,113	119	ISS 15A
ELM-ES	2,645	1,200	127	ISS 2J/A
JEM-EF	8,329	3,778	127	ISS 2J/A
ICC-VLD	9,038	4,100	127	ISS 2J/A
MPLM	27,509	12,478	128	ISS 17A
ELC1	14,034	6,366	129	ISS ULF3
ELC3	14,014	6,357	129	ISS ULF3
Node 3 Tranquility	40,000	18,144	130	ISS 20A
Cupola	4,144	1,880	130	ISS 20A
MPLM 1 Leonardo	27,273	12,371	131	ISS 19A
ICC-ULD	8,329	3,778	132	ISS ULF4
MRM1 Rassvet	17,636	8,000	132	ISS ULF4
ELC4	7,674	3,481	133	ISS ULF5
PMM	28,353	12,861	133	ISS ULF5
ELC3	13,472	6,111	134	ISS ULF6
AMS	15,300	6,940	134	ISS ULF6
TOTAL	648,115	293,988		
34 ISS ELEMENTS				

(continued)

Table 7.3 (continued)

		(continued):Canadarm ISS Payload Handoffs To Canadarm2 2002–2011		
STS	ISS Element	Mass (lb)	Mass (kg)	Hand offs	Date
113	P1	27,507	12,477	SRMS >>> Canadarm2	2002 Nov 26
115	P3/P4 Truss	34,885	15,824	SRMS >>> Canadarm2	2006 Sep 11
116	P5 Truss	4,109	1,864	SRMS >>> Canadarm2	2006 Dec 11
117	S3/S4 Truss	35,679	16,184	SRMS >>> Canadarm2	2007 Jun 10
118	S5	3,999	1,814	SRMS >>> Canadarm2	2007 Aug 10
118	ESP-3	5,676	2,575	SRMS >>> Canadarm2	2007 Aug 14
123	JEM ELM-PS	18,492	8,388	SRMS >>> Canadarm2	2008 Mar 14
119	S6	31,113	14,113	SRMS >>> Canadarm2	2009 Mar 18
127	JEM-EF	8,329	3,778	SRMS >>> Canadarm2	2009 Jul 18
127	ICC-VLD	9,038	4,100	SRMS >>> Canadarm2	2009 Jul 19
127	ELM-ES	2,645	1,200	SRMS >>> Canadarm2	2009 Jul 21
129	ELC1	14,034	6,366	SRMS >>> Canadarm2	2009 Nov 18
129	ELC2	14,014	6,357	SRMS >>> Canadarm2	2009 Nov 21
132	MRM1 Rassvet	17,636	8,000	SRMS >>> Canadarm2	2010 May 18
133	ELC4	7,674	3,481	SRMS >>> Canadarm2	2011 Feb 26
134	ELC3	13,472	6,111	SRMS >>> Canadarm2	2011 May 18
134	AMS	15,300	6,940	SRMS >>> Canadarm2	2011 May 19
TOTALS		263,602	119,572		17 ISS ELEMENTS

Information courtesy of MDA Corporation, Brampton, Ontario, Canada.
With special thanks to Lynn Vanin, Manager, Public Affairs.
[Ref: Email from L. Vanin to AIS December 4, 2015]

that was mounted on Unity and the P6 truss that was placed on Z1 were handled by the Shuttle RMS alone. The SSRMS transferred the S0 and S1 segments from the payload bay onto the station. It also moved the P6 truss in order to complete the final configuration of the ITS.

Installing Modules

The Destiny laboratory was installed onto Node 1 (Unity) by the STS-98 RMS in 2001. The Quest airlock, Node 2 (Harmony), ESA Columbia laboratory, Node 3 (Tranquility), and the Cupola were installed using Canadarm2. The various parts of the Japanese Kibo laboratory were installed using both the Shuttle and ISS arms.

EVA Support

The Shuttle RMS was used extensively in support of EVAs during the assembly of the ISS until 2003. Later the RMS was not needed as much in direct support of spacewalks but it was employed alongside the SSRMS in activities such as relocation of hardware, evaluation of new procedures, and inspection of certain sites and equipment.

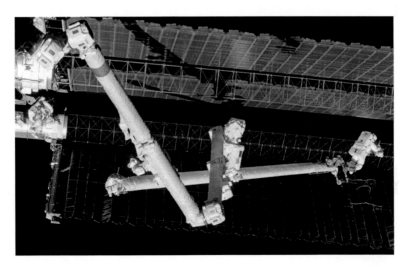

As the station grew in size, the reach of the Shuttle RMS was insufficient to support all EVA requirements, so the Orbiter and ISS arms were used in combination. There was also an evaluation of using the OBSS as an EVA support tool.

RMS LEGACY

Without a doubt, the development of the RMS for the Shuttle provided a bonus for mission operations. It served as a dispatcher and retriever of payloads and wayward satellites; it supported EVA astronauts; it captured telescopes and free-flying pallet payloads; it was

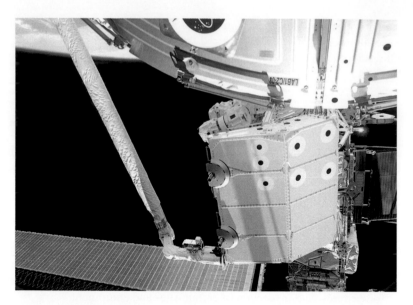

The RMS units on both the Shuttle and ISS were key elements in the transfer of large items of hardware from (and in some case back to) the payload bay during assembly missions.

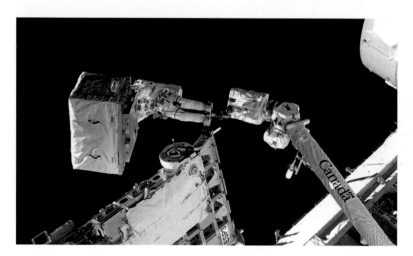

The experience of earlier EVA operations gave NASA confidence that it would be able to handle the large masses that would be involved in ISS assembly operations.

used as an inspection tool; it also provided a means of viewing the engine firings of deployed payloads to confirm their ignition (when the Orbiter was facing its belly towards the motor). For ISS operations, its use in EVA and logistics transfer was expanded by the OBSS inspection system and by joint operations with Canadarm2.

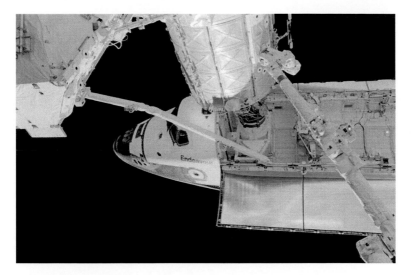

The Shuttle RMS and Canadarm2 work together as STS-111 installs the S0 truss segment on the Destiny module in June 2002.

The RMS design clearly delivered all that had been hoped for it, and much more besides. Although the Shuttle RMS flies no more, its descendants will find a worthy role on future vehicles that will return us to the Moon and later venture much farther into space.

We may have closed the Shuttle chapter of RMS operations at the ISS, but it was only the first part of the cargo delivery story.

SUMMARY

The capabilities of the Shuttle RMS and SSRMS enabled the assembly of the ISS to progress smoothly and efficiently. Inevitably there were small issues to address, but these were overcome. Today, the maturity of robotics in the space program is amply demonstrated by Canadarm2 and Dextre on the station.

Clearly the planning for an RMS in the Shuttle program in the early 1970s was an inspired decision, and together with the introduction of the RMS in 1981 permitted a gradual build-up of operating experience over the next seventeen years to make RMS operations at the ISS appear almost routine, although in reality they are very complex activities.

Together with the challenge of transferring tons of material first to Mir and then to the ISS, the evolution of station logistics that was pioneered by the Soviet Union with Progress vessels at the Salyut and Mir space stations and which continues with similar ships at the ISS, was taken to a new level with Shuttle flights to both stations. The key to extended duration human flight in space is one of sustained supply of consumables, propellant, food, and soft goods as well as new hardware, experiments and equipment. On the other side of the coin is the disposal of trash, unwanted material and hardware, and the recovery of experiment samples and of course the crew. The Shuttle certainly proved itself in these roles, and at the same time delivered large elements of hardware for ISS assembly and provided a re-boost capability to sustain the orbit of that facility.

The venerable Soyuz and Progress have been called 'universal' spacecraft because they demonstrated the capability to carry out a range of missions and tasks, as well as for proving the longevity of a basic design that was introduced fifty years ago.

For all its faults, doubters and critics, the Space Shuttle was able to demonstrate its flexibility in the delivery of hardware, consumables, cargo, crews and propellant to the two stations that it supported between 1995 and 2011. Its ability to undertake a variety of small secondary experiments and investigations facilitated multifaceted research by domestic, educational, military, and international parties. This broad capability of the vehicle allowed planners to extract the maximum benefit from each mission flown.

Notes

1. *Canadarm, The Space Shuttle's Helping Hand*, David J. Shayler, in preparation
2. ISS Assembly SRMS SSRMS Summary; Canadarm SRMS OBBS Summary documents courtesy Lynne Vanin, MDA Public Affairs, Brampton Ontario, Canada, December 3, 2015
3. STS-88 Mission Status Reports, various dates, December 1998
4. RMS, History, Evolution and Lessons Learned, Glenn Jorgensen and Elizabeth Bains, NASA Automation, Robotics and Simulation Division, NASA JSC, AIAA paper with PowerPoint notes, September 26, 2011, AIAA Space Conference and Exposition 27–29 September 2011, Long Beach California

8

Stepping Out

Many elaborate design studies [of space stations] are based
upon fitting together in orbit prefabricated sections. It is
fairly certain that these parts cannot be assembled by
remote control but will have to be man-handled
by crewmen in space suits.

J. G. Guignard, "Spaceman Overboard" in *Spaceflight*, vol. 1 #8, July 1958

Written four decades before any American astronaut left the airlock of a Shuttle to work at a space station – and a Russian one at that – these words would prove to be prophetic for what would unfold half a century later, in the first decade of the new millennium.

FROM DREAMS TO REALITY

It was soon evident to designers and planners that a large and complex space station could not be constructed without astronauts working outside. Officially called EVA,[*] this activity is more widely referred to as 'spacewalking' although it is something of a misnomer because there can be no 'walking' in the weightless environment of space, away from the solid surface of a celestial body. Nor can it be called 'space floating,' because humans on-orbit are actually moving or 'falling' through space in accordance with the laws of gravity. To speak of 'conducting a space fall' sounds rather peculiar, therefore for simplicity the term EVA is generally considered to be synonymous with spacewalking.

The illustrations made in the 1950s of space-suited astronauts undertaking space station assembly tasks often looked like a cosmic construction site in which several astronauts were depicted heroically bolting, welding and manipulating vast pieces of hardware, assisted by various small utility vehicles. In many of these iconic images a ferry vehicle

---

[*] Extra Vehicular Activity (EVA) is work conducted by astronauts outside a spacecraft, as opposed to that conducted inside, which is known as Intra Vehicular Activity (IVA).

© Springer International Publishing Switzerland 2017
D.J. Shayler, *Assembling and Supplying the ISS*, Springer Praxis Books,
DOI 10.1007/978-3-319-40443-1_8

could be seen in the background, the transportation system that operated between Earth and orbit for humans and cargo. Half a century later, this aspect of that vision has proved accurate with Shuttles supporting astronauts during EVA assembly tasks at the ISS. However, whilst the ISS is the largest construction project in the real world of space flight it is much smaller than those envisioned in the 1950s, the means of assembly are less visually dramatic, and the spacewalking teams are much smaller. Nevertheless, the principle is the same.

The benefits of the large Shuttle payload bay, the RMS, and the capabilities of the EVA astronauts were all essential to the decision to employ the Shuttle in the primary assembly role for the large elements of station hardware. Furthermore, the Orbiter also provided a platform to conduct a series of spacewalks that would be crucial to station construction. Starting in 1983, the roles and skills of EVA astronauts were designed, developed, and honed with great success on missions to recover stranded satellites, to repair Solar Max, and to service the Hubble Space Telescope. When the time came to start assembling a station fifteen years later, NASA had a cadre of highly experienced spacewalkers ready to tackle the most complex construction project in human history, several hundred miles in space and circling the globe sixteen times per day.

Simulations in the NBS tank at Huntsville, Alabama, in 1985 use a module provided by Boeing for Space Station Freedom. (Courtesy Bert Vis Collection)

EVA FROM THE SHUTTLE

As a back-up measure against the failure of specific equipment that might prevent an Orbiter from returning safely to Earth at least two astronauts on each crew trained for what was called a contingency or emergency spacewalk. Such contingencies included stowing and securing the remote manipulator arm, the K_u-band dish antenna, a variety of hardware in the payload bay such as tilt tables and non-deployed payloads, and the radiators and doors of the bay. If necessary, items would be jettisoned in order to shut the bay doors for re-entry.

Although not every Shuttle flight actually included an EVA, at least 135 teams of astronauts still had to prepare to venture outside as part of their mission training. This inevitably meant producing procedures and timelines, and training for and simulating activities that might never be performed on-orbit. Across the thirty-year history of the Shuttle program, only three flights were required to complete spacewalks that had not been planned as part of the mission. On STS-51D in 1985 astronauts attached a jury-rigged 'flyswatter' device to the end effector of the arm in an attempt to throw a stuck activation switch on a deployed satellite. STS-37 in 1991 required an EVA to deploy a jammed antenna of the Gamma Ray Observatory. And STS-49 in 1992 improvised an unprecedented (and never repeated) three-person EVA to manually capture an Intelsat satellite and put it on a cradle in the payload bay so that it could be fitted with a rocket motor.

All these experiences added to NASA's database of EVA and raised confidence in conducting spacewalks from the Shuttle. The methods that were developed to operate on a variety of satellites, in particular the Hubble Space Telescope, and during several demonstrations of space construction activities, all suggested that assembly of a space station supported by Shuttle EVAs would be feasible.

The main problem with Space Station Freedom was that its complex design would have made its assembly and maintenance heavily reliant on EVA. Although the design that emerged in early 1991 (and would become the ISS) addressed this issue by calling for less EVA, this work would still be considerably more than had been achieved in the past. Shuttle EVAs had been very successful but the sheer number required in the early years of ISS assembly was daunting. Further techniques would need to be developed to determine whether the remaining obstacles of the assembly program could be met, and overcome safely and successfully. This uncertainty gave rise to what became known as the 'Wall of EVA' at the ISS.

"THE WALL"

When NASA was given approval in 1984 to develop a space station, the design of the hardware to create a large, permanent facility on-orbit could begin. EVA was deemed integral to the assembly and maintenance program but as the complexity of the design increased so too did the magnitude of the spacewalking tasks it was thought would be needed to assure success. The issue of the investment in EVA, both in equipment and man-hours, would be a major factor in the ultimate demise of Space Station Freedom.

As the design evolved, it was realized that the effort needed to complete what was planned would be immense. Particularly alarming, was that the experienced people in the EVA Program Office at JSC thought the task daunting. The number of spacewalks required would increase with the missions, enhancing the chances of a serious failure threatening completion of the station. If just one mission were to fail or suffer serious setbacks, the entire effort could tumble like a pack of cards to create further problems from which it would be difficult to recover.

The overall program of EVA for the station appeared mountainous. A substantial obstacle in spacewalking activity would have to be surmounted before the next stage could be attempted. The fact that the plans left little, if any, margin for error or delay meant that the timeline pressure and risks would mount. This situation remained well past the transformation of the Freedom concept into that for the ISS.

In 1985 the planning for the EVA program estimated that by the early 1990s, if all the planned and potential hours of EVA required by every aspect of the program were added together, the annual EVA time for assembly work would have surpassed 2,800 hr. This would increase to over 4,000 hr for maintenance tasks by the start of the new millennium. This staggering number did not even take into account all the effort that would be required beforehand in planning, training and simulations. These figures are all the more astonishing in terms of overall American EVA experience. In total NASA has logged just over 12 hr of EVA during Gemini, 163 hr during Apollo, around 82 hr during Skylab, and a cumulative total of 700 hr during Shuttle missions. Summed over thirty-five years, all of this experience was less than half of what would be expected in a *single* year of the Freedom construction and operations.

Expecting a 100% success rate, with no delays, no set-backs, no crew sickness, and no failures of equipment was wildly optimistic. Even for the *minimum* estimate, EVA activity was projected at over 1,872 hr/year, which worked out at six EVAs per week, each lasting 6 hr, performed by a pair of astronauts (and thus 3,744 man-hours). At the *maximum* projected estimate the figure was 5,600 hr/year (11,200 man-hours). Such an exhausting schedule would impose a tremendous strain on the crews and would stretch hardware, consumables, timelines, and ground support to (and probably beyond) their limits. Although these calculations were only part of a paper study, they reflected both the complexity of the Freedom design at that time and the extent to which the program would be reliant upon EVA operations. Something had to be done to reduce this effort, and this concern fed into a review of the whole Freedom design.

Even after the design emerged as the ISS, it still called for a significant amount of EVA. In the first two years of on-orbit operations for this project, the EVA effort was estimated at 50 to 60 hr/year. This would go up to over 200 hr in years three, four and five, prior to declining to about 75 hr of assembly work in year six. In reality, the ISS EVA time never reached such a high rate. It averaged 80 hr because the peaks of over 140 hr in the years 2002, 2007 and 2009 were balanced out by the significant hiatus in activity between 2003 and 2005 caused by the loss of Columbia.

Gaining The Experience

Between 1965 and 1995, the only spacewalks by Americans at a space station were the ten made at Skylab between May 1973 and February 1974, which totaled 41 hr 46 min. The next opportunity came during Shuttle-Mir with two excursions from Atlantis, one during STS-76 in 1996 and the other during STS-86 in 1997. On the first occasion the astronauts affixed MEEP dust collectors to the exterior of the Mir docking module and evaluated foot restraints and tether hooks intended for ISS EVA operations. These two EVAs totaled 11 hr 3 min. This was useful up-to-date experience but was still far short of what would be needed to guarantee (as much as possible) that so many EVAs at the ISS would succeed.

Underwater simulations in 1986 involving a suited subject in a mock-up airlock. (Courtesy Bert Vis Collection)

If a difficulty on an EVA at the ISS prevented the installation of a large hardware element, requiring its return to Earth, this would risk the Orbiter being damaged by a heavy landing. Even if the landing was uneventful, the program would take a substantial budgetary hit in relaunching the payload once the problem had been resolved and NASA was ready to try again.

The Decision To Build

Prior to NASA being authorized to build a station many studies were conducted, both by the agency and by industry to evaluate the concept of constructing the station using robotics, automated methods, spacewalking, and a combination of all three. Numerous simulations were performed and assessed, all long before the design of the station was selected. That was just as well because, unlike the vast structures envisaged in the late 1960s, the 'stations' evaluated in the 1970s and early 1980s changed so many times it was a challenge just to follow the design process, let alone work out how to fabricate a structure.

Whilst the Freedom project authorized in January 1984 clearly had the Shuttle and its EVA capabilities at the heart of the assembly process, it is easy to forget that at that point only two astronauts had made a Shuttle-based spacewalk, and that was only 4 hr. But this capability expanded through 1984 and 1985, with the first demonstrations of space servicing and repair tasks and the first flights of the Manned Maneuvering Unit, the free-flying backpack that was believed to be an important asset to EVA capability and potentially useful in assembling a station.

However, the design of the station remained in flux at this point and any decisions regarding how it should be constructed, how much EVA time would be required, and what equipment might be utilized, were largely unknown.

Applying Human Factors Engineering

The issues involved in structural assembly in space were addressed as early as 1978 by the Massachusetts Institute of Technology (MIT). This was different to earlier studies, which had centered upon the productivity of an EVA crew for station assembly in the context of human factors engineering. It was apparent that time would be a significant factor – not from the Presidential decree requiring completion of the station within ten years but from the limitations to and frequency of the EVAs that predictions suggested would be required to complete the program. MIT then initiated a program of seventeen tests using the 75 ft (22.8 m) diameter 39 ft (12 m) depth Neutral Buoyancy Simulator in Building 4705 at the Marshall Space Flight Center in Alabama. These tests between March and September 1979 expanded upon earlier work that had tested the hypothesis that the assembly of erectable trusses during EVA was the best way to construct large structures in space.

The upshot of these studies, and others, was a program of flight tests for STS-61B in November 1985. Called the Experimental Assembly of Structures for EVA (EASE) and Assembly Concept for Construction of Erectable Space Structures (ACCESS), the tests were to provide data to compare the ability of astronauts to assemble truss structures in space with the data from the underwater exercises. Several important differences would of course have to be taken into account, namely: significant differences in gravitational forces; the viscosity of the water; and the differences between wearing and working in a pressure suit in water versus doing so in space. But from the underwater tests it seemed that such work could be conducted comfortably in space.

Astronauts Jerry Ross and Sherwood ('Woody') Spring made two EVAs that totaled 12 hr during which they assembled and evaluated the two systems. While the ACCESS girder design worked well, and was promising for future development, the three-sided pyramidal structure of EASE involved too much free-floating activity which was tiring and made the task more complicated than if the men had been able to work from a firm base. In the post-flight report, the astronauts commented that ACCESS had confirmed the feasibility of using EVA to assemble space structures and could play a key role in the assembly of Space Station Freedom, which at this point featured a series of 16.4 ft (5 m) erectable structures, but one limiting factor would be the time taken to assemble these lengths.

A follow-up evaluation of a large truss structure was proposed for a Shuttle mission planned for 1991, which was about 18 months in advance of the planned start of Space Station Freedom assembly. This Structural Assembly Verification Experiment (SAVE) was based on the ACCESS design, but was to be erected in stages until it stood 263 ft (80 m) above the payload bay. However, the task of erecting such a large structure and the likely resulting dynamics raised concern in the Astronaut Office about the safety of the astronauts performing the construction work and the integrity of the vehicle. It was an investigation that never left the drawing board.

Experiments for handling large structures in space continued with the Assembly of Station by EVA Methods (ASEM) experiment. Originally assigned to the second and fourth EVAs of STS-49 in May 1992, this test was reduced to just a single spacewalk by the need to undertake a three-person EVA to manually capture a satellite when the plan for that task went awry. Developed by McDonnell-Douglas, the prime contractor for the Space Station Freedom truss structure, ASEM was to evaluate truss assembly in space in preparation for assembly of Freedom. However, around the same time that the ASEM experiment was manifested NASA decided to launch the station trusses in pre-assembled and pre-integrated form. The experiment remained on STS-49 and provided useful data on handling large objects in space, even though the station trusses wouldn't be assembled in this way. Between them, Tom Akers and Kathy Thornton used struts and connectors (called 'sticks and balls') to assemble a pyramidal structure some 15 ft (4.6 m) wide in the payload bay. In a sense it was a step backwards. The narrow necks on the ends of the beams of the EASE-ACCESS experiments on STS-61B had proven easy to handle when wearing the bulky EVA gloves but they weakened the assembled structure. Over the subsequent seven years, the engineering requirements for Freedom had altered and the ASEM hardware was "not designed for optimum EVA handling."[1] This seemed to be at odds with the purpose of the experiment in the first place and led to Thornton's remark that the experiment was "an exercise in frustration" that required excessive arm work and a lot of free floating.

With the STS-49 mission falling behind schedule as a result of the work required to capture the satellite, other station-related work was either truncated or abandoned. This demonstrated the difficulty of keeping to the timeline when working with challenging equipment on-orbit.

An underwater simulation in the 1980s of assembling truss elements in preparation for Space Station Freedom. (Courtesy, British Interplanetary Society)

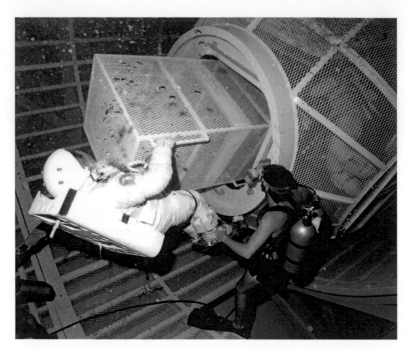

Further tests during 1986 in preparation for Space Station Freedom, here evaluating the manipulation of a large mass through a mock-up airlock and hatch. (Courtesy, Bert Vis Collection)

EVA AT SPACE STATION FREEDOM

After months of deliberation, it was decided to construct the erectable truss for Space Station Freedom from 16.4 ft (5 m) segments. All of the planned pressurized modules, system facilities, and utilities would be mounted on this framework. EVA was seen as key to assembling the station but the number of spacewalks that would be needed, the time required, and the number of astronauts assigned to these tasks changed as often as the design of the station.

It was apparent that more in-depth studies and simulations would be required for the EVA load expected at Freedom. To help with this effort, in part, several former Skylab astronauts (the only American astronauts with any space station experience) were hired as consultants and test subjects by leading industrial partners involved in the Freedom contracts. In December 1985 a four day conference was held at Ames Research Center to review the development and progress of Freedom. The Human Factors Engineering studies had identified several areas in which EVA would play a significant role and in which astronauts lacked skills and experience. This discrepancy would have to be fully addressed before progress could be made towards flight operations.

In reviewing just this first year or so of development it became clear to NASA how significant EVA would be to Freedom as designed, and how frustrating it was that the design was constantly changing. One thing was obvious, however. For whatever plan was chosen to remain on track, the productivity and performance of each EVA would be critical to ensuring that the mission sequence ran smoothly. Each assembly mission would require a thorough EVA training program, with a cadre of qualified astronauts available to receive detailed specialized training, both for the payload assigned to their flight and also in support of other EVA operations. The assembly program would also require extensive use of the robotic manipulator systems and specialist training would be needed to enable robotic arm operators to coordinate with spacewalkers. All of this preparation would involve several months of training. It was clear that a large team of astronauts would be required. This effort would add considerable strain to the already tight workloads for the astronaut training teams and support personnel assigned to the Shuttle flights that were not involved in space station work.

Several evaluations were undertaken to refine this workload, with the areas studied including the workstations on the Shuttle and the proposed station, as well as the tools and support equipment for contingency Shuttle EVAs and the specialized facilities that the station demanded. Studies were undertaken of the care and maintenance of pressure suits, as well as how much (or contrariwise how little) housekeeping should be done on-orbit between EVAs. Past EVAs were re-examined, including the work undertaken by Russian cosmonauts at the Salyut and Mir stations and the problems and setbacks that were acknowledged during those impressive programs. A plan emerged for a range of EVA apparatus and the time to be assigned to a single EVA (or series of EVAs) which would be necessary for every individual element of the station. For instance, different EVA techniques, procedures and apparatus would be needed in attaching the truss and solar panels versus installing pressurized modules. EVA training was also assessed. A review of the facilities highlighted a requirement for a much larger water tank than that at Marshall, preferably sited at the Johnson Space Center. Studies were also conducted into the likely requirements for EVA at the station *after* its construction was complete, in order to maintain and upgrade it.

Rise Of The Robots

In parallel with the EVA studies of the assembly of Space Station Freedom a series of studies examined the potential role for extensive robotic systems on the station. These considerations of fixed, portable and mobile workstations would eventually lead to the development of the Canadian systems that were installed on the ISS.

In 1987, Boeing and a Government-Industry Advisory Group produced the four-volume Man-Systems Integrations standard document. This focused on man-system integration in the station programs and investigated a variety of human engineering factors, including EVA operations. In July 1990, a two-volume report written by the External Maintenance Task Team (EMTT) reviewed how the evolving design of the station was leading to ever more EVAs of increasing complexity. The report noted a difference between the official estimate issued by NASA for EVA at the station and what the NASA study group had been requested to evaluate. This discrepancy raised further doubts regarding being able to deliver the projected EVA program safely, on time and efficiently. The role of robotics did little to reduce concerns about the EVA requirements for assembling the station.

Prior to the EMTT report (and highlighting the problem facing EVA planners) yet another review was launched in October 1989 that considered the maintenance of the station after its assembly was complete. The study estimated approximately 432 hr of EVA time would require to be devoted to maintenance per year, but this assumed that everything was working as designed. If contingencies and other unplanned activities were factored into the equation, this could increase to 1,732 hr each year. That would mean, on average, 2.8 EVAs *every week* throughout the operational life of the station. Worse, even that figure did not include any additional assembly and construction that might be required as the station was improved or upgraded owing to age. This amount of EVA would require a significant amount of the crew's time because, including pre-EVA preparations and post-EVA clean-up, a typical 6 hr EVA would occupy at least two astronauts for three days every week. Then there was the problem of training on-orbit for all of the EVA work, perhaps with apparatus recently sent up from Earth that they had not even seen before, as well as preparing and supplying the ground teams to develop each EVA. Other issues were the wear and tear on the pressure suits and EVA equipment, and the logistics required to launch EVA hardware and possibly return it to Earth. And the list went on… Remarkably, when this report was published, NASA had assigned just 132 hr *per annum* to station maintenance after its assembly was complete. This equated to a single two-man EVA per month. Even in the most optimistic scenario this was insufficient. It was evident that for some time to come there would be difficulty scheduling or estimating the EVA effort that would be needed to keep the station flying.

During qualification of the new NBL tank at the Sonny Carter Training Facility in 1996, a team of scuba divers guide a large element of the space station to be used in the training of astronauts. (Courtesy, Bert Vis Collection)

Scuba divers maneuver the pre-integrated truss in the NBL for use by astronauts in EVA train-ing. This design of the truss replaced the Space Station Freedom plan to have spacewalkers assemble the truss segments on-orbit. (Courtesy Bert Vis Collection)

Slamming The Airlock On Space Station Freedom

By December 1989 the External Maintenance Task Team had started its review of the issues relating to EVA at Freedom. Headed by Charles R. Price, Chief of the Robotic Systems Branch at JSC, the team comprised thirty-three specialists and included seven astronauts: Michael Foale, Mike Gerhardt, Linda Godwin, Tammy Jernigan, Dave Low, Dave Leestma, and Jerry Ross. They assessed the design and assembly of each element of the proposed space station. They then reviewed the likely failure rates of the various items, the variety and number of Orbital Replacement Units, and the predicted levels of mainte-nance. Finally, they looked at the plans for assembling the station and the EVAs required to support that effort. The numbers were staggering.

The investment in EVA for assembly and maintenance would be a vast undertaking, especially during the proposed assembly period from 1995 to 1999. When the station's expected lifetime of thirty-five years was factored in, the group found that, on average, over 500 EVA maintenance tasks would be required *every year*. This equated to about 3,276 hr of EVA time. In terms of crew time, at least 273 two-person EVAs would be required every year throughout that period – a staggering rate of 5.3 two-person EVAs *per week*. This was far more than had been predicted by an official NASA report just a few months earlier and therefore called the official figures sharply into question.

The Price report also raised issues about how such a program of spacewalks could be trained for, supported, and kept on schedule and within budget, simply in order to keep the station functioning. If it required more than five spacewalks per week, then it would cer-tainly need more than one two-person EVA team to sustain the workload week after week.

This forecast soon led to studies of the EVA workload that a Shuttle crew could be expected to satisfy. These calculations were backed up by hard data from a series of demonstration tests that were carried out by several Shuttle missions during the 1990s.

Although they were not the only reason for curtailing development of Freedom and transforming the concept into what became the ISS, these EVA studies and projections certainly helped to inform that decision. Freedom was an overly complex concept that lacked a clear appreciation of the balance between what was desired and what could actually be achieved.

Perhaps the clearest example of this is evident from the projected assembly time for Freedom, estimated at fifty-four months. Assembling the ISS using mostly the Shuttle system took almost thirteen years to complete. This included the downtime caused by the loss of Columbia, but even without this delay the process would have taken over a decade. The assembly of Freedom would inevitably have suffered its own delays, and these would likely have exacerbated the pressure on the EVA work.

When the design that became the ISS emerged from the confusion of Freedom this required a simpler, but still substantial, EVA program to support the assembly; the so-called Wall of EVA.

Revisiting Freedom's basic EVA plans for the years of assembly suggests it would have been more of a mountain than a wall. The five-year process would have required, it was estimated, more than 1,700 maintenance actions. This would have necessitated some 11,517 hr of EVA, corresponding to a staggering 950 two-person spacewalks.

What did this really mean in terms of crew time? To assemble Freedom a cadre of EVA specialists (excluding RMS-trained astronauts) would have to perform 4.1 two-person EVAs per week for over 4.5 years. This was utterly impractical even on paper. The Shuttle itself would never achieve anything like the twenty-six flights per annum that were originally envisaged throughout its thirty-year operational lifetime, without the entire fleet being called upon to support the intensive five-year assembly. None of this even makes allowance for launch delays, miscellaneous failures, or the notorious Florida weather. Although it was estimated that the task of maintaining Freedom after its assembly would lighten the EVA load, even in a 'low year' there might be as many as 189 EVAs. Yet as Freedom aged and it became necessary to replace ailing systems, those figures would rise again. It was estimated there would have to be 3.7 EVAs per week in a low EVA maintenance year and an alarming 10.4 EVAs per week in a high maintenance year.

This was not what the Freedom planners wanted to hear, especially coming from an internal NASA team, but the data was clear and could not be ignored. In hindsight, the EVA plans for Freedom were unattainable on paper, let alone in terms of training and support – even at the most conservative estimates. The actual amount of EVA that has been achieved since Alexei Leonov made the first excursion in 1965 readily bears this out. Between NASA, the Soviet-Russian program and the nascent Chinese effort, just over two hundred people have conducted fewer than four hundred EVAs over a period of fifty years.

Reviewing the EVA records of the Shuttle at the ISS (excluding those undertaken only by residents) the data shows the absurdity of the Freedom projections. Between December 1998 and July 2011 NASA astronauts, sometimes assisted by international partners or ISS crewmembers, undertook one hundred and nine EVAs totaling 723 hr 51 min. Of these, twenty-eight (187 hr 49 min) were from an Orbiter and eighty-one (536 hr 2 min) were from the airlock of the ISS. In total, the one hundred and eleven station related spacewalks at Mir and the ISS, which were made primarily by Shuttle astronauts or were at least under a Shuttle mission designation, totaled 734 hr 54 min. That is *less than half* of the original *minimum annual* rate that was naively envisaged for Freedom thirty years previously. Whilst improvements in equipment, techniques, training, and hardware over the years would probably have contributed to easing the burden of the Freedom EVA plan, as would occur with the ISS, even this would not have overcome the fundamental flaws inherent in the Space Station Freedom design. The major rethink that led to the development of the International Space Station was probably the only sensible way forward.

BUILDING A FOUNDATION FOR SCALING THE WALL

Although the development of Freedom itself had stalled, there was progress in EVA techniques. The operations envisaged for the station may have been well beyond the capabilities of humans and machines but significant work had been accomplished in defining just what would be feasible for conducting a significant EVA program from the Shuttle. One of the earliest objectives of the Shuttle program was to service and maintain satellites by EVA, and as the program evolved during the 1970s and 1980s procedures and hardware for spacewalking were developed in parallel.

Having The Ability

The ability to undertake an EVA from the Shuttle was made possible by locating the airlock on the rear wall of the mid-deck, although if necessary it could be transferred into the payload bay as part of a transfer tunnel system. The airlock could enable two space-suited astronauts to prepare for and undertake a spacewalk without needing to depressurize the entire habitable compartment of the Orbiter. This system functioned well for over thirteen years, then the outer hatch of the airlock jammed, preventing a planned spacewalk on STS-80. The crew had no suitable tools to address the problem. Subsequent Shuttle crews carried a set of EVA hatch tools that would allow access to the handle mechanism in order to overcome such a failure. Fortunately this was never necessary.

Although the two EVAs undertaken at Mir and the first ISS EVAs were conducted from the airlock of the Orbiter, the new Quest airlock was installed on the ISS in 2001 to allow spacewalks to be performed from the station by astronauts wearing American Extravehicular Mobility Units (EMU) irrespective of whether an Orbiter was docked. When the Russian Pirs airlock was added a few months later for use with the Russian Orlan suits, it allowed the station residents to make spacewalks from either the US or Russian segments of the station. Once Quest had been installed, the only other EVAs from the Shuttle airlock on a space station flight were by STS-114 in 2005. But these were not the final Shuttle-based EVAs because those were made during STS-125, the final servicing of the Hubble Space Telescope.[2,3] They brought to a close the era of Shuttle-based EVAs that began in April 1983 with the STS-6 mission.

The pressure garment for spacewalks using the Shuttle and the US segment of the ISS was developed by Hamilton Standard (Hamilton Sundstrand after 1999) based at Windsor

Locks, Connecticut. The company had initiated work on the EMU design in 1974, two years ahead of the formal request for proposals by NASA, and the contract was issued in 1977. The suits were available for contingency EVA operations during the inaugural mission in April 1981 but were not needed. In November 1982 the first planned EVA was aborted before it could start due to problems with the suit systems. The test was reassigned to STS-6 in April 1983 and successfully accomplished using the baseline configuration of the suit.[4,5]

The EMU used a new approach for American spacesuits. It had multiple layers for both orbital protection and comfort, and combined a hard upper torso and fabric arm assembly with a fabric lower torso and legs with integral boots. The upper and lower torso were locked by a waist ring. Underneath the suit, the astronaut wore a web-like liquid cooling garment and a communications cap. There were biomedical sensors to enable the flight surgeon on the ground to monitor the astronaut's performance. And there was a small tube for liquid refreshment during a long and exhausting EVA. An integrated life support system supplied air and coolant and dealt with excess heat. On the chest was a computer display that gave readings of the consumables and the status of the pressurized garment. The gloves were locked onto the arms, and a helmet with Sun visors completed the basic ensemble. Unlike previous programs, the suit was not custom built for each astronaut. Instead, it incorporated sizing adjustment sections to enable it to accommodate a span from fifth percentile female to ninety-fifth percentile male astronauts.[*]

The EMU functioned at a nominal internal pressure of 0.3 atmospheres (4.3 psi or 29.6 k/pa) with 100% oxygen. This necessitated a period of pre-breathing prior to the EVA to purge nitrogen from the blood and prevent the 'bends' during the EVA. In the early days of the Shuttle program, this required the EVA crew to wear the clam-shell-like helmet that they used during the launch and re-entry phases of the flight and pre-breathe pure oxygen for several hours to purge their blood. In parallel, the pressure of the oxygen-nitrogen atmosphere of the cabin was gradually cut from 14.5 psi to 10 psi. After the switch to full pressure suits following the loss of Challenger in 1986, it was necessary for spacewalkers to wear face masks during pre-breathing. With the Quest airlock installed on the ISS, astronauts would 'camp out' inside the sealed airlock at a reduced pressure overnight prior to venturing outside.

The EMU life support system included a 30 min air supply as a back-up in case the primary supply failed. This would give the astronaut time to get back to the airlock of either the Shuttle or Quest and plug into its consumables. The suit had a safe operating

---

[*]A percentile is a statistical measurement where a variable for a population is divided into 100 groups with equal frequency. Hence the ninetieth percentile is the value of a variable such that 90% of the relevant population is below that value. In this case, NASA included information about the human body, its size, posture, movement, surface area, volume, and mass. According to NASA Man-Systems Integration Standards Volume 1, Section 3, Anthropometry and Biomechanics, Revision B, July 1995, NASA STD-3001, these measurements are "limited to the range of personnel considered to be the most likely to be space module crewmembers or visiting personnel." Assumptions were made that the candidates would be in a good state of health, an average age of forty years, come from wide ethnic and racial backgrounds, and be either male or female. In this case the data reflected an example of a smaller frame female Japanese and a larger frame American male crewmember.

limit of 8 hr. The average fell between 6 and 7 hr, but there was one station spacewalk which was almost 9 hr. Additional safety came from a system of tethers and restraints designed to ensure the astronaut remained attached to the spacecraft at all times. This used the 'make before break' strategy in which the next attachment was made prior to breaking the extant one. And there was the Simplified Aid For EVA Rescue (SAFER), a nitrogen-powered backpack that would allow independent maneuvering in the event of an astronaut becoming detached from the Shuttle, station, or tethers. It was strictly for use in an emergency. On an independent mission the Orbiter would have been able to chase after a loose spacewalker, but this would have been impossible when docked with a station. All astronauts who were to wearing the Shuttle EMU needed to qualify to use the SAFER during training that included simulations of recovery scenarios.

An extensive range of tools and apparatus was available to enable an astronaut to undertake EVA tasks, including hand-held tools, tethers, restraints, storage containers, and attachments for the RMS such as fixed and maneuverable foot restraints. There were sunshades on the helmet to protect the wearer from the harshness of the Sun, in addition to the gold-coated pull-down visor. The helmets also had lights installed for use during the night side of an orbit. These were usually used in conjunction with the payload bay floodlights.

An EVA was filmed by a variety of cameras, not so much to provide footage for the world's news media (although this dissemination was never discouraged) but to record the progress of the EVA and the performance of the equipment for post-flight analysis and revision of future training. The outer layer of the EMU was mainly white in order to help to reflect heat, but every suit bore a variety of marks to identify the astronauts taking part. These included:

- Solid white stripes on an all-white suit meant it was essentially unmarked.
- Solid red stripes in the form of a red band around the upper legs of the suit.
- Vertical broken stripes, in essence the solid red line divided up into red and white dashes.
- Diagonal broken stripes in a pattern similar to a vintage barber pole design.
- Combinations of two stripes if more than four different crewmembers were scheduled to conduct EVAs during a mission.

SAFETY AT ALL TIMES

By 1997, Shuttle EVA operations had, on the whole, been very successful, although there was always more to practice and learn, or new equipment and techniques to try. EVA is one of the most hazardous and most exciting activities that an astronaut can experience during their career and is one which nobody can afford to be complacent about. Consequently, a great deal of effort was devoted to ensuring that EVAs at the ISS would be safe for the astronauts involved. Safety requirements were incorporated into the development of each EVA plan in terms of the use of the EMU, the tools and the ISS hardware. Safety specialists, experts from industry and the NASA JSC Safety, Reliability and Quality Assurance Branch were all involved in the early evolution of EVA concepts for the space station. They participated in the step-by-step approach to developing each EVA that was to be undertaken, starting with the Preliminary Design Reviews (PDR) though to the Critical Design Reviews (CDR), during which pertinent safety requirements were assessed in relation to the design of all hardware. The safety teams were also involved in underwater testing, in the development of procedures for spacewalks, and in the final certification for all ISS EVA hardware for each flight and its likely influence on subsequent missions. For example, the segmented truss system had to be assembled over several flights, therefore the activities of one EVA might impact upon another planned for several missions later.

Coordinating with the JSC EVA Office, the safety team would not only ensure the safety of the on-orbit tasks but also that the design, development, fabrication, testing, and training phases produced the best chance of achieving all the objectives set for a given mission. Of course, safety did not end with the addition of the hardware on the ISS, because every item had to be re-evaluated to verify that it would remain safe for future EVAs. Constant examinations and evaluations were carried out on equipment returned after a mission, such as the EMUs, EVA tools, and other hardware to ensure their safety prior to being reused. Finally the timeline, procedures, training and flight documentation were regularly reviewed, amended and updated. The results of all this effort were passed on to the people who were next in line.

All hardware requirements were reviewed and each safety document checked to ensure it met current standards. Operationally, the EVA safety support team created effective risk assessments for all aspects of training, including bench reviews, one-g simulations, NBL simulations and Virtual Reality (VR) training. They also compiled flight rules, identified and mitigated potential risks and verified all aspects of an EVA before a Shuttle left the ground. During the mission, members of the team gathered in the Mission Evaluation Room (MER) in Mission Control, Houston, to offer real-time safety support. They also served on the investigative Tiger Team of Shift Team 4 for each flight.[6]

There were several notable incidents related to Shuttle assembly flight EVAs that invoked safety precautions devised by the safety team in order to address the situation and devise a plan to avoid a recurrence of the incident during subsequent spacewalks. For example:

- Prior to STS-97 there were no plans for an EVA astronaut to work on the ISS solar arrays because they were so fragile. In fact this was listed as a Keep Out Zone. However on December 3, 2000, when one of the arrays being deployed by this crew jammed, a new plan of action had to be rapidly devised. A Tiger Team from representatives of the safety team, the developers of the hardware involved, operations engineers and astronauts, drew up a plan by which EVA astronauts would avoid the sharp edges and other hazards on the blanket box. Their work led to the successful deployment of the solar array and created the baseline for planning future contingency EVAs in case of a recurrence of the situation.
- On the first EVA of STS-98 on February 10, 2001, a small amount of frozen ammonia leaked from a coolant line when links were being made to the new Destiny laboratory. The coolant soon dissipated in the vacuum of space, but standard contingency decontamination procedures were followed by the two spacewalkers. These actions had been devised by the safety team in the event of an incident involving handling hazardous material during assembly. In this case, the astronaut whose suit was most

affected (Robert Curbeam) remained in sunlight for a total of 30 min while his colleague (Tom Jones) brushed off the suit. During the re-pressurization of the airlock, it was partially purged to help flush out any lingering contaminant. In addition the other crewmembers Ken Cockrell, Mark Polansky and Marsha Ivins, each wore an oxygen mask for 20 min after the airlock hatch was opened to the mid-deck. There were no reported ill effects from the crew.

- On October 30, 2007, the STS-120 crew encountered a different incident with the solar arrays to that which faced STS-97 seven years earlier. This time, the solar array snagged and tore during deployment. With the STS-97 experience in the bank, the safety team soon defined safe working processes for this new problem, allowing the astronauts to continue with the planned tasks. With the spacewalkers protected from the risk of electrical shock and able to avoid the sharp edges of the torn array, they managed both to complete the EVA and to avoid further damage to the ISS.
- During STS-124 in June 2008, the EVA Safety Controller who was on duty in the MER realized the required electrical inhibitors had not been applied for the installation of cameras prior to the second EVA. This was flagged to the MER manager, who in turn informed the Flight Director (FD) of the oversight. The FD then instigated the necessary safeguards to ensure that the inhibitors were applied in advance of the EVA, thereby reducing the risk of the spacewalkers getting an electrical shock. Such fine attention to detail and the advantages of having multiple back-up systems (and personnel) often went unreported.

Any such incident could potentially push the teams on the ground and crew on-orbit to the limits of their skills and overly stress the available equipment. NASA policy has always been to avoid such occurrences and to re-evaluate any EVA that was deemed to lie beyond the training of the astronauts or the capability of their equipment. But it was also recognized that, "The possibility of an EVA occurring which is at or beyond the bounds of sufficiency will increase substantially with the start of ISS construction."[7]

Preparing For The ISS

Space Station Freedom was to have been assembled on-orbit using a combination of the RMS and EVAs over a long series of Shuttle flights. In the spring of 1984, when NASA was authorized to create a space station, there had only been eleven Shuttle flights with just three of them supporting the five EVAs made. By the loss of Challenger in January 1986 the EVA log book had still only recorded a little over 67 hr on eleven excursions. After the Shuttle resumed flying in September 1988, there was a five-year gap between the spacewalks of STS-61B in December 1985 and those of STS-37 in April 1991.

The initial Shuttle EVAs were of considerable importance, being required to prove that the vehicle was a suitable platform for a spacewalk. Before the loss of Challenger, astronauts tested and evaluated the EMUs, restraints, tools, the Manned Maneuvering Unit, and the capabilities of the RMS in supporting an EVA crew. They also captured, repaired, and redeployed or returned to Earth a number of satellites. These excursions successfully proved the concept of space servicing that had been a cornerstone of the Shuttle's rationale since the early 1970s. But it was also apparent that before NASA attempted a task as large and complicated as assembling a space station, it needed to develop a lot more skills and operational techniques.

After the Shuttle resumed flying in 1988 the first opportunity for an EVA came on STS-37 in April 1991. One of the mission's EVA astronauts was Jerry Ross, who had participated in the STS-61B spacewalks. He and Jay Apt were to carry out an EVA to assess methods of moving crew members and apparatus around the future station. But first they were called upon to make an unscheduled EVA to release an antenna of the Gamma Ray Observatory that had not deployed when commanded. Since this was a contingency that they had prepared for, the task was rapidly achieved. The success of the spacewalks on this mission reaffirmed the Shuttle's capabilities, but it was readily acknowledged that much more remained to be done.

By now, however, the hiatus after Challenger and the escalating budget concerns with Freedom had already pushed the projected first assembly missions into the late 1990s. By the time of STS-37, Freedom was six years into its ten-year mandate and there was still no flight hardware ready to launch, let alone any confirmed assembly flights on the manifest.

In their post-flight debriefing the STS-37 astronauts remarked on the significant attrition rate, not simply from the Astronaut Office but also Mission Control, other NASA Directorates, field centers and industrial partners. A considerable amount of talent, experience and skills had been lost. Given the ongoing delays, the numerous changes to the design, and the lack of training and firm flight opportunities, this was perhaps unsurprising. The astronauts highlighted their concern that in addition to the obvious problems with Freedom there was an absence of clear mission planning and preparation for the extensive EVA tasks that would be required. It was evident that a focused series of EVAs would have to be assigned to forthcoming Shuttle missions in order to carry out preparatory work well in advance of any assembly missions. It was further suggested that test procedures and flight rules would need to be formulated in advance of their operational use by assembly missions, to expose any difficulties and devise and test alternative methods. It was also evident that while back to back EVAs by the same crewmembers would be possible, alternative schemes must be devised to set a less stressful pace on each mission.

The Apollo lunar missions had shown that making EVAs on three consecutive days was exhausting for the astronauts and it left little time for them to take stock at the end of one excursion to prepare for the next. One-sixth gravity was certainly a real help in undertaking the often strenuous moonwalks but the dust that was kicked up coated the suits and clogged zippers and connecting rings. This posed a problem because most of the work required suit mobility of the legs and arms and imposed severe variations of temperature between light and shade. Assembling a station would also require mobility and significant physical exertion, exposure to day-night lighting and resulting thermal variations. The spacewalkers would require retention devices to remain in place at the work site, and their tasks would have to be designed to minimize tiring movements of the arms and upper torso.

Hence a plan was instigated to create EVA teams for both the station assembly and HST service missions on which pairs of astronauts would alternate spacewalks so that EVAs could be conducted on a daily basis without exhausting the crew. When one pair was outside, the other pair would remain inside to rest, carry out support activities, and prepare for their next outing.

EVA Operations Procedures/Training DTO 1210

To help to address these concerns NASA organized Development Test Objective DTO 1210. The role of this new evaluation of EVA training and operations was to "broaden EVA procedures and training experience bases and proficiency in preparation for future EVAs, such as the Hubble Space Telescope (HST) and Space Station Freedom (SSF)." The announcement of DTO 1210 was made in November 1992, some six months after STS-49 astronauts performed the ASEM experiment.

The decision to go with pre-assembled truss segments for the ISS was borne out by the difficulties that Kathy Thornton and Tom Akers encountered while assembling the ASEM hardware. In particular, the excessive forearm movements that were necessary had seriously tired the astronauts, working as they were in the confines of the pressure garments. Surprisingly, the catalyst for the development of the EVAs for servicing the HST, which inadvertently assisted in developing spacewalking procedures for the ISS, was the unscheduled and improvised three-person EVA to capture the Intelsat satellite because that shone a light on the need to better plan and prepare for external activity.

As mentioned, the ASEM test was scheduled for two EVAs on the STS-49 mission but was cut to a single EVA by 'stealing' time to capture the Intelsat satellite so that it could be stowed in the payload bay, mated with a rocket, and then released to climb to geosynchronous orbit. The decision to undertake an unprecedented three-person EVA to manually grab the errant satellite gave rise to headlines which praised the mission's drama. This was particularly helpful to plans to repair the flawed optics of the Hubble Space Telescope the following year.

NASA Administrator Daniel ('Dan') S. Goldin had already authorized a number of reviews into the planning, content and details of the unprecedented five EVAs that had been identified as necessary to repair the HST. Although he publicly heaped praise on his agency for its determination to save Intelsat, Goldin was apparently not very happy with the improvisation.

According to former astronaut Thomas P. Stafford's 2002 autobiographical memoir *We Have Capture*, written in conjunction with author Michael Cassutt, the concept for the Intelsat retrieval had not been properly simulated and the documentation about the satellite had failed to match the hardware. The biggest problem in the process was the lack of a singular authority figure to oversee the entire mission. Stafford urged that the Apollo-style Mission Director role be reinstated for the HST repair. This worked well, and was carried forward for all of the subsequent Hubble service missions. There were also direct implications for the longer term and intensive planning and preparations for assembling the space station.

Developing ISS EVA Procedures And Testing Them In Flight

By 1993, DTO 1210 was ready to be tested in space. Nine Shuttle flights contributed to EVA evaluations for DTO 1210, DTO 671 (EVA Hardware for Further Scheduled EVA Missions) or the EDFT (EVA Development Flight Test) program. The three that flew in that year concentrated on developing techniques. STS-54 explored the task of moving a large mass (another astronaut) around in the payload bay. This was related to the use of large tools and the ability to transport bulky objects. STS-57 then tested moving a large mass (still another astronaut) while riding on the RMS. In addition to obtaining data for direct comparison between training in the water tank in Houston and performing a task in space, the STS-51 mission performed assessments of high and low torques on safety tethers and the usefulness of a provisional stowage assembly in the Shuttle payload bay. In December 1993, many of the lessons learned from DTO 1210 were put into practice by STS-61 making the triumphant Hubble repair. The program then concluded in 1994 with STS-64 testing the Simplified Air for EVA Rescue (SAFER).

The series of five EDFT experiment missions kicked off with STS-63 in 1995, the 'Near-Mir' rendezvous mission. In addition to yielding further experience of working with mass loads the experiments involved a series of cold soak tests. Then on STS-69 thermal sensors were installed on the RMS along with a micrometeoroid shield of the type planned for the ISS. The astronauts evaluated their ability to perform a variety of repetitive tool handling exercises, cold soak tests, small maintenance tasks and simple assembly tests when subjected to differing thermal conditions. The next EDFT was on STS-72 in 1996, with station assembly techniques being assessed by installing a rigid umbilical in the payload bay. Additional thermal soak tests were performed and a new portable work platform was evaluated. Developed by Lockheed Martin, this new tool was a combination of a restraint aid, a stowage facility for transporting equipment, a workstation, and an articulating portable foot restraint.

The fourth EDFT was set for STS-80 but had to be canceled when the outer airlock hatch jammed. This fault proved to have been caused by a loose screw that had found its way into the handle mechanism. Tools and procedures were developed to prevent a recurrence on future missions.

The final EDFT, made by STS-87 in 1997, involved assessing a telescoping crane boom in a simulation of replacing solar arrays batteries on the ISS. The spacewalkers also tested a cable caddy, a body restraint tether, a multiple-use tether and some other handling apparatus. This flight also included a test of the Autonomous EVA Robotics Camera Sprint (AERCam Sprint). This 13.97 in (35.5 cm) diameter, 34.8 lb (15.8 kg) sphere had twelve small nitrogen powered thrusters, a pair of TV cameras and its own avionics system. It was deployed in the payload bay and controlled from the aft flight deck. The idea was being promoted as the possible forerunner of future robotic EVA inspection and observation tools which could enter locations that were too small, too difficult, or too dangerous for a EVA astronaut to venture. It was also suggested they might be used to remotely document EVA activities, though at the time of writing this application is still to appear.

By the conclusion of this series of tests and evaluations, some 155 hr of EVA had been accrued and valuable lessons were learned relevant to future EVA operations in general and the assembly and maintenance of the ISS in particular.

Creating An EVA Cadre

In June 1997 NASA addressed the EVA challenge of assembling the ISS by naming a cadre of fourteen astronauts who either already possessed spacewalking experience or had been training for some time to undertake EVAs at the ISS. Jim Newman and Jerry Ross had started training in August 1996 for tasks required on the first assembly flight. They were now supplemented by Leroy Chiao, Robert Curbeam, Michael Gernhardt, Canadian Chris Hadfield, Tom Jones, Mark Lee, Michael Lopez-Alegria, Jim Reilly, Bill McArthur, Carlos Noriega, Joe Tanner, and Jeff Wisoff. Their training addressed specific tasks for the early assembly flights.

Table 8.1 Early ISS EVA Assignments June 1997

STS Mission	ISS Flight	Planning Date	EVA Crew members	Payload/ EVA tasks
STS-88	2A	1998 July	Ross and Newman	Node 1, connection of power cables
STS-92	3A	1999 January	Chiao, Wisoff, Lopez-Alegria and McArthur	Integrated Truss, Pressurized Mating Adapter-3
STS-97	4A	1999 March	Tanner and Noriega	Photovoltaic Module
STS-98	5A	1999 May	Lee[1] and Jones	US Laboratory module
STS-99[2]	6A	1999 June	Hadfield and Curbeam[3]	US Lab racks, Station Remote Manipulator System
STS-100[4]	7A	1999 August	Gernhardt and Reilly	Joint Airlock, High Pressure Gas Assembly

NOTES:

1. Lee was stood down from the mission on September 7, 2000
2. STS-99 was re-manifested as STS-100
3. Curbeam moved to replace Lee on flight 5A (STS-100)
4. STS-100 was re-manifested as STS-104

This would prove to be the first in a long series of EVA crew assignments over the next few years. However, as Director of Flight Crew Operations Dave Leestma noted, "It is important for us to begin work now to train the crews who will support station assembly flights. They will be exceptionally busy preparing for some challenging and demanding tasks from initial assembly right through the installation of the robotic arm and an airlock for station-based spacewalks."[8] However, by starting so far in advance of the planned flights, it was inevitable there would be changes to the assignments. As Leestma put it, "We expect that these assignments may be refined in the future." They certainly were, in part due to delays in producing the station hardware, particularly the two Russian elements, but also due to faulty electrical wiring and fuel lines aboard the Shuttle that were identified during routine ground processing and the inclusion of extra logistics flights.

Space Station Program Manager, Randy Brinkley greeted the announcement of the spacewalking cadre with, "The assignment of these crewmembers is a critical element in our ability to bring together the elements of the space station on the ground and then successfully assemble them in orbit. I am highly pleased with the level of expertise and dedication these crewmembers bring to the program."

THE ERGONOMICS OF EVA

Dozens of reports have been issued over the years regarding the development, theory, and practice of EVA. Many focused on the amount of EVA time that would be devoted to the assembly phase of a space station, initially Freedom and later the ISS. Since the pioneering spacewalks of the mid-1960s, the planning and execution of EVA has been greatly refined in order to make the best possible use of the limited time that would be available 'outside.' These efforts addressed the amount of consumables, the apparatus that would be used, and the physical attributes of the astronauts performing such tasks. Despite such improvements and the corresponding development of robotics, many of the studies concluded that EVA at the ISS would become a major issue.[9]

In addition, a number of studies and programs were performed in order to define the 'science' of EVA. One such type of study is known as Human Factors Engineering and Ergonomics (HFEE). It aims to enable humans and their machines to function as safely and efficiently as possible. One of the core tools employed in these particular studies is the time and motion program. It creates a database from which a variety of EVA tasks can be defined, along with the expected time each task should require. From such data, the expected and optimal durations of any given mission's EVA tasks can be predicted.

Some of the earliest work in this field was preparing the spacewalkers of STS-61 to repair the flawed optics of the Hubble Space Telescope in December 1993, which was some five years before the first elements of the ISS were launched.

A Role Model: STS-61

Researchers in the field of HFEE familiarized themselves with analyses of EVAs right back to the Gemini era. They identified a preliminary set of tasks for STS-61 and their predicted durations were plotted to produce a baseline database. The film of the actual spacewalks of that flight was then compared with this baseline in order to measure the performance of the astronauts in executing those tasks. Over the next two years, more EVA data was added to the database from several Shuttle EVAs and DTOs. Feedback from the astronauts provided further references to expand the database.

Traditionally, American EVAs were planned by people who had worked in the field for years and used their judgment to evolve an EVA plan that specified how any given task should be addressed and how long it should take to execute. Then data from many hours of simulations, both in one-g and underwater, would be used to refine these plans and a margin in the range 10–30% added to the estimated duration of the EVA to allow for unexpected developments and failures of apparatus. Despite meticulous attention to detail during planning and training cycles, this was not necessarily the most efficient or productive means of planning an EVA because it often failed to exploit the opportunity to its fullest.

Back in the 1970s, it was worked out that the best way to relieve potential hurdles to EVA was to standardize as much of the apparatus, tasks, and training as possible. Frank ('Ceppi') Cepollina of the NASA Goddard Space Flight Center in Maryland, pioneered this effort by applying to satellite servicing his idea of efficiently overcoming the issues of spacecraft reliability by creating a basic spacecraft framework. This framework, also called a bus, would feature a number of exchangeable boxes called Orbital Replacement

Units (ORU). It was really an early example of the 'plug in and play' method of repair that is widely employed today. By replacing the whole box, the time required to fix the problem would be reduced. There would always be the likelihood of something failing or needing replacement that wouldn't fit within a given ORU design, but incorporating more exchangeable units into the bus would free up time to focus on non-ORU activity. This could easily apply to EVA planning. Predicting the probable failure rate of ORUs would make it easier to predict how much time would be required to deal with them, to calculate the number of EVAs that it would involve, and to plan other tasks around this framework of activity.

However, experience gained over the years established that there were variations to such predictions due to using different methods to estimate EVA time requirements. It didn't help that the number and design of ORU varied as widely as did the time that it would take to exchange them. It was partly in response to this uncertainty that HFEE was used in planning the EVAs for the ISS, with the servicing of Hubble providing a baseline and offering projections for planned activities across the station both during and after its assembly.

Adapting Time And Motion Parameters To EVA

Usually for time and motion studies, a task is observed (either in real-time or in a film playback) and timed to identify how long it takes to perform the basic steps needed to achieve that task. Such investigations have been applied in industry for over a century and progressively refined. Each task to be performed by a worker is divided into either productive or non-productive actions, and the analysts identify and eliminate wasteful time and actions within each operation. Such studies are very useful for planning and analyzing an objective as a baseline from which to develop more extensive operations or activities. The parameters for such guidelines need to include worker-specific traits, training, physical abilities and motivation, and environmental effects. When the basics were applied to EVA, it was apparent that wearing a pressure suit, gloves, and helmet limited many of the visual and physical cues or activities. This meant the basics had to be adapted to the EVA environment, particularly the weightlessness of space and huge variations of temperature and illumination during an orbital cycle.

The first step in increasing the efficiency of EVAs while assembling the ISS was to review previous experiences and documentation to draw up a suitable database. Since its formation in October 1958, NASA had gained a well-earned reputation for creating veritable mountains of documents, reports, studies, memos and other materials. By the 1990s the analysts could draw upon a wide array of documents to assist in defining the time and motion database for EVA. One difficulty was that none of the documents was a single comprehensive resource on the topic; instead several references had to be used, relying mainly on personal experience, recollections, and the intuition of people deeply involved with EVA tasking. Interestingly, one of the most comprehensive sources was the study by William F. Fisher and Charles R. Price in 1990 which was so skeptical of the EVA proposals for Space Station Freedom.[10]

There remained the task of integrating all of the knowledge and experience from a range of sources into a singular, common source of reference for EVA. The data from previous EVA activities and underwater simulations enabled the analysts to establish a good representation of the expected timelines for a variety of tasks. The training library of the Mission Operations Directorate at JSC yielded the two volumes of *EVA Lessons Learned*. This was drawn from all previous American spacewalks from Gemini through to the Shuttle. There was also comprehensive data from post-flight reports of flights on which EVAs did not achieve the estimated timeline, detailing how much time each task really took. Whilst these reports were helpful, they lacked the pre-flight estimate of the timeline for each given task.

More useful to the analysts were the post-flight reports of STS-51A in November 1984, where spacewalkers retrieved the Westar and Palapa satellites which had been released nine months earlier by STS-41B, only to be stranded in low orbit when their motors failed to kick them into geosynchronous transfer orbit. During the recovery of these satellites, difficulties were encountered that revised the actions of the astronauts and the time to complete tasks. This mission report listed the expected durations from training simulations and the actual durations, and summarized the issues encountered and recommendations for the future.[11] Table 19 illustrates the depth to which that this type of research can go.

Table 8.2 Breakdown Of EVA Analysis Circa 1995

I: Effective versus Ineffective Tasks
  A. Effective Tasks
      1. Physical basic division
         a. Reach
         b. Move
         c. Grasp
         d. Release
         e. Pre-position
      2. Objective basic division
         a. Use
         b. Assemble
         c. Disassemble
  B. Ineffective Tasks
      1. Mental of semi mental basic division
         a. Search
         b. Select
         c. Position
         d. Inspect
         e. Plan
      2. Delay
         a. Unavoidable delay
         b. Avoidable delay
         c. Rest to overcome fatigue
         d. Hold
II: A breakdown of EVA basic tasks with examples
  • Adjust – the action of modifying a setting of a tool, PFR, or MFR
  • Assemble – the action that occurs when two or more components are joined together
  • Delay – any time during which no work is occurring
  • Disassemble – the action that occurs when two or more components are taken apart

---

(continued)

Table 8.2 (continued)

- Grasp – the action of closing the fingers around an object
- Hold – the action of supporting an object that is not being (dis)assembled or transported
- Inspect – the action of visually checking an object to determine something about it
- Operate – the action that occurs when work is being performed with a tool
- Plan – the action of deciding what operation or task is to occur next
- Position – the action of properly orientating and aligning an object
- Reach – the action moving a hand to or from an object
- Rotate – the action of moving a PFR or MFR about its axis
- Search – the action of locating an object
- Select – the action of selecting one object from among a large number of objects
- Tether – any action involving the use of a tether
- Transfer – the action of transporting an object that is within the grasp of the astronaut
- Translate – the action of moving from one place to another
- Translate command – the action of communicating the desired RMS translation by the IV RMS operator
- Verify – the required action of checking with the IV operators to ensure the proper tool setting or procedure
- Wait – a delay due to someone (or something) else.

In addition these categories were subdivided in a number of classes such as Translate: manual; Translate: via RMS; Tether: select; Tether: open etc. which were to be constantly modified as experience on EVA was gained via the Hubble servicing missions and experience from the DTO 1210, DTO 671 and EDFT programs.

III: Information incorporated in an EVA spread sheet database
- A reference number within the data base
- The astronaut performing the task
- Panned time of the task
- Actual time of the task
- Timeline procedure from the Flight Data File (FDF)#
- Task baseline
- Class of the beeline task
- Task description containing supplemental information
- Starting time of the task
- Stopping time of the task
- Elapsed time of the task
- Hand used for the task (left, right, both)
- The task
  - Direction (clockwise, counter clockwise, left, right etc.)
  - Value (measurement, revolutions, etc.)
  - Encumbrance
  - Obstacles
  - Rate (distance/sec, revolution/sec, etc.)
- View
- Light (day, night, helmet lights, helmet and sport, etc.)
- Restraint (free, holding with left hand/right hand/both hands, PFR, MFR, etc.)
- Posture (neutral, twisted, hunched forward, etc.)
- Metabolic rate
- Error committed
- Comments.

[Information courtesy *A Human Factor Analysis of EVA Time Requirements*, Final Report, NASA/ASEE Summer Faculty Fellowship Program-1995, JSC, Dennis W. Pate, Texas A&M University Engineering Technology Department, Contract Number NGT-44-001-800, August 8, 1995]

One of the difficulties in ensuring the accuracy of time and motion data for EVA is that it relies on the ability to observe the entire task and the actions taken to achieve it. But there are restrictions, not least the fact that the researcher is unlikely to be there to complete the study in person and thus must rely on recorded observations. Also, those undertaking the task are doing so in a unique environment. The EMU suit restricts the vision of its wearer. This is even more evident during the daylight portion of an orbit, when direct sunlight has to be diffused either by deploying sunshades or lowering the gold visor. The night pass poses a different limitation, with helmet lights and payload bay floodlights offering a rather limited perspective of the work station and task being performed. It is also difficult to recognize when a spacewalker is taking a refreshment break or resting. All these factors need to be taken into account in compiling the basic guidelines.

EVA Governing Factors

The fundamental objectives of a time and motion analysis are to improve productivity and to reduce waste. To achieve this, it is necessary to understand the specific factors which affect the actions being undertaken. Whilst documentation, studies, and ground simulations are useful for EVA tasks, they do not replicate the precise environment in which the final task is to be executed. Aside from the limitations arising from wearing the EMU, the most significant factor is the weightlessness in space which provides the sensation of 'free floating.'

One discovery by the Gemini spacewalkers in the mid-1960s was that to attempt a task while floating freely requires a much longer time than doing so while tethered in position or using a foot restraint. When preparing for the final Gemini mission, Buzz Aldrin simulated spacewalking activities underwater, employing neutral buoyancy to approximate some aspects of weightlessness. Over the years this has proven to be the most effective means of training for long periods of EVA. However, even though an astronaut wears the full EVA gear, they are governed by the laws of gravity on Earth, meaning that if they let go of an item in the water tank it will fall, whereas in space it will not. In addition, the viscosity of the water acts to slow their movements.

One of the primary factors to be considered is, of course, the astronauts themselves. The physical attributes of the individual, their motivation to undergo training, whether they possess prior EVA experience, and their mental abilities and reasoning all have a direct effect on their productivity while spacewalking. The anthropometric differences of stature, stamina, and strength between individuals and between genders can produce different approaches to the same task that can influence the time required to complete a specific operation. Even the issue of left- or right-handedness will have a bearing, with individual proficiency favoring a dominant side which will be faster, more precise, and more powerful. An ambidextrous individual (one who can use either hand effectively) would offer the broadest range of options in the case of a task where accessibility was limited.

The optimal positioning of an individual to address a given task would also have to be evaluated: whether they were free floating, tethered, using foot restraints, riding the RMS, aided by a colleague or performing the task alone.

The sequence of tasks would have to be laid out in a logical manner and understood in relation to the broader objectives of the EVA. Because of the uncertainties of space flight, most flight objectives are 'front ended' or 'front loaded,' meaning that the most important tasks are scheduled early in the mission to increase the chance of their being accomplished if a problem obliges the flight to be cut short. The same rationale applies to the majority of planned EVAs, with the tasks that have the highest priority coming first, then the secondary tasks and, if there is time, what are usually referred to as get-ahead or housekeeping tasks. Contingency, emergency, or unplanned spacewalks are not necessarily dealt with in this manner, since they are not part of the plan.

As an EVA progresses, the astronauts will naturally tire. If they encounter a task which proves more strenuous than expected, the sequence might be revised so that a lighter task will allow them to recover whilst maintaining the overall timeline.

The longer the astronauts spend on EVA or the more frequently they participate in EVAs, the more likely they are to slow down, making each task take longer on-orbit than it did during training or if it were to be undertaken separately. Motivation is the key to overcoming this. If the astronauts find themselves behind schedule, they may push a little harder to catch up. Care has to be taken in such a situation to avoid over-tiring the astronauts and thus creating a medical situation that might result in an early (emergency) termination of the EVA. Fortunately, astronauts are naturally motivated. Once they have secured a coveted flight assignment, sometimes after years of waiting and training, the desire to achieve mission success is intense.

One thing that the Americans had to learn for the ISS, after years of short-duration missions in space, was that work on a space station must be addressed in a completely different manner. From Mercury through Gemini and Apollo, then on to the Shuttle, a standard mission duration was several days to just a couple of weeks. The flight plan could be worked out in fine detail to ensure that all of the objectives were at least attempted, if not completed, before the mission was over. On space station expeditions however, as the Americans had discovered on Skylab and would re-learn at Mir, the tasks are more paced because there is usually another week or another crew to continue any work that remains. This approach was eventually adopted for the assembly of the ISS. Specialist EVA training was confined to the Shuttle crewmembers who were assigned a specific task. One of the lessons from the HST missions that was carried over to the ISS was a level of generic, contingency, and cross-training designed to allow station residents to pick up work left unfinished by a visiting EVA crew.

The amount of EVA training a crew receives can have a dramatic effect on the live performance of the spacewalk during a mission. The ability to maintain EVA training proficiency is also important, especially if there is a significant time interval between the training and the actual EVA, such as owing to launch postponements. The Shuttle assembly missions lasted typically 10–12 days, so the EVAs for which the astronauts trained occurred shortly after the mission launched. For station crewmembers though, several months might pass between training for and conducting an EVA. They might receive refresher training in space using virtual reality (VR) or video instruction. VR training came online at JSC for STS-61 in December 1993, which was the first HST service mission. Its role increased in the later years of Shuttle EVA training, and it is now the standard aid for station EVA training due to the complexity of the EVA tasks and the changing layout of the station.

The water tank of the Neutral Buoyancy Laboratory (NBL) had long been used for EVA training, but this was another area where the relationship with an EVA in space had to take into account several differences. For example, the positioning of apparatus in the tank is influenced by the drag that the viscosity of water imposes on movement. This is not encountered in space. Water resistance also influences the amount of force required to start, move, and stop an object in the tank compared to in space. A major aspect of training in the tank proved to be working 'upside down.' In the tank, gravity pulls everything downwards. If a suited astronaut is upside down, their body weight is supported by the collar bones pressing against the EMU bearing ring. Because this can be very tedious, training in the tank tends to occur in the 'head up' position whereas in space 'head down' isn't a problem and this has led to differences in the way that some tasks are performed on-orbit compared to in training.

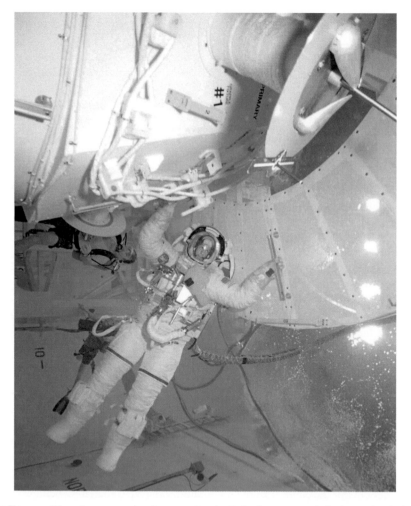

Scott Parazynski evaluates gasping between handrails in the water tank during a simulation in March 1996 for an ISS spacewalk.

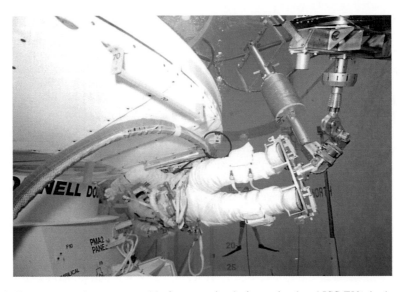

Wendy Lawrence evaluates a portable foot restraint during a simulated ISS EVA in the water tank in March 1996.

The external factors that must be taken into account when planning an EVA include lighting and thermal conditions. Temperature regulation is especially important, as was demonstrated by several of the EDFT 'cold soak' tests. Spacewalkers found that when the temperature drops sharply on exiting direct sunlight, so too does body temperature, especially in the extremities of the arms, hands, legs and feet. Heating elements in the gloves help, but the risks of frostbite are very real. Indeed, while training in a vacuum chamber for STS-61 Story Musgrave suffered severe frostbite in his fingers, which his colleague Jeff Hoffman said were "black and purple." The items astronauts work with can rapidly vary in temperature during transitions between the intense heat of sunlight and the chill of darkness. For example, on the early Hubble flights it proved difficult to ingress the foot restraints in full sunlight because the heat expanded either the restraint or the EMU boot. Seemingly innocuous operational issue like this can impact the EVA timeline.

Despite the most meticulous of planning, even the smallest things can influence an EVA. When a task is delayed or goes awry it can impose a significant knock-on effect on immediately following tasks, later spacewalks, or indeed the operation of hardware and systems. The 'front loading' of EVAs can help but unforeseen issues, unplanned activities, and emergencies can easily require a second EVA crewmember to come to the assistance of a colleague, thus impacting the schedule. If an EVA was terminated early, there had to be a plan to make the work site safe and secure, so that neither the vehicle nor the crew would be compromised by the incomplete task.

Story Musgrave has likened EVA to an orchestrated ballet, but one that involves a more restrictive costume. If all goes well, the movements are well rehearsed, and the equipment performs as intended, then the tasks flow seamlessly from one to the next and almost without conscious thought. But if there are interruptions it can take some time to get back into that flow. For every EVA, the clock is always ticking away the amount of time the crew can remain outside with the consumables in the life support system. It is another factor that can add to the stress felt. Wishing to get the job done quickly and efficiently is natural, but safety is paramount.

If an EVA crew is not conducting a spacewalk they do not have a 'day off.' Space flight is expensive and the maximum must be gained from the mission. In the Apollo days, it was discovered that working using the same crew on back-to-back EVAs was tiring. Nevertheless, on early Shuttle flights a single EVA team would be sent out on successive days. For the record-breaking STS-61 mission, two pairs of spacewalkers would go out on alternating days. This worked very well. For station assembly flights the entire Shuttle crew was on an intensive timeline, not merely those assigned EVA. Inside the spacecraft other crewmembers operated robotics, controlled the vehicle(s), documented the activities, and otherwise aided the EVA crew along the timeline. And there were the pre-EVA and post-EVA chores of preparing and cleaning the suits and apparatus for the next excursion. When there was more than one EVA team, those not actually outside were fully occupied by support roles, leaving them little time to relax and wind down until all the spacewalks and other planned tasks had been wrapped up safely and successfully. It was often at this point, after the excitement and intensity of the work was over, that tiredness would catch up to them. This was why a day was set aside at the end of a mission to enable the astronauts to get some decent sleep prior to returning to Earth.

For The Records

The patterned stripes helped to identify which spacewalker was conducting which task when still and motion picture imagery was being studied post-flight. Together with the air-to-ground commentary and the post-flight debriefing, this enabled the flight crew to inform future crews, trainers, managers, controllers, planners, and hardware makers of those things that worked well and, just as importantly, those that did not. Whether any recommended changes could be implemented immediately or be phased in later would depend upon the importance and complexity of the amendments.

It is noteworthy that most of the EVAs conducted at the ISS between 1998–2011 progressed smoothly, with relatively few failures, delays, or setbacks. This was due mostly to dedicated teams on the ground diligently working out the EVA plans and sequences, making available effective tools and equipment, and conceiving back-up plans and alternative scenarios. The training included integrated sessions where the astronauts in the water tank worked closely with their trainers and the flight control team that would be on duty during the EVAs.

The smooth way in which the ISS was assembled over a thirteen-year period was greatly assisted by NASA building up a cadre of astronauts experienced in EVA and use of the RMS, then either reassigning them to missions farther down the assembly sequence or passing them through management posts on the way to receiving further flight assignments to ensure that the lessons learned in space filtered upward through the organization.

Handling Mass During EVA

In June 1997, just over a year before assembly of the ISS began, NASA published a four volume report *Understanding Skill in EVA Mass Handling*.[12] It derived from a three-year project to create a better understanding of whole-body skills required for mass handling during EVA and was funded by NASA Research Announcement 93-OLMSA-07, rather mundanely entitled "Environmental Constraints on Postural and Manual Control." A series of reports are regularly updated, reassessed and routinely referenced by many areas of space flight. A general search of the web will generate scores of such papers devoted to the theory, operations, history, and development of EVA, including spacesuits, EVA apparatus, techniques for spacewalking, and likely future developments.

The 1997 study evaluated research into postural control and its relationship to the manipulation of the Orbital Replacement Units (ORU) that were critical to servicing satellites such as the Hubble Space Telescope and would be required in operating the ISS. Achieving a clear understanding of mass handling during EVAs would improve planning, mitigate safety issues, training procedures and training simulators.

As with many such reports, the authors highlighted the challenges, "The nature of EVA is such that it remains one of the most dangerous of all operations during a space mission."

The challenges cited for a crewmember on an EVA included:

- Reduced visibility due to illumination, contrast, field of view, and clutter.
- Reduced sense of orientation because of inadequate visual stimulation.
- Reduced proprioception as a consequence of inadequate stimulation of the skin, joints, and muscles.
- Reduced range of motion owing to the EMU and the limitations of the suit's joints.
- Compromised strength as a result of fatigue, hardware design, and adaptation to microgravity.
- Reduced body support due to inadequate rigidity, extent, friction, orientation, and location of services.

Completing an EVA successfully is testament to the skills and adaptability of those who accept the challenge. The 1997 report conservatively estimated that at least 10 hr were spent in mission-specific ground-based EVA training by each assigned astronaut for every hour spent in space on the actual EVA. Many more hours were also spent on contingency training for events that would likely never happen. Prior to the ISS, EVA training in America was intensive, very task-orientated and extremely detailed. Every spacewalk was firmly set into well-known, tried and tested scenarios and was planned down to the minute with highly detailed timelines. For the ISS this degree of intensity was impractical, chiefly because of the accelerated programs required by each Shuttle mission to achieve the EVA objectives, but also the limited time that was available to any given crew with the facilities available for EVA training.

The report raised several specific points in regard to ISS assembly:

- Constant construction did not allow extensive task training. Instead, the Shuttle EVA crew would receive skill-based training.

- Decisions concerning work site techniques would have to be made by the EVA crewmember.
- Simulations of the ISS controls were limited in fidelity due to the scope of the project.
- High fidelity simulations were also not yet available for EVA training and rehearsal (including that for resident crews).
- Activities at the ISS would increase the number of EVA hours and frequency, thereby presenting a further burden on the already tight training facilities and procedures.

There were concerns that orbital assembly work would be compromised by the lack of proficiency of inexperienced EVA crewmembers, but this research (strongly backed by experienced astronauts and engineers) emphasized that the skill of mass handling during EVA firmly depended on managing whole body stability with a high degree of coordination. The emphasis in EVA training should therefore be on the acquisition of knowledge and skills, rather than focusing on a specific set of procedures. This would maximize flexibility to tackle any situation that might arise. It would also benefit even the least experienced EVA astronaut. During training, which included positioning and restraint at a range of different work sites, crews were taught to limit their motion and momentum to maintain control. The tests performed on STS-54 showed that the EMU was far more stable in the WETF than in space; something that future crews needed to be aware of. The difference between small and large object handing in the tank and in space was one of many factors that required to be addressed, but the main conclusion was that body position and stability were essential to completing a task on schedule in space. Achieving and maintaining the optimal body position would be greatly assisted by tethers and foot restraints. In fact, the study concluded that stable body positioning was 90% of the solution to an EVA task because, "A crewmember could not expect to control something else if they cannot control themselves."

This conclusion was a potentially key factor when dissecting the latest estimates for the planned five-year assembly of the ISS. By March 1997 there had been about 453 hr of Shuttle-based EVA since 1983, yet the projected total for ISS assembly was 910 hr. Spread over the five-year program this averaged 182 hr/year. Even at its peak in 1993, Shuttle EVA had achieved only 106 hr. The basic training for a Shuttle crew required over 400 hr. The STS-61 EVA crew required 10 hr of training for every hour that was assigned to EVA in space. With that ratio the typical assembly mission with four 6 hr EVAs assigned to two pairs of astronauts would require 480 hr of additional training.

The 1997 report strongly reinforced this fact. "ISS construction will require double the number of EVA hours completed in the Shuttle program to date; moreover it will require these EVA hours to be performed in half the time [i.e. ten years of Shuttle EVA compared to five years of ISS construction] and it will exceed the peak year hours to date by a minimum of 60%." And this did not even take into consideration the likely contingency and/or maintenance EVAs that "provide a [further] reasonable cause for wanting to understand the characteristic skills during EVA mass handling."

SUMMARY

With actual assembly of the ISS commencing the following year, there still remained much to be learned regarding the manipulation of weightless objects with large mass, either using the RMS or manually by spacewalkers. The series of EVA development exercises helped, as indeed did water tank simulators and virtual reality, but the most beneficial experience comes with practical experience, which meant handling large payloads on-orbit in order to learn lessons.

Notes

1. *Walking to Olympus*, p. 90, entry for May 14
2. *Hubble Space Telescope, From Concept to Success*, David J. Shayler with David M. Harland, Springer-Praxis, 2016
3. *Enhancing Hubble's Vision, Service Missions That Expanded Our View of the Universe*, David J. Shayler with David M. Harland, Springer-Praxis, 2016
4. *US Spacesuits*, 2nd Edition, Kenneth S. Thomas and Harold J. McMann, Springer-Praxis, 2012
5. STS EVA Report No. 1: STS-5, David J. Shayler with Keith T. Wilson, AIS Publications, January 1984; and STS EVA Report No. 2: STS-6, David J. Shayler with Keith T. Wilson, AIS Publications, April 1984
6. Climbing the Extravehicular Activity (EVA) Wall – Safely, a PowerPoint presentation by Jose Fuentes (SAIC) and Stacy Green (GHG Corp.), GHG Corporation Project Management Challenge 2011, February 9–10, 2011
7. Understanding Skill in Mass Handling, G.E. Riccio et al., NASA Technical Paper 3684, vol. 1, p. 7, 1997
8. NASA News Release #97-126, June 9, 1997
9. A Human Factor Analysis of EVA Time Requirements, Final Report, NASA/ASEE Summer Faculty Fellowship Program-1995, JSC, Dennis W. Pate, Texas A&M University Engineering Technology Department, Contract Number NGT-44-001-800, August 8, 1995
10. William F. Fisher and Charles R. Price, Space Station Freedom External Maintenance Task Team, Final Report, vol. 1, parts 1 and 2, NASA, July 1990
11. *Walking In Space*, David J. Shayler, STS-51A section, pp. 265–267, Springer-Praxis, 2004
12. Understanding Skill in EVA Mass Handling, Gary E. Riccio, P. Vernon McDonald, Brian T. Peters, Charles S. Layne and Jacob J. Bloomberg, NASA Technical Paper TP-3684, NASA JSC June 1997. Volume 1: Theoretical & Operational Foundations; Volume II Empirical Investigations; Volume III Data and Results from Volume II; Volume IV Summary

9

From Inner To Outer Space

*What we're doing is we're really learning to take
spacewalks from a special, rare feat into a production
mode...[to] get the station built. It's a large task, but
they're already planning out how to do it.
I think [we] will certainly be able to do the job.*

From a 1998 pre-flight interview with
Jim Newman, Mission Specialist STS-88

There was a long-term concerted program to prepare for extensive EVA activities while assembling a space station but these were primarily in the form of studies, simulations, and the definition of requirements. As the start of ISS assembly loomed, NASA figured some hands-on experience in space was necessary.

In terms of training, the enormous size of the proposed ISS meant there was no way that all the components expected to be handled during EVA could fit into the available emersion pools, and that a new purpose-built facility was required near to the Johnson Space Center in Houston where the Astronaut Office was located. With this facility in place, the time would come to put all the theory and practice to the test and get on with the job.

GETTING WET

Since the mid-1960s, training in water tanks has proven to be an effective method of simulating spacewalking techniques. In addition to the influence of Earth's gravity on the apparatus and test subject and the viscosity of the water, its limitations include the fixed dimensions of the tank and the restricted time available to each training session due to other evaluations or training programs and the sheer number of crews vying for tank time.

© Springer International Publishing Switzerland 2017
D.J. Shayler, *Assembling and Supplying the ISS*, Springer Praxis Books,
DOI 10.1007/978-3-319-40443-1_9

Neutral Buoyancy Laboratory

The focus of EVA training for the ISS is the Neutral Buoyancy Laboratory (NBL) at the Sonny Carter Training Facility near Ellington Air Force Base, just north of the Johnson Space Center and some 30 miles (48.3 km) south of downtown Houston, Texas. By far the largest of NASA's neutral buoyancy facilities it was designed to be large enough to accommodate training on spacecraft mock-ups for the same duration (7 hr) as was being experienced on EVA at the time. Measuring 202 ft (61.56 m) long and 102 ft (31.08 m) wide, the laboratory has a 40 ft (12.19 m) deep tank, half of which stands above ground level and the rest below ground level with a volume of 6.2 million gallons (28.1 million liters).[1]

Despite its impressive size the facility is not large enough to house a complete, full-size mock-up of the ISS. Nevertheless, all of the pressurized modules, elements of the truss segment, various pallets, and the SSRMS Canadarm2 were installed, configured and repositioned as the station expanded in order to satisfy the changing requirements of EVA training.

Management and Organizational Structure

The program of EVAs was designed by the EVA Development and Verification Test (EDVT) team that was responsible for developing the spacewalks and evaluating the tools and procedures for Shuttle operations at the ISS from the mid-1990s through to the retirement of the Shuttle in 2011. After that, the team continued to evaluate EVA requirements at the space station and investigate possible future programs.

The EDVT team functioned under the direction of the Tools and Equipment Branch (EC7) at JSC to evaluate the tools and hardware certified as Class III EVA-related. The team was also responsible for coordinating, developing, and verifying hardware testing in the NBL, working alongside the test conductors, aerospace human factors engineers, and the team of support divers. The Tools and Equipment Branch was part of the Crew and Thermal Systems Division (CTSD) at NASA and worked in conjunction with both the EVA Projects Office (XA) and the Space Suit and Crew Survival Systems Branch (EC5) within CTSD. In addition, EC7 coordinated activities with the EVA, Robotics and Crew Systems Operations Division (DX), the Astronaut Office (CB), a number of primary contractors and subcontractors, NASA field centers, particularly the Goddard Space Flight Center in Greenbelt, Maryland and partner agencies such as the Japanese Aerospace Exploration Agency (JAXA), the Italian Space Agency (ASI), the Canadian Space Agency (CSA), and the European Space Agency (ESA).

Approach to Testing

The development of accurate flight hardware simulations to train in an environment arranged to resemble both the EVA site and the items that are to be manipulated and assembled on-orbit is a significant advance over physical one-g and wire-suspended simulations. Advances with virtual reality and computer software have enhanced the fidelity of the tasks performed. This was started by the need to develop tools for the HST service

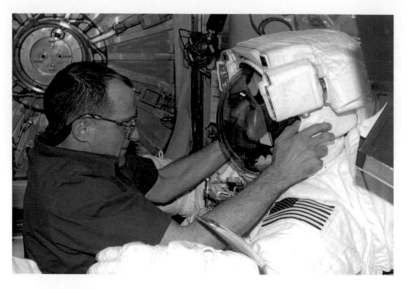

Intravehicular activity includes helping the EVA crew to don their suits.

All suited up and ready to go.

missions, and was carried through to the ISS assembly missions. It has permitted the astronauts and the designers of their tools and hardware to develop the practicalities of mass handling, operational safety, body restraints, and efficient and effective working practices, as well as to develop and test contingency and back-up procedures for many situations.

The lessons learned in servicing the HST based on the "test, re-test and test again" rationale of 'Ceppi' Cepollina at Goddard gave the ISS EVA planners the opportunity to adapt similar principles to the assembly of the largest and the most complex vehicle yet created off-planet. The one hundred and nine ISS assembly EVAs made by Shuttle astronauts constituted the most intricate timelines yet devised for human space flight. Many activities were performed by spacewalkers with extensive support from robotic systems in a closely coordinated combined effort. Over a period of thirteen years each spacewalk was repeatedly rehearsed in the NBL before the entire thing was performed in space. The dedication and hard work of a handful of divers, engineers and planners, trainers, controllers, and especially the astronauts who trained for and then performed EVA tasks on-orbit, enabled the ISS assembly to pass off remarkably smoothly.

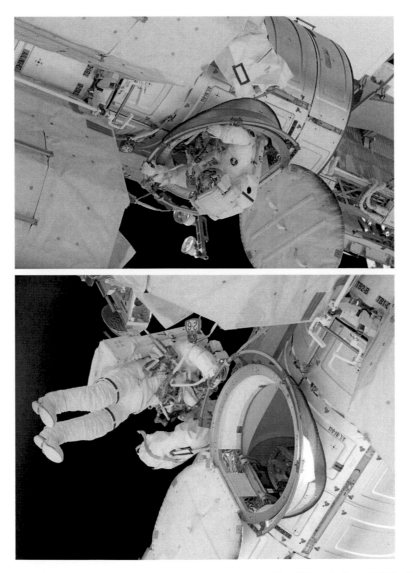

Installed on the ISS in 2001, the Quest Joint Airlock was used by all Shuttle-based EVAs from 2002 except those of STS-114 at the ISS in 2005 and of STS-125 in servicing the Hubble Space Telescope in 2009.

PREPARING THE HARDWARE

The system installed at NASA for certifying hardware for EVA followed the long-established tradition of Requirements and Design reviews, supported by NBL tests. There were three phases in this system, beginning with the Pre-Preliminary Design Review that determined and developed the requirements. This was followed by the Preliminary Design Review (PDR) in which full-scale mid-fidelity mock-ups of the modules and main elements of the ISS that required human EVA intervention were placed into the NBL to evaluate particular EVA tasks.

Next came the period between the PDR and the Critical Design Review (CDR) in which items that were nearer to that of the final flight hardware were placed into the NBL to assess whether what the EVA was expected to accomplish could actually be done. The areas investigated at this point included the availability and usefulness of handling and restraint aids to enable a specific task to be done efficiently and safely; contingency plans in the event of equipment failure; whether the tools supplied were adequate for the task; whether the crew could complete the assembly or disassembly comfortably and in safety by using those tools; whether the training for new elements was adequate and whether the necessary tools and fixings were available; the balance between human and robotic activities involved in a given EVA task; and the level of cooperation that would be required to complete a task.

In addition to the planned EVA tasks, there were also unplanned and contingency EVA simulations with hardware to feed better data into training based on simulation software. As the process of ISS assembly developed across the years, the program of external maintenance and housekeeping increased and began to include systems that had not originally been designed to be accessed during EVA, such as the previously designated Keep Out areas such as electrical connectors. As always, prior to tackling these tasks, EVA simulations were undertaken in the NBL to assess the risks and the time limits imposed by the ever changing EVA constraints.

After the CDR most of the hardware was fully defined, but the NBL continued to support any issues that arose during nominal integration of the hardware into a flight payload. Such issues included changes to the hardware, material replacements, issues with structural loads, launch delays, and the evolving flight rules. All this monitoring and testing was costly, but it saved money by identifying potential difficulties which, left uncorrected, could have led to serious failures in space with potentially disastrous and certainly more costly consequences.

THE SMALL STEPS THAT MAKE A WALK

An old Chinese proverb suggests that every journey, however lengthy, begins with the first small step.[2] This is certainly the case with EVA. The requirement for an EVA to support an operation can be identified many years prior to the event occurring in space. This was particularly true for the HST service missions, for which work on each EVA began months before the *previous* servicing mission had even flown. As soon as a task or a series of tasks are determined that will require spacewalking, a crew is assigned to train. From that point on, the various elements of the EVA plan, the contingencies, the safety rules, and the integration of the EVA into the mission progressively evolve.

The first decision is to estimate what will be possible and practical for the EVA over a standard duration of 6.5 hr. Time may become available to tackle get-ahead tasks that are of a lower priority at that moment but would assist with future EVAs. The tasks are prioritized in the time constraints for the nominal EVA duration. Data is collected from the makers of the hardware and a detailed procedure list is prepared for each EVA task or operation. It is at this point that the crew is given a crib sheet that will allow them to become familiar with any contingencies associated with a given piece of hardware that they might have to address on-orbit although they would not train extensively for these situations.

With the EVA plan and necessary mock up hardware provided, the EVA astronauts and their colleagues in support roles such as the RMS operator and the IV crewmember responsible for the checklist and timeline, would undertake end to end simulations with scale models, one-g mock-ups, and the NBL. There would be debriefings and reviews after each phase or test, with suggestions from the crew and trainers being recorded. In due course the EVA plan would be modified to iron out any problems. After all of this preparation a program of acceptable tasks that were achievable within the timeline of a single two-person EVA would be created.

Depending on the complexity of the EVA task and the time available before launch, an additional two to six runs would be executed in the NBL to further hone the crew's actions. A significant amount of cross-training would also be completed with the EVA crew assisting with and rehearsing one another's tasks to generate a strong framework of redundancy and better chance of mission success. Finally, the flight crib sheets and EVA timelines would be written and a last round of simulations performed with RMS operators and other relevant crewmembers. The next phase would be to open the hatch in space and complete the assigned tasks.

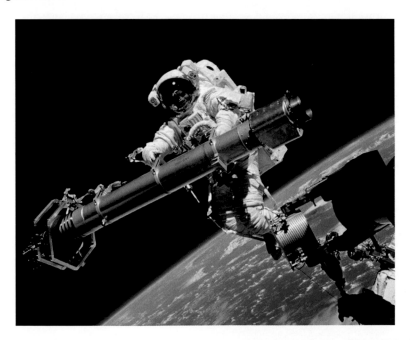

One of the STS-101 astronauts relocating hardware outside the embryonic ISS in 2000, prior to the arrival of the first resident crew.

The Devil is in the Detail

Having a set of objectives to complete in 6.5 hr is all very well, but there remain many factors to be considered. What are the correct steps in the procedure? Is it a one-person or a two-person task? Will it include robotics? Are restraints needed and if so of which type, how many, and for how long? What tools will be needed? And what is needed to ensure the hardware will be left in a safe state if the task is unable to be finished in the allotted time?

Planners will bear in mind the potential need to reassign tasks, either to minimize the amount of crew movement or to bring a higher priority task ahead in the EVA. It is also important to match the abilities of the crew to each task, so that two single-person tasks can be addressed simultaneously. Equally, any two-person task ought not to be so time consuming and complex that it cannot be achieved safely by a single person if the need should arise. Efficient use of tethers and restraints, motions of the robotic arms, and the distance that a crewmember needs to travel in order to reach the safety of the airlock in the event of a suit malfunction or emergency situation must all be considered, planned, and rehearsed. Astronaut input into the development of equipment and procedures or support equipment is helpful but has an impact in terms of costs, time and practicality; therefore such work must be considered very carefully by the crew, their trainers, and the mission planners.

The choice of the crewmembers to conduct and support an EVA is a vital one. It is clear that previous experience helps but this is not always the key factor. When Steve Smith and Joe Tanner were assigned as EVA crewmembers to STS-82, which was the second HST service mission, they were overjoyed, presuming their past achievements had earned them this assignment. They were brought down to Earth, as it were, when Mark Lee, Chief of the EVA Office at that point, told them that they had been selected because they were tall and some of the tasks at Hubble required long arms. Sometimes crew assignments were simply a matter of genetics.[3]

Early in the Shuttle era, the EVA pairs tended to be a veteran and a rookie. For the early ISS assembly missions they were usually both veterans, with first-timers being assigned later in the program. Continuing the style pioneered by the Hubble missions the EVA veteran was the 'free floater' who used tethers, while the rookie rode a foot restraint on the end of the robotic arm. It was necessary to ensure that any given task was well within the personal work envelope of each of the astronauts, lest they had to switch tasks.

SCALING THE WALL

When the time came to assign astronauts to the first EVAs for the assembly of the ISS, experience was a valuable commodity. Three of the astronauts who participated in this early work were interviewed for this book about their experiences at that time.

Table 9.1 Shuttle Crewmember EVA Assignments On ISS Assembly Missions

STS	EV#	EVA Crewmember	Suit Stripe Identification	EV#	EVA Crewmember	Suit Stripe ID	IV Crewmember(s)/Backup EVA astronaut
88	1	Ross	Solid Red	2	Newman	Solid White	Sturckow
96	1	Jernigan	Solid Red	2	Barry	Solid White	Payette; Husband
101	1	Williams J.	Solid White	2	Voss J.S.	Solid Red	Horowitz (also served as BUp EVA astronaut)
106	1	Lu	Solid Red	2	Malenchenko	Solid White	Burbank (also served as BUp EVA astronaut)
92	1	Chiao	Solid Red	2	McArthur	Solid White	Wisoff
	3	Wisoff	Vertical Red	4	Lopez-Alegria	Diagonal Red	Chiao
97	1	Tanner	Solid Red	2	Noriega	Solid White	Garneau
98	1	Jones T.	Solid White	2	Curbeam	Solid Red	Polansky
102	1	Voss, J.S.[1]	Solid Red	2	Helms[1]	Solid White	Richards, P.
	3	Thomas A.	Vertical Red	4	Richards P.	Diagonal Red	Helms
100	1	Hadfield	Solid Red	2	Parazynski	Solid White	Phillips; Lonchakov
104	1	Gernhardt	Solid Red	2	Reilly	Solid White	Hobaugh (also served as BUp EVA astronaut)
105	1	Barry	Solid Red	2	Forrester	Solid White	Sturckow (also served as BUp EVA astronaut)
108	1	Godwin	Solid Red	2	Tani	Solid White	Gorie (Walz served as BUp EVA astronaut)
110	1	Smith S.	Solid Red	2	Walheim	Solid White	Ross; Morin
	3	Ross	Vertical Red	4	Morin	Diagonal Red	Smith; Walheim
111	1	Chang-Diaz	Solid Red	2	Perrin	Solid White	Lockhart
112	1	Wolf	Solid Red	2	Sellers	Solid White	Melroy; Yurchikhin
113	1	Lopez-Alegria	Solid Red	2	Herrington	Solid White	Lockhart; Wetherbee
114	1	Noguchi	Solid Red	2	Robinson	Solid White	Thomas, A.
121	1	Sellers	Solid Red	2	Fossum	Solid White	Kelly, M.

(continued)

Table 9.1 (continued)

STS	EV#	EVA Crewmember	Suit Stripe Identification	EV#	EVA Crewmember	Suit Stripe ID	IV Crewmember(s)/Backup EVA astronaut
115	1	Tanner	Solid Red	2	Stefanyshyn-Piper	Solid White	Jett (assisted by T. Reiter from the ISS resident crew)
	3	Burbank	Vertical Red	4	MacLean	Diagonal Red	
116	1	Curbeam	Solid Red	2	Fuglesang	Solid White	Oefelein
	3	Williams, S.	Solid White				
117	1	Reilly	Solid Red	2	Olivas	Solid White	Forrester; Williams S.
	3	Forrester	Vertical Red	4	Swanson	Diagonal Red	Reilly; Williams, S.
118	1	Mastracchio	Solid Red	2	Williams, D.	Solid White	Caldwell
	3	Anderson, C.	Vertical Red				
120	1	Parazynski	Solid Red	2	Wheelock	Solid White	Nespoli (EVA 1-4); Tani (EVA 5)
	3	Tani	Vertical Red	4	Whitson	Solid Red & Diagonal Red	
	5	Malenchenko	Diagonal Red				
122	1	Walheim	Solid Red	2	Schlegel	Solid White	Poindexter
	2/3	Love[2]	Vertical Red				
123	1	Linnehan	Solid Red	2	Behnken	Diagonal Red	Foreman (EVA 1, 3); Behnken (EVA 2); Linnehan (EVA 4, 5)
	3	Foreman	Vertical Red	4	Reisman	Solid White	
124	1	Fossum	Solid Red	2	Garan	Solid White	Ham
126	1	Stefanyshyn-Piper	Solid Red	2	Bowen	Solid White	Kimbrough (EVA 1, 3); Boe (EVA 2, 4)
	3	Kimbrough	Vertical Red				
119	1	Swanson	Sold Red	2	Arnold	Solid White	Acaba (EVA 1); Arnold (EVA2); Swanson (EVA 3); Acaba (EVA 4)
	3	Acaba	Vertical Red				

127	1	Wolf	Solid Red	2	Kopra	Solid White	Cassidy (EVA 1, 2); Marshburn (EVA 3); Kopra (EVA 4); Wolf (EVA 5)
	3	Marshburn	Diagonal Red	4	Cassidy	Vertical Red	
128	1	Olivas	Solid Red	2	Stott	Vertical Red	Forrester
	3	Fuglesang	Solid White				
129	1	Foreman	Solid Red	2	Satcher	Solid White	Bresnik (EVA 1); Satcher (EVA 2); Foreman (EVA 3)
	3	Bresnik	Vertical Red				
130	1	Behnken	Solid Red	2	Patrick	Solid White	Robinson
131	1	Mastracchio	Solid Red	2	Anderson, C.	Solid White	Metcalf-Lindenburger
132[3]	1/2	Reisman	Solid White	1/2	Good	Diagonal Red	Antonelli
	1/2	Bowen	Solid Red				
133	1	Bowen	Solid Red	2	Drew	Solid White	Stott
134	1	Feustel	Solid Red	2	Chamitoff	Solid White	Fincke (EVA 1); Chamitoff (EVA 2, 3); Feustal (EVA 4)
	3	Fincke	Vertical Red				
135[4]	1	*Fossum*	Solid Red	2	*Garan*	Solid White	Walheim

Notes:

[1] Voss and Helms performed their EVA prior to transferring as ISS-2 crew

[2] Love replaced Schlegel on the missions first EVA

[3] As all three crewmembers were experienced spacewalkers they had elected not to designate one person as lead spacewalker for the mission; instead they each took turns in the role on a different EVA.

[4] As this mission was manifested at short notice the EVA was accomplished by STS EVA experienced ISS resident crewmembers, allowing the four person Shuttle crew to focus on other training matters, and contingency EVA activities only

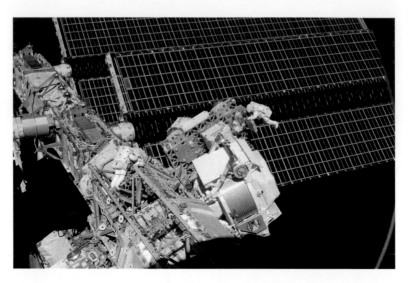

A typical view of EVA operations at the ISS, in this case during STS-134, the penultimate Shuttle mission.

Spacewalker Jerry Ross

Jerry Ross became involved with the EVA requirements for creating a space station very soon after becoming an astronaut. "When I first entered the Astronaut Office in 1980, an EVA was considered something to avoid if at all possible because the crew would be, 'One failure away from certain death.' When we started to think seriously about building and maintaining a station we had numerous EVA-related questions… There was the role of EVA in the station's assembly and operations, the number and duration of EVAs, enhancements to space suits and gloves, crew-hardware interface designs, tools, EVA aids, labels, expendable versus rechargeable life support systems, hardware development and testing, and the need for facilities to do the testing."[4]

In his notes, Ross explained that over the years he had participated in many forums which discussed the space station and its design. He offered suggestions but not all of his ideas were adopted. "I strongly argued for a reduced station cabin pressure to help minimize EVA pre-breathing requirements, but lost." He had presumed that the data obtained from the station would provide the life sciences community with hard facts to start compiling a database at cabin pressures relevant to bases on the Moon and Mars, but at that time this was apparently not so.

In his excellent 2013 memoir *Spacewalker,* Ross went into detail about some of the important things the Astronaut Office had to fight for in order to ensure that the EVA work to assemble the ISS would go to according to plan. In a 2015 E-mail, he added, "Thermal vacuum tests of certain hardware and flight hardware-to-flight hardware fit checks were two of the other areas that unbelievably we had to fight for even though our management knew full well that these were the correct things to be doing."

A spectacular view of Steve Robinson riding the RMS foot restraint during the third EVA of STS-114 in 2005.

Ross was Chief of the EVA and Robotics Branch of the Astronaut Office through 1996 and early 1997. "In that time we accomplished many activities that we felt were critical to the future success of the ISS assembly EVAs. We established an EVA team. Its membership came from the Astronaut Office, the Mission Operations Directorate, and the EVA Project Office. We visited all three of the Station Product Groups. These visits were to allow our EVA team to gain a better understanding of the types of EVA tasks and hardware interfaces the engineers were designing, and to enable our team to help the engineers better understand the capabilities and limitations of EVA astronauts and their tools and equipment. These visits were very beneficial for both the engineers and the EVA team, and helped to establish good working relationships.

"We convinced the ISS program management to add Human Thermal Vacuum (HTV) testing of exterior station mechanisms that were judged to be susceptible to failure or malfunction under the vacuum and/or thermal environment of space. Our management also agreed to add On-Orbit Constraints Tests (OOCT), to verify that electrical cable and fluid line lengths and routings were properly engineered and to verify their connectors could physically be mated. Both of these series of tests have identified problems that would have caused serious difficulties during the assembly process if not corrected. All of this testing was in addition to the more traditional fit checks, sharp edge inspections, and go/no-go gauging of all of the bolt heads on the station EVA hardware. We also ensured that adequate labeling on the exterior of the ISS would be provided for the EVA operations. This took an incredible amount of effort, but it eventually paid big dividends for the assembly crews."

Ross also noted that during a ten-year period he made at least two trips per year to NASA Headquarters in an effort to gain approval for the construction of the NBL near JSC. "I was charged with chairing the Operational Readiness Inspection team for the NBL and I was privileged to be the first crewman to do a suited run in this incredible facility. This NBL permits us to meet the very great demand for training that the ISS assembly and operations dictate." The new water tank could support two parallel but independent simulations by pairs of suited crewmembers, each run lasting up to 6 hr, and operate five days/week. It was the introduction of an increased oxygen content (NITROX) breathing gas mixture which allowed 6 hr sessions at depth while safely avoiding bends problems. As Ross admits, "I don't see how we could have properly trained for the ISS challenges without this important facility."*

When the NBL became operational in 1997 it enabled full scale EVA development for the ISS to get underway. This program included each of the planned assembly and maintenance tasks on each of the US station elements and some of those provided by the international partners. As Ross, a former Air Force Flight Test Engineer recalled, "We developed a testing methodology based upon aircraft flight test principles with a modified Cooper-Harper scale to judge the feasibility of each EVA task. We used six crewmembers to perform evaluations on each of these tasks. The crewmembers were selected to yield a wide spectrum of EVA experience and physical sizes. This helped ensure that the on-orbit EVA tasks would be doable by anyone eventually assigned to perform these tasks on a given mission."

In summarizing his thirty-two year career as an astronaut, Ross outlined some of his personal experiences from nine EVAs performed across seven Shuttle missions. All of his

---

* Even with this new facility, training for EVAs to assemble Space Station Freedom would have been challenging with its originally conceived structure.

spacewalks were associated in some way with the space station concept: the EASE-ACCESS construction experiments by STS-61B; being instrumental in instigating the series of DTOs to allow NASA to accumulate EVA experience prior to ISS assembly; evaluating a prototype CETA Cart on STS-37, together with tests on an instrumented pallet to measure EVA force applications and loads induced by EVA motion, with the data helping in the development of EVA loads design criteria; working to improve the fit of EVA gloves; evaluating the production variant of the SAFER during the STS-88 EVAs, where problems were discovered that needed fixing; evaluating the new EMU helmet mounted camera and pre-breathing protocol during STS-110, in the latter case using a period of exercise to help quicken the de-nitrogenation process. It was a solid record of achievement.

Joe Tanner: From Hubble To ISS

Joe Tanner flew two ISS missions, STS-97 and STS-115, both of which included his participation in assembly EVA tasks. On coming off STS-82, which was the second Hubble service mission, Tanner had no prior warning of his next assignment, which was with STS-97 several months later. It was probably that experience which helped towards his first ISS crew assignment. "One was seldom assigned to another mission before completing the current mission. That would be an unnecessary distraction. And they had to assess how I would perform on STS-82. The basics of EVA are the same, but the fine details between HST and ISS are rather different. They wanted as much experience as possible on the first seven assembly flights. I had experience, and had demonstrated at least some level of competence."[5]

As Tanner recalled of STS-97, "The main challenge was that we were doing things we had never done before. We had the Space Vision System to guide Marc [Garneau, RMS operator] in bringing P6 close enough to Z1 for the capture latch to bring it into full mate, but it was basically a manual RMS movement. We were also de-mating and mating ammonia lines for the first time. We were to perform the first deploy of a solar array and remove an MMOD shield. All of these procedures had to be developed. We had a tremendous amount of input, working closely with EVA controllers and trainers and the hardware people. Carlos [Noriega, his EVA partner] and I traveled all across the country to see the real hardware first hand and make any appropriate inputs. It was the ideal way to prepare for the tasks ahead, because we ultimately knew the hardware almost as well as the designers knew it. The basic timeline was developed by the EVA training team and the flight director before we were even assigned. One of the driving factors was the health of P6 after its removal from the relatively warm environment of the payload bay. There was no thermal control of P6 available until it was mated and connected electrically to Z1. Adjustments to the timeline were made as the procedures were refined and we got a better idea from NBL training for how long it would take to complete each task."

Tanner agreed with Ross concerning the NBL. "We really needed that large training facility to prepare for all of the ISS assembly missions. It was an advantage over STS-82 HST training because our fellow crewmembers could come to the NBL to participate in our training and still be able to train for their duties on the same day. We didn't have that luxury for STS-82, where we did most of our EVA training in the Huntsville NBS.

It was good to be home with the family every night. The advantage of Huntsville was that we were totally immersed [deliberate pun] in the EVAs for the two weeks of each training period. Being together away from home was good for team building, not only with fellow crewmembers but also with our EVA controllers and the Goddard [Space Flight Center] engineers.

"We also trained with virtual reality a great deal, rather more so than on STS-82, to include some mass handling. That was a very good tool for situation awareness. With translation distances longer than previous ISS assembly missions, we wanted to make sure that we didn't become lost when a long way from home, particularly in darkness. The VR experiences helped. We also used VR for the P6 mating and SAFER training. The operators would cause us to tumble off of P6 and it was our job to deploy SAFER and find our way back. That was actually a lot of fun! And the helmet cameras were a tremendous new tool for all aspects of EVA operations. Those views helped situation awareness for the IV and RMS operator crewmembers and allowed the ground team a real-time view of what was happening when we had K_u-band coverage. We recorded both helmet cameras during every EVA to enable future crews to learn from what we did. These views also made for more interesting post-flight presentations. I must say it was a little intimidating, knowing potentially millions of people were watching you at work."

Tanner and his fellow crewmembers were well aware that installing the P6 array on top of the Z1 truss was critical to the mission. "We had thoroughly studied and trained for every procedure. Marc would be using the SVS for the first time. Carlos and I were at the interface between those modules, to verify the SVS was performing as it should. As it turned out, having us monitoring the approach was a good thing since there was a bias in the SVS solution that resulted in P6 being slightly closer to Z1 than indicated by the cockpit display. We called for an early stop. An interesting event occurred after we called for a stop. Carlos was relocating to a position where he would have a better view of the other corner, for which he was providing clearance. I couldn't see him, but I saw the entire P6 move a little laterally. That startled me, to say the least. I asked Carlos if he had seen it move. He said he had inadvertently bumped into it, causing the motion. We were surprised that an EVA crewmember having about 600 pounds of mass could make a mass of 35,000 pounds move by just bumping into it. Welcome to space! The capture latch and RTAS bolts engaged perfectly [to produce a rigid connection] and so did the electrical and data connectors which Carlos engaged. P6 was safely installed."

The difficulties in deploying the solar arrays added tasks to a third EVA that Tanner would clearly recall some fifteen years later. "The solar array problem was a surprise to everyone. No one expected the panels to stick together in the way that they did. An on-orbit test of a similar design had shown that the panels should deploy successfully with the retract reels in the low-tension mode. As we realized after the flight, the specified amount of force was insufficient to release the eleven panel pairs after their storage at high compression. It all turned out well in the end, but we ought to have stopped the deployment when the tension bar was pulled away from the top of the blanket box. The ground did not have coverage in the K_u-band at the time, so had no live video. The bar was pulled to the max length of the two tension cables before those stuck panels finally released. The resulting 'crash' of the tension bar was faster than the tension reels could retract, throwing both cables off their reels.

"That malfunction drove the need for our third EVA. The ground people set to work Apollo 13 style and came up with a procedure to fix the problem. The major credit goes to [astronaut] Dave Wolf, who flew with some of the EVA team to Lockheed Martin in Sunnyvale, California, and they all basically didn't sleep until they had come up with a fix that they developed and tested on another flight array. Carlos and I thought a little about what we could do, but mostly focused on what we needed to do on our second EVA too. The deployment procedure for subsequent arrays was revised to do it in the high-tension mode, and allow the array to bake in the Sun for twenty minutes in order to loosen the bonds between the panels. After the mission Carlos went into training for another flight but I was able to participate in the team work to understand the problem and modify the procedure. It was great work, and I had a lot of fun with the solar array experts."

As Tanner said, "The start of EVA number three was dominated by the re-spooling of both tension reels on the blanket box. No one knew how long that would take. As it turned out, the procedure was so well developed that we completed both repairs in no time. Carlos and I had studied the video Dave made in Sunnyvale until we knew exactly what had to be done and how to do it. The repair was one of the more rewarding EVA experiences for both of us. That left us plenty of EVA time to tackle the other tasks for which we were prepared." These other tasks were preparatory work to enable the next mission to install the Destiny laboratory. "I would not say we took our time, but there was certainly less sense of urgency. I was very comfortable with EVA by this time and really didn't want to come back inside. We had taken some time to look around during the first two EVAs, so we already had an appreciation for how large the station was at that time, and how far we were at times from the Shuttle airlock. The HST is small by comparison."

The STS-97 astronauts were on a different sleep and work cycle to the residents of the ISS. Owing to the lower pressure used on the Orbiter in preparation for the EVAs, the visitors did not enter the station properly until the EVA program was finished. This made things seem strange, knowing their colleagues were on the other side of the hatch inside the ISS during the EVAs. "It was a little strange being that close to your friends and not being able to properly greet them until after the EVAs," Tanner reflected. "We did use PMA-2 as an airlock, to transfer equipment to the ISS crew after docking. The other reason [for not entering the station immediately] had to do with electrical power. Until the new solar arrays were deployed there was not enough power to run the wall heaters or ventilation system in the Unity module. There would have been no way to control the humidity generated by human activity in the node."

Finally, Tanner spoke of the physical effects of making three challenging EVAs. "I don't recall that Carlos and I were very tired afterward. We had a day off between each EVA. And the third EVA was only four and a half hours. Carlos is a very fit guy and I was too pumped up waiting to go into the ISS to worry about being tired." Clearly their physical conditioning and training had paid off.

Tom's Destiny In Space

Following STS-80, Tom Jones was assigned as a deputy branch chief for space station operations. "I was aware of all the assembly missions and the major milestones to get the

One of the STS-113 EVA astronauts maneuvers across the handrails on the central elements of the ITS.

station off the ground, but I had only a passing awareness of my particular flight assignment to STS-98, the 5A assembly mission."

Mark Lee, at that time the CB EVA Branch Chief, had assigned Jones to a series of underwater tests for generic task development; specifically how they should approach working on the Destiny laboratory, work on the truss, replace some of the components, and hook up utilities. So when Lee asked Jones if he wanted to continue doing the test work, he was very eager to accept because it complemented the desk work that he was doing and the meetings he was attending. When he was assigned to STS-98 to help to develop the plans for three spacewalks he really didn't know much about that mission other than it was going to deliver the laboratory.

"For this assembly mission, we visited the laboratory once or twice a month during almost three years. At first it was an aluminum shell on the factory floor at Huntsville. By the time it was finished it was a beautiful, self-contained spacecraft module that was going to be connected to the ISS. That transformation was really encouraging. It was remarkable to see how the technicians invest their heart and soul into building it. As an astronaut I knew what the interior would be like, because we had run test procedures in it. When it transferred to the processing facility in Florida we thoroughly inspected the finished exterior. We donned spacesuit gloves and we manipulated the very things that we were going to work with on-orbit. We got to be very familiar with the lab."[6]

When Jones was asked for this book how closely the training for STS-98 matched the work in space – whether it was confusing or difficult to work outside on the station surfaces with so many appendices protruding out, with the changing lighting conditions and his orientation while working – he said, "You have raised a good point. There are a couple of things that you have to cope with out there. The laboratory and the rest of the station structure was mocked up on the floor of the NBL pool, so we got very familiar with the layout during our hundreds of hours in the tank. Firstly Mark [Lee] and myself and then Bob [Curbeam] spent hours underwater addressing this point. I think my total time under-water for STS-98 was about 230 hr in the suit, working on the station, so you get this very good mental map of the station and how you move around it in the pool, as well as the specific choreography of the task that you're supposed to do on each of your spacewalks.

"So when I got on-orbit, I knew where all the big pieces were. But what training in the pool does not give you is – as you say – all the little appendages and the wires and antennas that stick out. We studied photographs before we left. We did that by flicking though books of pictures of the station taken by the most recent crew. When you get out there you really have to take an extra level of care. You have to take all of your caution and knowledge up a notch in order to ensure you do not ding something while working on something else. The lighting conditions were totally new. We did not simulate that anywhere. However, we are all used to turning on a light switch when a room is dark, and vice versa. So when the Sun came up we would lower our visor and switch off the helmet lights. Mark Polansky, inside the Orbiter, would warn us when the sunrise was coming so that we wouldn't get any glare. And then we'd do the opposite when it got dark."

Jones was next asked whether the viscosity of the water in the training tank was an issue. "It is noticeable when you make your first spacewalk," Jones agreed. "You are very aware of how easy it is to manipulate your space suit in three dimensions using just a couple of finger tips on the hand rails. You notice there is no damping from the water. It takes maybe thirty minutes to figure out how your suit moves ever so slightly differently than in the water. Those initial minutes of your first EVA are fun. They are programmed – I think it's just called mobility familiarization or something like that, or translational famil-iarization. Once you discover the difference from training, you don't notice it anymore. It isn't a dramatic difference. It is a very fast learning curve and you very quickly internalize it and incorporate it into your motions."

As an example of how you cheat in the water, Jones said, "If you are swinging your body around in the water, you get used to the fact that you are swinging your torso or your legs and the motion stops after you cease using your wrist muscles. In space I was doing the same kind of thing. I would swing my body around and of course I wouldn't stop as quickly as there was no drag from the water, or I would overshoot my motions. During the second EVA, Marsha Ivins [operating the RMS] was moving PMA-2 up to the front of the laboratory, and I was working very close by on some thermal covers. I swung my body out, overshot, and my thighs slammed right into the shaft of the robot arm above the wrist joint. So I banged into it and the entire PMA-2 vibrated and shook several inches back and forth. They saw that immediately on the TV cameras and said, 'Wow, what's going on.' When they realized I had bumped the arm, they said, 'Tom, calm down.' So I had to back off and apologize. I actually said, 'OK, I'll go and work somewhere else while you guys are finish-ing.' Because I had thrown off the targeting from the Space Vision System they had to lock onto those targets again."

Jones agreed the ISS assembly EVAs were tiring. "I was really beat after the first spacewalk, which lasted for more than seven hours. In particular your forearms and shoulder muscles get really fatigued – everything in your arms and fingertips – and it takes a while for that to come back. I found during the second spacewalk I had gotten almost complete relief from that and was ready to go again. It builds up cumulatively, so that on the third spacewalk, when I went on the outside and started moving around again, I noticed that my arms were already somewhat tired at the start. But fortunately for us there wasn't a lot of hard and intensive work that really taxed us. We had ample reserves of strength. I think that was because Bob [Curbeam] and I did a lot of physical training beforehand, knowing we were going to be challenged like that. I didn't notice being tired on the inside, on the in-between days, because, again, you are not doing the demanding work to put those muscles to the test, so it was not something that affected anything we did inside. I just noticed, by the third EVA day, I was struggling to pay a penalty in recovery for all of the intensive work. It's just a fact of life. I don't think it's an obstacle that cannot be overcome. But you better know – and we got this from our predecessors who had worked on three EVAs – that you had better prepare physically for the triple spacewalk."

It was during the initial EVA that Jones and Curbeam experienced a small leak of frozen ammonia crystals while umbilicals were being connected to Destiny. "I didn't think it was going to be a problem, because it happened fairly early in the spacewalk and I brushed Bob off. As I was brushing him, I did not see any crystals of ammonia coming off because they had already sublimed. We parked him in the sunlight for one entire daylight pass around our orbit, to 'bake' him clean. My judgement was that we had decontaminated him adequately and he had several more hours in vacuum before we came back inside. But they stuck to the pre-flight rules and procedures that say we have to do a decontamination process. So we got back inside, closed the airlock hatch, partially re-pressurized it, then we went to vacuum again hoping that would flush out any fumes that were still emanating from the suit. Then when we came back to cabin pressure, they tossed in wet towels and a zip lock bag. Bob and I swabbed each other off with the towels, then wiped the walls and floor of the airlock and bagged the towels so that any ammonia gas would be contained in the plastic bag. Then Mark Polansky came in wearing an oxygen mask, took a sniff of the air and didn't detect any odors so we were fine. I think that was overkill. I complained about it to [CDR] Ken Cockrell. I told him I was tired and wanted to get out of the suit. He said, 'I understand where you are coming from, but let's do this for just ten minutes and then we will get you out of there.'"

During their third and final EVA, Jones and Curbeam paid tribute to the people who had helped to plan the mission, helped the crew to train, and prepared the hardware. It had been an intense period of team work both prior to and during the flight. Reflecting later he said, "You work with the divers on a day to day basis in all of your training. It took about three years of underwater training to get ready for that mission. You see the divers and you know them on a first name basis. It is the same with the instructors and the flight controllers that work with you. Then you've got the tool developers. And of course you have the payload crew who worked with the technicians in Huntsville who built the laboratory, and also at the Cape. They are all counting on you to do justice to their payload; if you will, to their handiwork. So you feel a great kinship to them. But they are not in touch with you in space except by way of the Capcom. You hear advice from the EVA flight controllers and instructor people, and so on, but you're separated from them emotionally. The tribute that we made on the third spacewalk was a chance to recognize the fact that we had all these hundreds of people helping us. It was nice to step back from the intense work on the spacewalk and acknowledge them. I think they were all listening in because we had known that the opportunity was coming up, and I think a lot of them had a chance to listen in. I would say we were very much working together on the ground leading up to the mission."

These three examples of hands-on experience during the early ISS assembly EVAs, and indeed the years of preparation and training that went into every EVA there, reflect the dedication and focus that was required to ensure the station was safely assembled as planned.

Table 9.2 Shuttle Crew EVAs At ISS Annual Summary

Year	Visited Space Station	Total Shuttle Missions with station EVAs	Crew Members who conducted STS Station EVAs	Total STS Station EVAs by Flights	Total Annual STS Station EVA Duration (HH:MM)	Annual Accumulative Shuttle based Station EVAs
1998	ISS	1	2	3	21:22	21:22
1999	ISS	1	2	1	7:55	29:17
2000	ISS	4	10	9	59:37	88:54
2001	ISS	6	14	13	82:23	171:17
2002	ISS	4	10	13	87:29	258:46
2003	ISS	0	Shuttle fleet grounded due to loss of Columbia			
2004	ISS	0	Return to Flight program post Columbia tragedy			
2005	ISS	1	2	3	20:05	278:51
2006	ISS	3	11	10	67:33	346:24
2007	ISS	3	12	12	78:27	424:51
2008	ISS	4	12	15	102:49	527:40
2009	ISS	4	13	14	88:16	615:56
2010	ISS	3	7	9	59:51	675:47
2011	ISS	3	7	7	48:03	723:50
TOTALS		37	102	109	723:50	723:50

STS-SUPPORTED EVAS AT THE ISS

In March 2010, just over a year before the final Shuttle mission flew, NASA created a PowerPoint presentation that briefly looked at the lessons learned from assembling the ISS.[7] This observed that the external assembly was far more complex and required more EVAs than had been predicted by the plan that was in force when the assembly started. It was argued that if the assembly of a complex spacecraft was planned in the future then EVA should be used "only when absolutely necessary." But in the case of the ISS, "The investments made in EVA over many years made EVA assembly tasks the preferred mode of operation." The robotic arms of the Shuttle and the ISS proved critical to the assembly of the larger elements of the station, but these operations were complemented by well over 1,000 hr of EVA.

The Primary Activities

Shuttle EVA crews began external activity at the ISS on the very first assembly flight, STS-88, in December 1998, when the Unity Node was mated with the Zarya Control Module. At least one EVA was included on each Shuttle assembly mission from that date until the end of the program, although the final 'Shuttle' ISS assembly EVA was conducted in July 2011 by members of the resident crew owing to the late decision to fly STS-135; the previous EVA experience of the residents removed the need for the minimal four-person crew of the final Shuttle mission to undergo additional training.

Below is a summary of the work conducted by Shuttle EVA crews at the station in flight order. For further details see Table 9.3 here, the *Praxis Manned Spaceflight Log 1961–2006*, and *Manned Spaceflight Log II 2006–2012*. For clarity the STS numerical is abbreviated to the mission designation and the ISS assembly designation. Therefore 88/2A refers to the EVAs of the STS-88 mission on ISS Assembly Flight 2A. A more detailed account of space station EVAs forms part of a larger work in preparation.[8]

Get-Ahead Tasks And Deferred Objectives

In addition to the primary tasks of an EVA, wherever possible and if there was time in hand, smaller, lower priority tasks would be carried out. These get-ahead tasks were a valuable tactic for planning future EVAs, reducing workloads, and ensuring timelines were conducted at a reasonable pace. Conversely, there were times when tasks could not be completed as planned and therefore had to be deferred either to a later EVA on the same mission or to another mission. A setback could result from faulty hardware, fixtures and/ or fittings, delays in achieving higher priority tasks, problems with EVA apparatus, or simply running out of time.

A review of EVA activities during the ISS assembly missions indicates the sheer complexity of sequencing the preparation of such a large amount of completely new hardware and assembling it on-orbit by a combination of robotic arm work and often very intense manual EVA over a period of thirteen years.

STS-116 MS Robert Curbeam works with the P6 solar array prior to its retraction for relocation by a later crew to complete the ITS.

Table 9.3 Space Shuttle Crewmember EVA's at ISS 1998–2011

Year	Date of EVA	Mission	Mission EVA #	EVA Time	Orbiter/Quest Airlock used	Orbiter crew EVA at station	STS/ Station EVA	EVA Crewmembers	Mission EVA Total	Shuttle at station EVA Accumulative Total
1998	Dec 7	STS-88	1	7:21	OV 3rd	Endeavour 1st	ISS 1st	J. Ross; J. Newman	21:22 (3)	21:22 (3)
	Dec 10	STS-88	2	7:02	OV 4th	Endeavour 2nd	ISS 2nd	J. Ross; J. Newman		
	Dec 13	STS-88	3	6:59	OV 5th	Endeavour 3rd	ISS 3rd	J. Ross; J. Newman		
1999	May 30	STS-96	1	7:55	OV 6th	Discovery 1st	ISS 4th	T. Jernigan; D. Barry	7:55 (1)	29:17 (4)
2000	May 22	STS-101	1	6:44	OV 7th	Atlantis 3rd	ISS 5th	J. Williams; J.S. Voss	6:44 (1)	36:01 (5)
	Sep 11	STS-106	1	6:14	OV 8th	Atlantis 4th	ISS 6th	E. Lu; Y Malenchenko	6:14 (1)	42:15 (6)
	Oct 15	STS-92	1	6:28	OV 9th	Discovery 2nd	ISS 7th	L. Chiao; W. McArthur	27:19 (4)	69:34 (10)
	Oct 16	STS-92	2	7:07	OV 10th	Discovery 3rd	ISS 8th	M. Lopez-Alegria; J. Wisoff		
	Oct 17	STS-92	3	6:48	OV 11th	Discovery 4th	ISS 9th	L. Chiao; W. McArthur		
	Oct 18	STS-92	4	6:56	OV 12th	Discovery 5th	ISS 10th	M. Lopez-Alegria; J. Wisoff		
2001	Dec 3	STS-97	1	7:33	OV 13th	Endeavour 4th	ISS 11th	J. Tanner; C. Noriega	19:20 (3)	88:54 (13)
	Dec 5	STS-97	2	6:37	OV 14th	Endeavour 5th	ISS 12th	J. Tanner; C. Noriega		
	Dec 7	STS-97	3	5:10	OV 15th	Endeavour 6th	ISS 13th	J. Tanner; C. Noriega		
	Feb 10	STS-98	1	7:34	OV 16th	Atlantis 5th	ISS 14th	T. Jones; R. Curbeam	19:49 (3)	108:43 (16)
	Feb 12	STS-98	2	6:50	OV 17th	Atlantis 6th	ISS 15th	T. Jones; R. Curbeam		
	Feb 14	STS-98	3	5:25	OV 18th	Atlantis 7th	ISS 16th	T. Jones; R. Curbeam		
	Mar 11	STS-102	1	8:56	OV 19th	Discovery 6th	ISS 17th	J. S. Voss; S. Helms[2]	15:17 (2)	124:00 (18)
	Mar 13	STS-102	2	6:21	OV 20th	Discovery 7th	ISS 18th	A Thomas; P. Richards		
	Apr 22	STS-100	1	7:10	OV 21st	Endeavour 7th	ISS 19th	C. Hadfield; S. Parazynski	14:50 (2)	138:50 (20)
	Apr 24	STS-100	2	7:40	OV 22nd	Endeavour 8th	ISS 20th	C. Hadfield; S. Parazynski		
	Jul 15	STS-104	1	5:59	OV 23rd	Atlantis 8th	ISS 21st	M. Gernhardt; J. Reilly	16:30 (3)	155:20 (23)
	Jul 18	STS-104	2	6:29	OV 24th	Atlantis 9th	ISS 22nd	M. Gernhardt; J. Reilly		
	Jul 21	STS-104	3	4:02	Quest[3]	Atlantis 10th	ISS 23rd	M. Gernhardt; J. Reilly		
	Aug 16	STS-105	1	6:16	OV 25th	Discovery 8th	ISS 24th	D. Barry; P. Forrester	11:45 (2)	167:05 (25)
	Aug 18	STS-105	2	5:29	OV 26th	Discovery 9th	ISS 25th	D. Barry; P. Forrester		
	Dec 10	STS-108	1	4:12	OV 27th	Endeavour 9th	ISS 26th	L. Godwin; D. Tani	4:12 (1)	171:17 (26)

(continued)

Table 9.3 (continued)

Year	Date of EVA	Mission	Mission EVA #	EVA Time	Orbiter/Quest Airlock used	Orbiter crew EVA at station	STS/Station EVA	EVA Crewmembers	Mission EVA Total	Shuttle at station EVA Accumulative Total
2002	Apr 11	STS-110	1	7:48	Quest	Atlantis 11th	ISS 27th	S. Smith; R. Walheim	28:22 (4)	199:39 (30)
	Apr 13	STS-110	2	7:30	Quest	Atlantis 12th	ISS 28th	J. Ross; L. Morin		
	Apr 14	STS-110	3	6:27	Quest	Atlantis 13th	ISS 29th	S. Smith; R. Walheim		
	Apr 16	STS-110	4	6:37	Quest	Atlantis 14th	ISS 30th	J. Ross; L. Morin		
	Jun 9	STS-111	1	7:14	Quest	Endeavour 10th	ISS 31st	F. Chang-Diaz; P. Perrin	19:31 (3)	219:10 (33)
	Jun 11	STS-111	2	5:00	Quest	Endeavour 11th	ISS 32nd	F. Chang-Diaz; P. Perrin		
	Jun 13	STS-111	3	7:17	Quest	Endeavour 12th	ISS 33rd	F. Chang-Diaz; P. Perrin		
	Oct 10	STS-112	1	7:01	Quest	Atlantis 15th	ISS 34th	D. Wolf; P. Sellers	19:41 (3)	238:51 (36)
	Oct 12	STS-112	2	6:04	Quest	Atlantis 16th	ISS 35th	D. Wolf; P. Sellers		
	Oct 14	STS-112	3	6:36	Quest	Atlantis 17th	ISS 36th	D. Wolf; P. Sellers		
	Nov 27	STS-113	1	6:45	Quest	Endeavour 13th	ISS 37th	M.Lopez-Alegria; J. Herrington	19:55 (3)	258:46 (39)
	Nov 28	STS-113	2	6:10	Quest	Endeavour 14th	ISS 38th	M.Lopez-Alegria; J. Herrington		
	Nov 30	STS-113	3	7:00	Quest	Endeavour 15th	ISS 39th	M.Lopez-Alegria; J. Herrington		
2003	Due to the loss of Columbia on February 1 (STS-107) the shuttle fleet was grounded pending an inquiry									
2004	The Return to Flight program recovering from the loss of Columbia continued but the shuttle fleet remained grounded									
2005	Jul 30	STS-114	1	6:50	OV 28th	Discovery 10th	ISS 40th	S. Robinson; S. Noguchi	20:05 (3)	278:51 (42)
	Aug 1	STS-114	2	7:14	OV 29th	Discovery 11th	ISS 41st	S. Robinson; S. Noguchi		
	Aug 3	STS-114	3	6:01	OV 30th [4]	Discovery 12th	ISS 42nd	S. Robinson; S. Noguchi		

Year	Mission	EVA	Date	Orbiter	ISS	Duration	Airlock	Crew	Subtotal	Total
2006	STS-121	1	Jul 8	Discovery 13th	ISS 43rd	7:31	Quest	P. Sellers; M. Fossum	21:29 (3)	300:20 (45)
	STS-121	2	Jul 10	Discovery 14th	ISS 44th	6:47	Quest	P. Sellers; M. Fossum		
	STS-121	3	Jul 12	Discovery 15th	ISS 45th	7:11	Quest	P. Sellers; M. Fossum		
	STS-115	1	Sep 12	Atlantis 18th	ISS 46th	6:26	Quest	J. Tanner; H. Stefanyshyn-Piper	20:19 (3)	320:39 (48)
	STS-115	2	Sep 13	Atlantis 19th	ISS 47th	7:11	Quest	D. Burbank; S. MacLean		
	STS-115	3	Sep 15	Atlantis 20th	ISS 48th	6:42	Quest	J. Tanner; H. Stefanyshyn-Piper		
	STS-116	1	Dec 12	Discovery 16th	ISS 49th	6:36	Quest	R. Curbeam; C. Fuglesang	25:45 (4)	346:24 (52)
	STS-116	2	Dec 14	Discovery 17th	ISS 50th	5:00	Quest	R. Curbeam; C. Fuglesang		
	STS-116	3	Dec 16	Discovery 18th	ISS 51st	7:31	Quest	R. Curbeam; S. Williams[5]		
	STS-116	4	Dec 18	Discovery 20th	ISS 52nd	6:38	Quest	R. Curbeam; C. Fuglesang		
2007	STS-117	1	Jun 11	Atlantis 21st	ISS 53rd	6:15	Quest	J. Reilly; J. Olivas	27:58 (4)	374:22 (56)
	STS-117	2	Jun 13	Atlantis 22nd	ISS 54th	7:16	Quest	P. Forrester; S. Swanson		
	STS-117	3	Jun 15	Atlantis 23rd	ISS 55th	7:58	Quest	J. Reilly; J. Olivas		
	STS-117	4	Jun 17	Atlantis 24th	ISS 56th	6:29	Quest	P. Forrester; S. Swanson		
	STS-118	1	Aug 11	Endeavour 16th	ISS 57th	6:17	Quest	R. Mastracchio; D. Williams	23:15 (4)	397:37 (60)
	STS-118	2	Aug 13	Endeavour 17th	ISS 58th	6:28	Quest	R. Mastracchio; D. Williams		
	STS-118	3	Aug 15	Endeavour 18th	ISS 59th	5:28	Quest	R. Mastracchio; C. Anderson[6]		
	STS-118	4	Aug 18	Endeavour 19th	ISS 60th	5:02	Quest	D. Williams; C. Anderson[6]		
	STS-120	1	Oct 26	Discovery 21st	ISS 61st	6:14	Quest	S. Parazynski; D. Wheelock	27:14 (4)	424:51 (64)
	STS-120	2	Oct 28	Discovery 22nd	ISS 62nd	6:33	Quest	S. Parazynski; D. Tani[7]		
	STS-120	3	Oct 30	Discovery 23rd	ISS 63rd	7:08	Quest	S. Parazynski; D. Wheelock		
	STS-120	4	Nov 3	Discovery 24th	ISS 64th	7:19	Quest	S. Parazynski; D. Wheelock		

(continued)

Table 9.3 (continued)

Year	Date of EVA	Mission	Mission EVA #	EVA Time	Orbiter/Quest Airlock used	Orbiter crew EVA at station	STS/ Station EVA	EVA Crewmembers	Mission EVA Total	Shuttle at station EVA Accumulative Total
2008	Feb 11	STS-122	1	7:58	Quest	Atlantis 25th	ISS 65th	R. Walheim; S. Love	22:08 (3)	446:59 (67)
	Feb 13	STS-122	2	6:45	Quest	Atlantis 26th	ISS 66th	R. Walheim; H. Schlegel		
	Feb 15	STS-122	3	7:25	Quest	Atlantis 27th	ISS 67th	R. Walheim; S. Love		
	Mar 14	STS-123	1	7:01	Quest	Endeavour 20th	ISS 68th	R. Linnehan; G. Reisman[8]	33:28 (5)	480:27 (72)
	Mar 15	STS-123	2	7:08	Quest	Endeavour 21st	ISS 69th	R. Linnehan; M. Foreman		
	Mar 17	STS-123	3	6:53	Quest	Endeavour 22nd	ISS 70th	R. Linnehan; R. Behnken		
	Mar 20	STS-123	4	6:24	Quest	Endeavour 23rd	ISS 71st	M. Foreman; R. Behnken		
	Mar 22	STS-123	5	6:02	Quest	Endeavour 24th	ISS 72nd	M. Foreman; R. Behnken		
	Jun 3	STS-124	1	6:48	Quest	Discovery 25th	ISS 73rd	M. Fossum; R. Garan	20:32 (3)	500:59 (75)
	Jun 5	STS-124	2	7:11	Quest	Discovery 26th	ISS 74th	M. Fossum; R. Garan		
	Jun 8	STS-124	3	6:33	Quest	Discovery 27th	ISS 75th	M. Fossum; R. Garan		
	Nov 18	STS-126	1	6:52	Quest	Endeavour 25th	ISS 76th	H. Stefanyshyn-Piper; S. Bowen	26:41 (4)	527:40 (79)
	Nov 20	STS-126	2	6:45	Quest	Endeavour 26th	ISS 77th	H. Stefanyshyn-Piper; S. Kimbrough.		
	Nov 22	STS-126	3	6:57	Quest	Endeavour 27th	ISS 78th	H. Stefanyshyn-Piper; S. Bowen		
	Nov 24	STS-126	4	6:07	Quest	Endeavour 28th	ISS 79th	S. Bowen; S. Kimbrough		
2009	Mar 19	STS-119	1	6:07	Quest	Discovery 28th	ISS 80th	Swanson; Arnold	19:04 (3)	546:44 (82)
	Mar 21	STS-119	2	6:30	Quest	Discovery 29th	ISS 81st	Swanson; Acaba		
	Mar 23	STS-119	3	6:27	Quest	Discovery 30th	ISS 82nd	Acaba; Arnold		
	Jul 18	STS-127	1	5:32	Quest	Endeavour 29th	ISS 83rd	D. Wolf; Kopra	30:30 (5)	577:14 (87)
	Jul 20	STS-127	2	6:53	Quest	Endeavour 30th	ISS 84th	D. Wolf; T. Marshburn		
	Jul 22	STS-127	3	5:59	Quest	Endeavour 31st	ISS 85th	D. Wolf; Cassidy		
	Jul 24	STS-127	4	7:12	Quest	Endeavour 32nd	ISS 86th	Cassidy; T. Marshburn		
	Jul 27	STS-127	5	4:54	Quest	Endeavour 33rd	ISS 87th	Cassidy; Marshburn		
	Sep 1	STS-128	1	6:35	Quest	Discovery 31st	ISS 88th	J. Olivas; N. Stott[9]	20:15 (3)	597:29 (90)
	Sep 3	STS-128	2	6:39	Quest	Discovery 32nd	ISS 89th	J. Olivas; C. Fuglesang		
	Sep 5	STS-128	3	7:01	Quest	Discovery 33rd	ISS 90th	J. Olivas; C. Fuglesang		
	Nov 19	STS-129	1	6:37	Quest	Atlantis 28th	ISS 91st	M. Foreman; R. Satcher	18:27 (3)	615:56 (93)
	Nov 21	STS-129	2	6:08	Quest	Atlantis 29th	ISS 92nd	M. Foreman; R. Bresnik		
	Nov 23	STS-129	3	5:42	Quest	Atlantis 30th	ISS 93rd	R. Satcher; R. Bresnik		

2010	Feb 12	STS-130	1	6:32	Quest	Endeavour 34th	ISS 94th	R. Behnken; N. Patrick	18:14 (3)	634:10 (96)
	Feb 14	STS-130	2	5:54	Quest	Endeavour 35th	ISS 95th	R. Behnken; N. Patrick		
	Feb 17	STS-130	3	5:48	Quest	Endeavour 36th	ISS 96th	R. Behnken; N. Patrick		
	Apr 9	STS-131	1	6:27	Quest	Discovery 34th	ISS 97th	R. Mastracchio; C. Anderson	20:17 (3)	654:27 (99)
	Apr 11	STS-131	2	7:26	Quest	Discovery 35th	ISS 98th	R. Mastracchio; C. Anderson		
	Apr 13	STS-131	3	6:24	Quest	Discovery 36th	ISS 99th	R. Mastracchio; C. Anderson		
	May 17	STS-132	1	7:25	Quest	Atlantis 31st	ISS 100th	G. Reisman; S. Bowen	21:20 (3)	675:47 (102)
	May 19	STS-132	2	7:09	Quest	Atlantis 32nd	ISS 101st	S. Bowen; M. Good		
	May 21	STS-132	3	6:46	Quest	Atlantis 33rd	ISS 102nd	M. Good; G. Reisman		
2011	Feb 28	STS-133	1	6:34	Quest	Discovery 37th	ISS 103rd	S. Bowen; A. Drew	12:48 (2)	688:35 (104)
	Mar 2	STS-133	2	6:14	Quest	Discovery 38th	ISS 104th	S. Bowen; A. Drew		
	May 20	STS-134	1	6:19	Quest	Endeavour 37th	ISS 105th	A. Feustel; G. Chamitoff	28:44 (4)	717:19 (108)
	May 22	STS-134	2	8:07	Quest	Endeavour 38th	ISS 106th	A. Feustel; E. Fincke		
	May 25	STS-134	3	6:54	Quest	Endeavour 39th	ISS 107th	A. Feustel; E. Fincke		
	May 27	STS-134	4	7:24	Quest	Endeavour 40th	ISS 118th	E. Fincke; G. Chamitoff[10]		
	Jul 12	*STS-135*	*1*	*6:03*	*Quest*	*Atlantis 34th*	*ISS 119th*	*M. Fossum; R. Garan[11]*	*6:31 (1)*	*723:50 (109)*

Total time of Shuttle based EVAs at ISS	723:50 (109)
Total time of EVAs at ISS from Shuttle airlock	191:22 (30)
Total time shuttle EVAs from Quest airlock	532:28 (79)
Total time of EVAs while docked to Mir	11:03 (2)
TOTAL SHUTTLE CREW EVA TIME AT SPACE STATIONS	734:53 (111)

(continued)

Table 9.3 (continued)

Year	Date of EVA	Mission	Mission EVA #	EVA Time	Orbiter/Quest Airlock used	Orbiter crew EVA at station	STS/Station EVA	EVA Crewmembers	Mission EVA Total	Shuttle at station EVA Accumulative Total

Notes:

[1] Jim Voss and Susan Helms were (together with Yuri Usachev) members of the recently arrived ISS-2 resident crew on STS-102.

[2] This was the first EVA from the Quest (Joint) Airlock on ISS. It represented the first time in 18 years that shuttle astronauts had conducted an EVA from an airlock other than one installed on the orbiter. It was also the first *U.S. EVA* from a space station hatch since Skylab in February 1974 (27 years earlier). NOTE: During 1997 and 1998 three American astronauts (J. Linenger, C. M. Foale and D. Wolf) had participated Russian EVAs from a Mir hatch during their long duration residencies.

[3] This EVA was the last EVA from a shuttle airlock whilst docked to ISS. The five EVAs conducted at the Hubble Space Telescope during STS-125/Service Mission 4 in May 2009 were the final EVA conducted directly from an airlock located on the orbiter.

[4] Sunita Williams was a member of the ISS-14 resident crew.

[5] Clayton Anderson was a member of the ISS-15 resident crew.

[6] Dan Tani was a member of the ISS-16 resident crew.

[7] Garrett Riesman was a member of the ISS-16 resident crew.

[8] Nicola Stott was a member of the ISS-20 resident crew.

[9] This was the final EVA performed by shuttle crewmembers completing 28 years of EVA operations since STS-6 in April 1983.

[10] This was the last EVA conducted at the ISS while a shuttle was docked to the station. It was conducted by resident ISS crewmembers due to the late scheduling of the final shuttle mission, restricted training time, reduction in Shuttle core crew number and urgency of planned tasks

A close up of Robert Curbeam using a specially prepared, tape-insulated tool to guide the P6 solar array back into its blanket box during STS-116.

External Fixtures And Fittings

- (88/2A) installed additional handrails and other equipment designed to support future EVA operations at the embryonic station; deleted launch restraints from Unity's hatches.
- (106/2A.2b) installed a magnetometer.
- (92/3A) released launch locks on PMA-3.
- (97/4A) installed a Floating Potential Probe and a centerline camera.
- (98/5A) relocated PMA-2 on the Destiny laboratory.
- (102/5A.1) installed Lab Cradle Assembly on Destiny.
- (104/7A) transferred two oxygen and two nitrogen tanks from Atlantis to the exterior of Quest.
- (105/7A.1) installed the Early Ammonia Servicer (EAS) on P6; installed six EVA handrails and relocated another two on the Destiny laboratory.
- (111/UF2) transferred power/data/grapple fixture to a solar array; attached space debris shields.
- (112/9A) attached cables for the Ammonia Tank Assembly; installed Spool Positioning Line on cables.
- (113/11A) made connections between the P1 and S0 trusses; installed Spool Positioning Devices; attached Ammonia Tank Assembly lines.
- (114/LF1) removed faulty CMG-1 from the Z1 truss and installed new unit.
- (116/12.1A) installed grapple fixture.
- (118/13A.1) installed a new CMG on the Z1 truss and stowed the failed CMG on an External Stowage Platform for later return to Earth.

- (122/1E) replaced nitrogen tank on the P1 truss; stowed older tank in Orbiter payload bay, stowed failed CMG in Orbiter payload bay.
- (124/1J) removed and replaced starboard Nitrogen Tank Assembly.
- (126/ULF2) transferred empty nitrogen tank from ESP-3 to Orbiter payload bay.
- (119/15A) relocated crew equipment cart.
- (128/17A) exchanged old ammonia tank for new unit.
- (129/ULF3) deployed S3 outboard Payload Attach Systems; installed a bracket for ammonia lines; deployed a pair of brackets to attach cargo to the station's truss; installed High Pressure Gas Tank on Quest; released bolts on Ammonia Tank Assembly.
- (131/19A) relocated new ammonia tank and removed old tank.
- (132/ULF4) retrieved spare PDGF from Atlantis and stowed in Quest airlock.
- (134/ULF6) installed grapple bars of port radiators; installed grapple fixture on Zarya to enable the SSRMS to translate to Russian segment; transferred OBSS boom extension to the starboard side of the ITS.

Truss Segments And Solar Arrays

- (92/3A) supported installation of Z1 truss and prepared for attachment of solar arrays.
- (97/4A) supported installation of P6 truss segment and deployment of solar array.
- (102/5A.1) made minor adjustments to solar array brace.
- (105/7A.1) strung heater cables for further installation of S0 truss segment.
- (108/UF1) installed insulation blankets on Beta Gimbal Assembly on top of P6; unsuccessful attempt to free stuck solar array cable.
- (110/8A) supported installation of S0 truss.
- (111/UF2) prepared P6 truss for future relocation.
- (112/9A) supported installation of S1 truss; released launch lock of radiators; added new Z1 and P6 junctions to install Spool Positioning Devices; installed two jumpers to facilitate flow of coolant between S1 and S0 trusses; released launch restraint on S1 and stowed it as a drag link.
- (113/11A) removed drag link on P1; installed fluid jumpers at S0/P1 point of attachment; removed P1 port and starboard keel pins.
- (115/12A) supported installation of P3/P4 trusses and radiators; configured Solar Array Rotary Joint (SARJ).
- (116/12.1A) supported installation of P5 truss; prepared P6 for relocation; assisted in deploying P6 solar array.
- (117/13A) supported installation of S3/S4 truss; assisted in retraction of P6 truss; partial failure discovered because of reversed wiring on S3/S4 SARJ; removed final SARJ launch restraints.
- (118/13A.1) supported installation of S5 truss; retracted forward radiator on P6 in readiness for relocation; secured gimbal locks on Z1; supported relocation of P6; disconnected P6/Z1 truss segment fluid lines and umbilicals; configured S1 radiator; inspected S4 SARJ; supported attachment of P6 to P5 segments of the truss; reconfigured S1 following redeployment; inspected the port SARJ.

- (123/1J/A) removed five covers from starboard SARJ; visually inspected and photographed area.
- (124/1J) replaced the trundle bearing assembly on SARJ.
- (126/ULF2) cleaned and lubricated troubled bearing assemblies on starboard and port SARJ.
- (119/15A) supported installation of S6 to S5 truss and deployment of the S6 radiator; deployed P3 nadir UCCAS.
- (127/2J/A) reconfigured Z1 patch panel.
- (130/20A) replaced final gyro on S0 truss.
- (132/ULF4) installed P4/P5 ammonia jumpers.

Spares

- (121/ULF1.1) delivered a spare pump for the station's cooling system.
- (123/1J/A) installed spare equipment on an ESP mounted on the exterior of Quest airlock.
- (126/ULF2) transferred a new flex hose rotary coupler from the Shuttle to ESP-3 for future use.
- (127/2J/A) transferred Orbital Replacement Unit from an ICC to ESP-3.
- (129/ULF3) installed a spare antenna on the truss.
- (131/19A) installed a spare K_u-band antenna on Z1 truss.
- (132/ULF4) stowed a spare PDGF in Quest airlock.

Modules

- (98/5A) supported installation of US Destiny laboratory.
- (104/7A) supported installation of Quest.
- (117/13A) installed hydrogen ventilation valve on Destiny; installed computer cable on Unity; opened hydrogen vent valve on Destiny.
- (120/10A) supported installation of Node 2 Harmony.
- (122/1E) supported installation of ESA Columbus module and installed keel pin thermal covers on that module.
- (123/1J/A) supported installation of the Pressurized Section of the Japanese Experiment Logistics Module on nadir of Harmony; removed port and nadir CBM launch locks from Harmony; installed ELM-PS trunnion covers.
- (124/1J) supported installation of the Pressurized Section on the Japanese Experiment Module (Kibo); installed covers and external TV equipment on Kibo.
- (126/ULF2) removed and replaced Kibo External Facility berthing mechanism insulation cover; installed handrails on Kibo.
- (119/15A) installed Kibo unpressurized cargo carried on P3 in preparation for STS-127.
- (127/2J/A) installed Japanese Exposure Facility.
- (128/17A) completed preparatory work in advance of delivery of Tranquility Node on STS-130.
- (129/ULF3) worked on docking adapter heater cables.

- (130/20A) delivered Tranquility Node; prepared nadir port for attachment of Cupola; installed Cupola; installed handrails on Tranquility; insulated cables on Unity and S0 truss.
- (131/19A) removed no longer required micrometeoroid shield from Quest and secured it in payload bay of the Orbiter.
- (132/ULF4) installed Russian Mini Research Module 1 called Rassvet.
- (133/ULF5) installed PMM.

Canadian Dave Williams during the first EVA of STS-118 when the S5 segment of the ITS was installed.

External Stowage Platforms And Experiments

- (102/5A.1) configured storage platform.
- (105/7A.1) installed Material ISS Experiment (MISSE) on Quest.
- (114/LF1) installed base and cabling for External Stowage Platform; retrieved two exposure experiments.
- (118/13A.1) retrieved MISSE containers 3 and 4.
- (122/1E) installed SOAR telescope and EuTEF facility on ESP of Columbus.
- (123/1J/A) failed to attach MISSE-6 experiments on to Columbus module due to faulty latching pins at first attempt (installed on subsequent EVA).
- (119/15A) installed unpressurized cargo carrier on P3 truss; failed to deploy a cargo carrier.
- (128/17A) retrieved MISSE and EuTEF facilities from exterior of Columbus and installed them in payload bay of the Orbiter.
- (129/ULF3) installed a range of small devices across station including the GATOR (Grappling Adapter to On-Orbit Railing) bracket on Columbus; installed MISSE 7A and 7B on ELC-2.

The closing moments of an EVA as both astronauts re-enter the Quest airlock.

- (130/20A) attached two micrometeoroid shields to ESP-2; retrieved Japanese seed experiment from exterior of Kibo.
- (132/ULF4) released the bolts holding the replacement batteries of ICC-VLD cargo carrier.
- (133/ULF5) installed ELC-4; removed Japanese vacuum museum educational experiment.
- (134/ULF6) retrieved two MISSE-7 experiments; installed new MISSE-8 experiments on ELC-2.

Robotics

- (98/5A) installed SSRMS base.
- (102/5A.1) installed SSRMS cable tray.
- (100/6A) installed Canadarm2 (SSRMS).
- (110/8A) routed power connections to SSRMS through S0 truss.
- (116/12.1A) installed thermal cover on SSRMS.

- (123/1J/A) installed Dextre assembly and removed covers; completed transfer of OBSS from Orbiter on ITS (temporary).
- (124/1J) transferred OBSS from ITS back to Orbiter.
- (126/ULF2) lubricated end effector snare bearing of SSRMS.
- (119/15A) lubricated SSRMS grapple snares.
- (129/ULF3) lubricated ORU Attachment Devices grapple mechanism I MBS and the hand snares on the Kibo RMS; fitted insulation covers on camera of the MSS and Canadarm2 end effector.
- (132/ULF4) installed new tool platform on Dextre.
- (133/ULF5) installed camera assembly on Dextre.
- (134/ULF6) attached grapple fixture to array to facilitate use of SSRMS on Russian segment; OBSS permanently transferred to the ISS to offer greater reach for the SSRMS.

Repair And Evaluation

- (88/2A) made a survey of station's exterior after a month in space in order to determine its condition early in the assembly and provide a baseline reference for later surveys to record the deterioration and condition of the early elements over time.
- (97/4A) inspected partially stuck solar array on P6 segment; repaired P6 solar array.
- (104/7A) inspected gimbal assembly on top of solar array truss.
- (120/10A) inspected S4 SARJ and port SARJ.
- (124/1J) replaced trundle bearing assembly of SARJ.
- (126/ULF2) photographed umbilical system cables and radiators.
- (119/15A) infrared images taken of P1 and S1 radiator panels.
- (127/2J/A) visually inspected equipment installation.
- (132/ULF4) repaired OBSS on Atlantis.

Plumbing And Power Utilities

- (88/2A) installed EVA power cables between Zarya and Unity.
- (106/2A.2b) installed electrical cables between Zvezda and Zarya.
- (92/3A) connected and reconfigured electrical cables on Z1; installed two direct current converters on Z1 truss.
- (97/4A) connected power and coolant lines as well as data cables from P6 truss.
- (98/5A) connected power and data cables and released cooling radiator.
- (102/5A.1) completed connections of cables begun on previous EVA.
- (100/6A) rewired and rerouted power and data cables for Canadarm2.
- (104/7A) connected heater cables.
- (110/8A) installed redundant power cable from S0 truss; routed power connections to SSRMS through S0 truss.
- (111/UF2) installed power, data and electronics cables to MBS.
- (112/9A) installed power, data and fluid lines between S0 and S1 truss segments.
- (113/11A) reconfigured external electrical harnesses which routed power through Main Bus Switching Units.

- (114/LF1) rerouted power to CMG-2.
- (116/12.1A) reconfigured electrical wiring, first bringing P3/P4 channel 2 and 3 on line and then channel 1 and 4.
- (117/13A) installed computer cable on Unity.
- (120/10A) supported relocation of umbilicals for Z1 and P6 truss segments.
- (122/1E) continued work on SSPTS.
- (130/20A) installed plumbing and connections for ammonia supply between Destiny, Unity and Tranquility modules and applied insulation.
- (131/19A) disconnected old fluid lines on S1 truss.
- (132/ULF4) replaced 6 of 6 batteries on P6 and stowed the old batteries for return to Earth; installed P4/P4 ammonia jumpers.
- (133/ULF5) connected power extension cables between Unity and Tranquility.
- (134/ULF6) installed back-up power cables to Russian segment.

Throughout an EVA, fellow crewmembers would monitor activities to ensure that the assigned tasks were completed as per the pre-flight EVA plan.

Maintenance

- (111/UF2) replaced failed SSRMS wrist roll joint.
- (112/9A) replaced Interface Umbilical Assembly on Mobile Transporter.
- (114/LF1) removed faulty CMG-1 and replaced with new unit.
- (116/12.1A) replaced broken video camera on S1 truss; attached three bundles of Zvezda debris panels for future installation on Russian segment.
- (117/13A) tethered two debris panels to Zvezda Service Module.
- (124/1J) removed TV camera due to failed power supply.
- (126/ULF2) performed maintenance on Kibo robotic arm grounding tab.

- (134/ULF6) vented nitrogen from ammonia servicer; refilled P6/P5 radiators with ammonia; completed servicing of Early Detection Ammonia Systems; lubricated port SARJ as well as parts of Dextre; replaced thermal insulator of spare gas tank on Quest.

Communications

- (88/2A) installed two antennas and released a stuck TORU antenna on Zarya.
- (101/2A.2a) replaced failed antenna on Unity.
- (106/2A.2b) installed communication cable between Zvezda and Zarya.
- (92/3A) deployed two antenna assemblies.
- (97/4A) replaced S-band assembly at P6 truss and installed a sensor on a radiator.
- (98/5A) attached S-band antenna.
- (102/5A.1) relocated early communications antenna from Unity to PMA attachment.
- (100/6A) installed UHF antenna and video command and power cables from the SSRMS to the robotic workstation on Destiny.
- (112/9A) deployed second S-band communications system; installed two external camera systems.
- (113/11A) installed two wireless video systems on External Transceiver assembly on Unity to support EVA helmet camera operations.
- (114/LF1) replaced faulty GPS antenna; installed PCSat2 ham radio satellite.
- (121/ULF1.1) installed zenith Interface Umbilical Assembly (IUA) to protect undamaged power, data and video cables.
- (115/12A) replaced S-band antenna and installed insulation on a second unit.
- (117/13A) moved TV camera from External Stowage Platform on Quest onto S3 truss.
- (118/13A.1) relocated antenna base from P6 to P1; installed upgraded communications equipment; installed External Wireless Instrumentation System antenna.
- (126/ULF2) installed video camera; installed GPS antenna on Kibo.
- (119/15A) installed GPS antenna on Kibo.
- (129/ULF3) installed an additional ham radio antenna and truss antenna for EMU wireless helmet camera.
- (134/ULF6) installed external wireless video communications antenna on Destiny.

EVA Tools And Equipment

- (88/2A) installed the first set of external EVA tools for later crews.
- (96/2A.1) installed a second set of external EVA tools; installed the US Orbital Transfer Device and elements of the Russian Strela crane.
- (101/2A.2a) completed installation of OTD and Strela.
- (92/3A) installed two EVA tool boxes; tested SAFER backpack.
- (98/5A) test flew SAFER backpacks.
- (110/8A) connected Mobile Transporter; installed work lights and airlock spur for future EVA work on S0 truss.

- (111/UF2) installed Mobile Base System.
- (112/9A) released launch restraints on CETA Carts; relocated CETA cart from P1 to S1 truss to permit MT to translate along P1 for future EVA assistance.
- (121/ULF1.1) evaluated RMS/OBSS system as a platform for EVA inspection and repair of damaged Orbiter; completed full operation of MT CETA carts.
- (116/12.1A) installed EVA tool bags for future use; relocated two CETA handcarts.
- (118/13A.1) relocated two CETA carts from the port side of the MT to the starboard side; installed OBSS Boom Stand.
- (120/10A) installed handrails on Harmony module.
- (126/ULF2) relocated two CETA carts from starboard side of MT to port side.
- (129/ULF3) relocated a foot restraint.
- (132/ULF4) installed new tool platform on Dextre; replenished EVA tools in toolboxes.
- (133/ULF5) replaced guard for railcar system; installed a work light on cargo cart.

Contingency Situations

- (98/5A) Robert Curbeam's suit became contaminated with leaking ammonia requiring cleaning via a 'bake-out' in sunlight.
- (114/LF1) Shuttle repair techniques on TPS demonstrated in payload bay; two gap fillers manually removed from underside of Discovery; documentation of Orbiter underside.
- (121/ULF1.1) Piers Sellers SAFER pack became loose and required Mike Fossum to secure using tethers; further evaluation of title repair systems; evaluated using infrared camera to image Shuttle wing to test capability of detecting potential damage.
- (116/12.1A) assisted in deployment of P6 solar array.
- (117/13A) repair of Shuttle OMS pod thermal blanket.
- (118/13A.1) slight damage to second layer in Richard Mastracchio's EVA glove; not serious damage but returned to airlock early.
- (123/1J/A) further testing of Shuttle TPS repair materials and techniques.
- (127/2J/A) high levels of carbon dioxide recorded in Christopher Cassidy's suit.
- (134/ULF6) faulty carbon dioxide sensor on Greg Chamitoff's suit delayed completion of some tasks to later EVAs.

Resident (EO-) Crew STS-EVA Support Tasks From Quest

- (EO-4) connected cables from Destiny to Z1 truss; removed tools and handrails used on earlier EVAs.
- (EO-6) continued outfitting and activating the P1 truss; removed debris from sealing ring on nadir docking port of Unity; completed several maintenance tasks; reconfigured ISS power system to provide secondary power source for one of the CMG's secured thermal control system quick disconnect fittings.

Table 9.4 Individual STS EVA Crewmember Experience 1998–2011

STS Crewmember	Nationality/ Agency	STS-Mission	Station Mission Designation	Year of Mission EVAs	Total Mission EVAs	Total EVA time hh:mm	Total Station EVAs	Total Station EVA Time hh:mm
Acaba	American/NASA	119	15A	2009	2	12:57	2	12:57
Anderson C.	American/NASA	118	13A.1	2007	2	10:30	5	30:47
		131	19A	2010	3	20:17		
Arnold	American/NASA	119	15A	2009	2	12:34	2	12:34
Barry	American/NASA	96	2A.1	1999	1	7:55	3	19:40
		105	7A.1	2001	2	11:45		
Behnken	American/NASA	123	1J/A	2008	3	19:19	6	37:33
		130	20A	2010	3	18:14		
Bowen	American/NASA	126	ULF2	2008	3	19:56	7	47:18
		132	ULF4	2010	2	14:34		
		133	ULF5	2011	2	12:48		
Bresnik	American/NASA	129	ULF3	2009	2	11:50	2	11:50
Burbank	American/NASA	115	12A	2006	1	7:11	1	7:11
Cassidy	American/NASA	127	2J/A	2009	3	18:05	3	18:05
Chamitoff	American/NASA	134	ULF6	2011	2	13:43	2	13:43
Chang-Diaz	American/NASA	111	UF2	2002	3	19:31	3	19:31
Chiao	American/NASA	92	3A	2000	2	13:16	2	13:16
Curbeam	American/NASA	98	5A	2001	3	19:49	7	45:34
		116	12A.1	2006	4	25:45		
Drew	American/NASA	133	ULF5	2011	2	12:48	2	12:48
Feustel	American/NASA	134	ULF6	2011	3	21:20	3	21:20
Fincke	American/NASA	134	ULF6	2011	3	22:25	3	22:25
Foreman	American/NASA	123	1J/A	2008	3	19:34	5	32:19
		129	1J	2009	2	12:45		
Forrester	American/NASA	105	7A.1	2001	2	11:45	4	25:30
		117	13A	2007	2	13:45		

Name	Nation/Agency	STS	Mission	Year	EVAs	Time	Total EVAs	Total Time
Fossum	American/NASA	121	ULF1.1	2006	3	21:29	7	48:32
		124	1J	2008	3	20:32		
		135	ULF7	2011	1	6:31		
Fuglesang	Swedish/ESA	116	12A.1	2006	3	11:36	5	25:16
		128	17A	2009	2	13:40		
Garan	American/NASA	124	1J	2008	3	20:32	4	27:03
		135	ULF7	2011	1	6:31		
Gernhardt	American/NASA	104	7A	2001	3	16:30	3	16:30
Godwin*	American/NASA	108	UF1	2001	1	4:12	1	4:12
Good	American/NASA	132	ULF4	2010	2	13:55	2	13:55
Hadfield	Canadian/CSA	100	6A	2001	2	14:50	2	14:50
Helms	American/NASA	102	5A.1	2001	1	8:56	1	8:56
Herrington	American/NASA	113	11A	2002	3	19:55	3	19:55
Jernigan	American/NASA	96	2A.1	1999	1	7:55	1	7:55
Jones T.	American/NASA	98	5A	2001	3	19:49	3	19:49
Kimbrough	American/NASA	126	ULF2	2008	2	12:52	2	12:52
Kopra	American/NASA	127	2J/A	2009	1	5:32	1	5:32
Linnehan	American/NASA	123	1J/A	2008	3	21:02	3	21:02
Lopez-Alegria	American/NASA	92	3A	2000	2	14:03	5	33:58
		113	11A	2002	3	19:55		
Love	American/NASA	122	1E	2008	2	15:23	2	15:23
Lu	American/NASA	106	2A.2b	2000	1	6:14	1	6:14
Malenchenko	Russian/RSA	106	2A.2b	2000	1	6:14	1	6:14
Marshburn	American/NASA	127	2J/A	2009	3	18:59	2	15:23
Mastracchio	American/NASA	118	13A.1	2007	3	18:13	6	38:30
		131	19A	2010	3	20:17		
McArthur W.	American/NASA	92	3A	2000	2	13:16	2	13:16
McLean	Canadian/CSA	115	12A	2006	1	7:11	1	7:11
Morin	American/NASA	110	8A	2002	2	14:07	2	14:07

(continued)

Table 9.4 (continued)

STS Crewmember	Nationality/ Agency	STS-Mission	Station Mission Designation	Year of Mission EVAs	Total Mission EVAs	Total EVA time hh:mm	Total Station EVAs	Total Station EVA Time hh:mm
Newman	American/NASA	88	2A	1998	3	21:22	3	21:22
Noguchi	Japanese/JAXA	114	LF1	2005	3	20:05	3	20:05
Noriega	American/NASA	97	4A	2000	3	19:20	3	19:20
Olivas	American/NASA	117	13A	2007	2	14:13	5	34:28
		128	17A	2009	3	20:15		
Parazynski*	American/NASA	100	6A	2001	2	14:50	6	42:04
		120	10A	2007	4	27:14		
Patrick	American/NASA	130	20A	2010	3	18:14	3	18:14
Perrin	French/CNES	111	UF2	2002	3	19:31	3	19:31
Reilly	American/NASA	104	7A	2001	3	16:30	5	30:43
		117	13A	2007	2	14:13		
Reisman	American/NASA	123	1J/A	2008	1	7:01	3	21:12
		132	ULF4	2010	2	14:11		
Richards P.	American/NASA	102	5A.1	2001	1	6:21	1	6:21
Robinson	American/NASA	114	LF1	2005	3	20:05	3	20:05
Ross	American/NASA	88	2A	1998	3	21:22	5	35:29
		110	8A	2002	2	14:07		
Satcher	American/NASA	129	ULF3	2009	2	12:19	3	20:05
Schlegel	German/ESA	122	1E	2008	1	6:45	1	6:45
Sellers	American/NASA	112	9A	2002	3	19:41	6	41:10
		121	ULF1.1	2006	3	21:29		
Smith S.	American/NASA	110	8A	2002	2	14:15	2	14:15
Stefanyshyn-Piper	American/NASA	115	12A	2006	2	13:08	5	33:42
		126	ULF2	2008	3	20:34		
Stott	American/NASA	128	17A	2009	1	N6:35	1	6:35
Swanson	American/NASA	117	13A	2007	2	13:45	4	26:22
		119	15A	2009	2	12:37		
Tani	American/NASA	108	UF1	2001	1	4:12	1	4:12
		120	10A	2007	1	6:33	1	6:33
Tanner	American/NASA	97	4A	2000	3	19:20	5	32:28
		115	12A	2006	2	13:08		

Name	Nationality/Agency							
Thomas A.	American/NASA	102	5A.1	2001	1	6:21	1	6:21
Voss J.S.	American/NASA	101	2A.2a	2000	1	6:44	1	6:44
		102	5A.1	2001	1	8:56	1	8:56
Walheim	American/NASA	110	8A	2002	2	14:15	5	36:43
		122	1E	2008	3	22:08		
Wheelock	American/NASA	120	10A	2007	3	20:41	3	20:41
Williams D.	Canadian/CSA	118	13A.1	2007	3	17:42	3	17:42
Williams J.	American/NASA	101	2A.2a	2000	1	6:44	1	6:44
Williams S.	American/NASA	116	12A.1	2006	1	7:31	1	7:31
Wisoff	American/NASA	92	3A	2000	2	14:03	2	14:03
Wolf	American/NASA	112	9A	2002	3	19:41	6	38:05
		127	2 J/A	2009	3	18:24		

Between December 1998 and July 2011, a total of 71 individuals (63 American and 8 international crewmembers) performed at least one EVA at ISS during Shuttle assembly missions. Of these 20 Americans performed EVA on two separate missions, and 2 conducted EVA on three different missions. The international crewmembers included 3 Canadians, 1 French, 1 Japanese, 1 German, 1 Russian and 1 Swede (who conducted EVA on two separate missions).

NOTE: *US astronauts Godwin and Parazynski also completed an EVA during the time their Shuttle orbiter was docked to the Mir station. Though they did not cross over the docking threshold onto Mir hardware their total EVA time at 'space stations' is:

Godwin:	Mir (STS-76)	1 EVA	06 hr 02 min
	ISS (STS-108)	1 EVA	04 hr 12 min
	Total	2 EVAs	10 hr 14 min
Parazynski:	Mir (STS-86)	1 EVA	05 hr 01 min
	ISS (STS-100 & 120)	6 EVAs	42 hr 04 min
	Total	7 EVAs	47 hr 05 min

Always time to enjoy the view…

…and the experience.

- (EO-12) installed camera on the P1 truss; retrieved failed rotary Joint Motor Controller; removed and replaced remote power controller module on MT; jettisoned Floating Potential Probe.
- (EO-13) installed a Floating Potential Measurement Unit on MISSE; repaired malfunctioning GPS antenna; installed starboard jumper and Spool Positioning Device of S1; replaced light on cart; tested infrared camera and took close up photography of ISS exterior.
- (EO-14) reconfigured two coolant loops on Destiny into permanent systems; connected up the SSPTS; removed Early Ammonia Services from P6; photo documentation for retraction of P6 during STS-117; removed sunshade from data relay devices; removed and discarded P3 shrouds on truss Rotary Joint Motor Controllers Bay 18 and 20; deployed the Unpressurized Cargo Carrier Assembly Attachment Point on zenith of P3; removed launch locks from P5; connected four SSPTS cables to PMA-2 at forward port on Destiny.
- (EO-15) replaced MT redundant power systems; cleaned CBM on nadir of Unity; jettisoned ammonia tank and flight support equipment.
- (EO-16) completed a series of five spacewalks from Quest to support the disconnection and rerouting of SSPTS cables; stowed PMA-2 umbilicals; temporary stowage of avionics umbilicals for Harmony, then retrieval and attachment; hooked up PMA-2 and Destiny; inspected and photographed starboard SARJ and Beta Gimbal Assembly; replaced one of the Bearing Motor Roll Ring Modules on a solar array; continued photo inspection of SARJ.
- (EO-24) replaced failed ammonia pump module on S1; quick disconnect would not release on first attempt but replacement completed on next two EVAs.
- (EO-28) retrieved failed ammonia pump module from ESP-2 and put it on Orbiter for return to Earth; installed Robotic Refueling Mission payload on station; retrieved a materials science package; released stuck wire of a power grapple fixture; installed thermal covers on PMA-3; conducted the final EVA while a Shuttle was docked, thereby completing Shuttle assembly program.

This listing conveys both the enormity of the assembly of the ISS and the manner in which a vast number of individual tasks had to be specified, sequenced, and performed over the years.

SUMMARY

The Shuttle-based EVA program at Mir and the ISS involved one hundred and eleven separate EVAs totaling over 734 hr, with only two of those sessions occurring at Mir. All of this was a far cry from the estimated 2,800 hr of EVA at Space Station Freedom projected during the assembly phase alone. Clearly the change of design and reduction of components helped to reduce the total figure to about one quarter of that which had been predicted two decades earlier. One major time-saving change was to provide pre-assembled truss elements instead of trying to assemble them on-orbit.

Putting aside the EVAs performed at Mir, the EVAs for ISS assembly, which were conducted primarily by Shuttle crews, were very impressive.

Just stepping out for the thrill of a lifetime.

Across the thirty-seven assembly missions a total of seventy-three individual Shuttle crewmembers (including the EVA of STS-135 that was actually done by ISS residents) accumulated 1,428.85 (crew) hr during 219 (crewmember) EVA sessions. Of this group sixty-four astronauts (six female) were Americans and nine were from the international partner agencies: three Canadians (CSA), two Russians (RSA), and one astronaut each from Japan (JAXA), France (CNES), Germany (ESA) and Sweden (ESA).

This program created a huge database of experience and knowledge on conducting EVAs at a structure as large as the ISS over many years, supplemented of course by a number of station-based EVAs by resident crews.

Furthermore the data from the planning, training, and simulation of all these EVAs enlarged the knowledge gleaned from analyzing earlier programs, both American and Russian.

This mass of talent and skills will be of tremendous assistance in deciding on and planning future human space exploration because the clear demonstration of what can be achieved on-orbit will be applied to the next step in space. Unfortunately, many of those who participated in the ISS EVA effort, both on the ground and on-orbit, have either retired from active involvement in the space program or have transferred to the aerospace industry. Hence by the time that we are ready to embark on the next major space project on a scale comparable to the ISS, a new generation will face the task of taking spacewalking skills to the next level because the achievements of the ISS will have become part of space history.

Notes

1. EVA Development and Verification Testing at NASA's Neutral Buoyancy Laboratory; Juniper C. Jairala, Jacobs Technology, Houston, Texas, Robert Durkin, NASA JSC et al, conference paper (JSC-CN-26179) 42nd International Conference on Environmental Sciences, (ICES) July 15–19,2012, San Diego, California, AIAA Paper [January 1, 2012] also JSC/EC5 U.S. Spacesuit Knowledge Capture Series, JSC/B5S/R3102, August 14, 2012, www.jsc.nasa.gov/history/spacesuits/presentations/Jairala-Durkin_ EVA.pdf; Last accessed May 7, 2016
2. "A journey of a thousand miles begins with a single step," Laozi (c 604 BC – c 531 BC), Chinese philosopher and founder of Taoism, in *Tao Te Ching* (The Classic of the Way's Virtues), 6th century, published in English in 1891
3. *Enhancing Hubble's Vision*, p. 5
4. E-mail from Jerry Ross, April 28, 2015, also notes from a presentation, reproduced with permission, intended for a World Space Congress but which he never got to attend
5. E-mail from Joe Tanner December 13, 2015
6. AIS interview with Tom Jones, August 3, 2006
7. International Space Station Assembly Lessons Learned, Briefing to International Workshop on on-orbit Satellite Servicing, 25 March 2010, Sam Scimemi, Deputy International Space Station, NASA Headquarters
8. *Opening The Door To The Universe, The Story of EVA*, David J. Shayler (a working title)

10

Getting Back

> *Although we got to take the ride,*
> *we sure hope that everybody who has*
> *ever worked on, or touched, or looked at,*
> *or envied, or admired a Space Shuttle was able*
> *to take just a little part of the journey with us.*
>
> Post-landing comments by STS-135 CDR Chris Ferguson,
> Shuttle Landing Facility, KSC, Florida, July 21, 2011

In all, 133 of the 135 Shuttle launches returned successfully; the crews of the other two missions were lost along with their vehicles: Challenger in 1986 and Columbia in 2003. Returning an Orbiter to Earth was as difficult, dynamic, and dangerous as getting it into orbit in the first place.

Though the Shuttle was roomier than the American spacecraft for Mercury, Gemini and Apollo it was still fairly confined, even when it carried a Spacehab or Spacelab. It therefore came as a welcome interlude on a busy flight, to explore the modules of the ISS as that grew over the years. Despite the enjoyment of floating through the various station components a Shuttle crewmember had, at the back of their mind, the sobering realization that the Orbiter that they could see through the windows of the station was their primary ride home. As a mission drew to a close, the satisfaction of achieving all the assigned tasks and the prospect of returning to loved ones combined into a feeling of pride and a longing to come home.

As Joe Tanner (STS-97 and STS-115) said, "I think we were all impressed with the relative roominess of ISS once we were able to ingress. Of course, we all got a guided tour by the crew. I spent most of my time there except for quick meals and sleeping. I wanted to sleep over there but we thought it best that everyone was on the right side of the hatches overnight just in case we had to leave in a hurry. We actually spent a lot of time with the ISS crew during STS-97, mainly because they did not sleep much while the hatches were open. They were on a tight schedule to film as much as possible using an IMAX camera

© Springer International Publishing Switzerland 2017
D.J. Shayler, *Assembling and Supplying the ISS*, Springer Praxis Books,
DOI 10.1007/978-3-319-40443-1_10

for the new *Space Station* movie before we left, since we were taking the exposed film home. I helped them all day with scenes and loading film until it was time for me to go back to the Shuttle, but they kept working on it."[1]

The mission emblem for STS-135, the final Shuttle flight, is added to those displayed on the ISS from previous missions, one of several traditions upheld by arriving and departing crews.

UNDOCKING THE SHUTTLE

The procedure for undocking an Orbiter from a space station started by closing the internal hatches between the two vehicles. Then the docking lights and TV cameras would be switched on and the vestibule airlock depressurized while leak tests were performed. The docking system would then be powered up in order to carry out the uncoupling. Because the APAS circuit incorporated inhibitors designed to preclude accidental undocking, certain button pushes were necessary to initiate the processes. These commands enabled the Undocking Open Hooks and Open Latches buttons. In normal circumstances the actual undocking then needed only the Undocking button to be depressed. The hooks would open and once they released, four spring plungers that were compressed between the mating surfaces imparted a force of 700 lb (317.8 kg) to gently push the vehicles apart. When the Orbiter had finished the separation burns, the docking system would be powered off and deactivated.

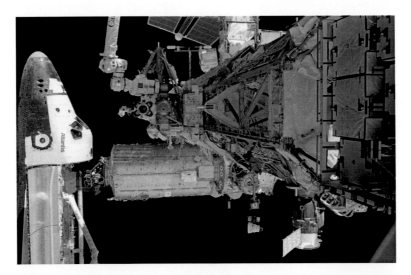

With all of their objectives achieved the Shuttle crew returns to the Orbiter, closes the hatches, and prepares for undocking.

Activity on the aft flight deck during the final moments prior to undocking from the ISS, showing PLT Doug Hurley (left) and MS Rex Walheim (right).

In the case of a contingency, the procedure would assume that the Orbiter hooks had 'failed closed' during an earlier undocking attempt. This situation would have required using the pyro system to initiate separation. If the failed hooks were on the station side, the pyro system for the passive hooks on the Shuttle side would be used to achieve the desired separation. There was a contingency plan for if the pyro bolts failed to fire, but this would have involved a contingency EVA where the astronauts undid the ninety-six bolts that attached the airlock to the docking system.

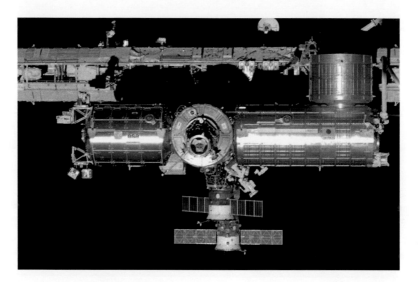

An STS-124 view of the ISS shortly after undocking. The Harmony Node 2 and the docking port are centrally located, the ESA Columbus laboratory is to the left and the Japanese Kibo laboratory is to the right.

The undocking was usually done by the PLT to allow them hands-on experience of controlling the Orbiter during the post-undocking maneuvering. Fly-around maneuvers at Mir and the ISS were initiated only if sufficient propellant was available. This phase of the flight helped to prepare a PLT for a future assignment as CDR, when they would be responsible for docking with the station.

Each undocking at Mir and the ISS was normally made in time to allow the resident crew to prepare to receive further visitors on the next Soyuz or Shuttle, or cargo on the next unmanned freighter. This was not always the case though. On STS-100 in 2001, a problem in qualifying the station's computer software prompted the ground controllers to keep Endeavour docked for an extra day in order to check and verify the data of the US Command and Control Computer. As a result the Shuttle didn't undock until a day before the arrival of the first Soyuz taxi (Soyuz TM-32), but there was no problem and the new arrival docked at the Russian segment as planned.

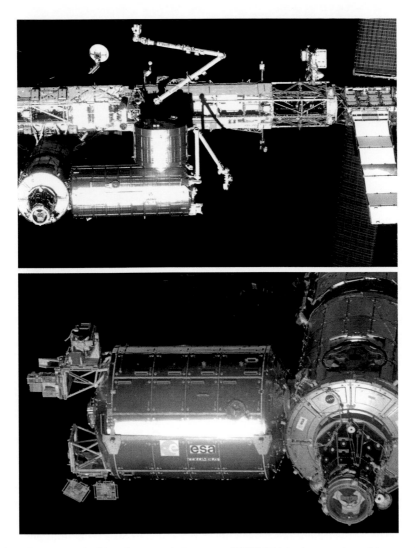

During the post-undocking fly-around the crew of STS-124 took a number of images that detailed the science modules, truss structure, and solar arrays as part of an ongoing program of documenting the expanding (and aging) station to add to the database on longevity of spacecraft materials and components begun in the 1960s. Details of the Kibo lab, truss, SSRMS, and solar arrays (top) and of the docking port on Harmony Node 2 and the Columbus lab and its external facilities (bottom).

An Orbiter's thrusters were inhibited from firing in close proximity to ISS elements such as the solar panels. At 2 ft (0.6 m), the jets were activated in a procedure that was basically the reverse of the sequence used in docking. Once the Orbiter had withdrawn to a safe stand-off distance, it would either depart immediately or perform a fly-around maneuver prior to departing.

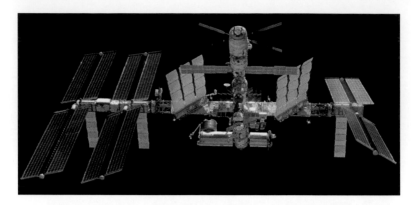

With its fly-around maneuver accomplished, Discovery moves away from the almost complete ISS to prepare for the return to Earth.

Flying Around The ISS

To undertake a station fly-around, the Shuttle's attitude control jets were put in Low Z mode and the PLT, working on the aft flight deck, would return the vehicle back to the 450 ft (137.16 m) point from the station. Initially the fly-around started directly behind the station but then the Orbiter was maneuvered beneath before circling over the top of the station. If there was sufficient propellant in reserve, more than one circuit might be flown.

A fly-around was an opportunity to document the station. Combined with comments by the Shuttle crew these pictures recorded the condition of the exterior surfaces of the station. For Mir, these updates came between nine and twelve years into its orbital life, but in the case of the ISS they ran from the very beginning of the program and through the next thirteen years until the Shuttle fleet was retired. On completing the fly-around maneuver the PLT would fire the control jets to execute the final separation burn. This put the departing Orbiter on a flight path that passed about 0.5 miles (0.80 km) behind and 1.5 miles (2.41 km) below the station prior to moving ahead (as a result of orbital dynamics) and increasing the separation distance.

Those Precious Final Orbits

During the final days on-orbit, the Shuttle crew resumed the normal operations which were a feature of every returning mission. Configuring the Orbiter for the descent took several hours and included stowing all unpacked apparatus as well setting up mid- and flight deck Mission Specialist seating.

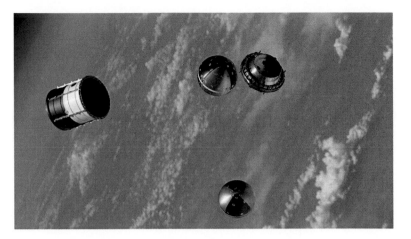

On several assembly missions, usually towards the end of the flight, small student or scientific satellites were deployed to increase the overall return on the investment that was put into each mission.

Not only were the Orbiter's systems checked for re-entry and landing, it was also necessary to pack away the equipment in the crew compartment, set up the MS seats, and don the 'pumpkin' launch and entry suits.

Several missions to space stations also deployed small satellites in the final days of the flight, and small mid-deck experiments and science investigations were conducted on board the Orbiter, independent of the equipment and experiments transferred to and collected from the station.

There were also a large number of DTOs and DSOs on each mission, undertaking research on a variety of subjects in order to maximize the results from a mission. On STS-66 in 1994 for example, DTO 680 (Recumbent Seat Systems Assembly) tested hardware and procedures for missions that were to recover crewmembers from space stations.[*]

Landing Day Minus One

While activities continued on the mid-deck, the flight deck crew would undertake a systems check, including the Orbiter control surfaces and, if necessary, firings of its RCS. During the final full day on-orbit, the crew normally received weather advice about primary, secondary and contingency landing sites across the globe. And there were press conferences with reporters and radio chats with families.

At least half of the final full day on-orbit was spent 'off duty' to recuperate after a busy and usually challenging flight. The flight plan was deliberately kept light, and as long as the crew was ahead of schedule and their craft was ready to carry out re-entry the astronauts tended to do what all space explorers enjoy most, which is look at Earth passing by. If the landing on the primary day was waved off, the crew would undo the re-entry preparations, called De-Orbit Backout, and prepare to spend another night in space by opening the payload bay to enable the radiators to dissipate heat. This would also be a great opportunity to enjoy the fun of weightlessness.

As there would be no Orbiter available for a potential rescue flight, the four-person STS-135 crew received Russian Sokol suit for use in the unlikely event of their having to return to Earth in Soyuz capsules.

---

[*]This is described in the companion volume *Linking the Space Shuttle and Space Station: Early Docking Technologies from Concept to Implementation.*

A unique image captured by the ISS-28 residents as Atlantis streaks through the atmosphere during re-entry to end the STS-135 mission and the Shuttle program.

The Shuttle carried a contingency of supplies to cope with delays to the landing but when the time finally came to return, the crew compartment was again prepared for re-entry, seats were installed, the crew donned their launch and entry suits and strapped in for the OMS burn that would start the long descent.

In a 2006 interview for this book, STS-98 MS Tom Jones reflected upon this final phase of an ISS flight. "It's a mixed bag of emotions. The first thing I remember when we undocked was I was so relieved that we were away from the station, that we could not mess up on the space station, we couldn't make any mistakes that would harm it. So I definitely was glad to be out of the 'exposure to error' which the station represented to me and my crew. We had done our work successfully and we were happy and satisfied. When we got away, my immediate emotion was I was so tired I simply wanted a break. On undocking day, after we cleared away from the station, we finally got to slow down and have a lei-surely afternoon of being able to get a look at Earth for the first time in a focused way. That was just a wonderful decompression feeling.

"And then, of course, we began to focus, even that first night we began to focus, on stowing the cabin and preparing for re-entry. I was eager to get home because I wanted to see my family again. But I was really aware, in contrast with my previous missions, we hadn't had any time to look at Earth while docking at the station. We had zero time for observation. Our windows were pointing in the wrong direction. So this final phase was our only chance during the mission, and so every day that we waved off we really took advantage, in terms of Earth, to shoot the rest of our film. Mark Polansky was a rooky on the flight and really wanted to learn something about geography from space. So we took the time to wake up in the middle of the night with him several times, and look at the Himalayas and Mount Everest for example. Those extra couple of days [by wave offs] was really a treat. We were joking about how we might remain on-orbit. My Commander Ken 'Taco' Cockrell and I together had this record of always waving off landings, so we were joking about we weren't ever going to get to land. Taco wrote a nice note to the flight director, talking about how things were getting desperate aboard and it was time to bring us home."[2]

COMING HOME

As the assembly and resupply of the ISS increased, so did the need to get each Orbiter back on the ground, ideally in Florida to ensure a prompt and smooth flow for its next flight and simplify the processing flow of the other Orbiters. As mentioned earlier, the sequence of assembling the station was similar to building a house of cards, where one slip would topple the stack. Any delay in the sequence would have knock-on influence not only in launching the Orbiters into space but also in returning them promptly to the ground for future processing.

Columbia was unable to participate in early missions to the ISS because its structure was so heavy that it couldn't carry a full payload. The availability of just three Orbiters for the ISS missions had always offered a very slim margin for delay. In fact it was one of the reasons for eliminating so many missions unrelated to the assembly of the station. When a retirement date for the Shuttle fleet was set following the loss of Columbia in 2003 the margin for delay was reduced even further.

Following the undocking and fly-around activities, the flight plan for an assembly mission was much lighter in order to allow the crew to stow their gear and prepare the vehicle to return to Earth.

The preferred landing option was the 15,000 ft (4,572 m) Shuttle Landing Facility runway at the Kennedy Space Center in Florida. An approach from the northwest was to Runway 15 and an approach from the southeast was to Runway 33. For landings at the primary back-up site at Edwards AFB in California there was a runway of similar length that was called Runway 4 or Runway 22 depending on the line of approach. By 2007, after fifty years, repairs were necessary to resurface this runway. In September 2009 a temporary runway was completed alongside the first. If an Orbiter approached from the northeast it could land at either the original Runway 22 now designated 22L (left) or at the temporary strip designated 22R (right). A southwestern approach would use either Runway 4L (the former temporary Runway 4) or 4R (the original).

Of the thirty-seven missions flown to the ISS between 1998 and 2011, all but eight successfully returned to the Cape, while those that were diverted went to Edwards. Of the twenty-nine missions that landed in Florida only seven (STS-110, -112, -113, -115, -120, -129, and -132) used Runway 33, the others landed on Runway 15. At Edwards six missions used the original Runway 22, one mission (STS-126) used the temporary landing strip and one (STS-128) used the new Runway 22L. During the ISS assembly period there were also thirteen night landings, twelve at the Cape and the other one at Edwards.

Table 10.1 End Of Mission Landing Summary ISS Assembly Missions 1998–2011

Mission	Orbiter	Date	Runway	Night Landing
KSC Landings				
STS-88	Endeavour	1998 Dec 15	15	Yes 22:53:29 EDT
STS-96	Discovery	1999 Jun 6	15	Yes 02:02:43 EST
STS-101	Atlantis	2000 May 29	15	Yes 02:20:19 EST
STS-106	Atlantis	2000 Sep 20	15	Yes 03:58:01 EDT
STS-97	Endeavour	2000 Dec 11	15	Yes 18:04:20 EST
STS-102	Discovery	2001 Mar 21	15	Yes 02:31:00 EST
STS-104	Atlantis	2001 Jul 24	15	Yes 22:38:55 EDT
STS-105	Discovery	2001 Aug 22	15	
STS-108	Endeavour	2001 Dec 17	15	
STS-110	Atlantis	2002 Apr 19	33	
STS-112	Atlantis	2002 Oct 18	33	
STS-113	Endeavour	2002 Dec 7	33	
STS-121	Discovery	2006 Jul 17	15	
STS-115	Atlantis	2006 Sep 21	33	Yes 06:21:00 EDT
STS-116	Discovery	2006 Dec 22	15	
STS-118	Endeavour	2007 Aug 21	15	
STS-120	Discovery	2007 Nov 7	33	
STS-122	Atlantis	2008 Feb 20	15	
STS-123	Endeavour	2008 Mar 28	15	Yes 20:39:08 EDT
STS-124	Discovery	2008 Jun 14	15	
STS-119	Discovery	2009 Mar 28	15	
STS-127	Endeavour	2009 Jul 31	15	
STS-129	Atlantis	2009 Nov 27	33	
STS-130	Endeavour	2010 Feb 21	15	Yes 22:20:31 EST
STS-131	Discovery	2010 Apr 20	15	
STS-132	Atlantis	2010 May 26	33	
STS-133	Discovery	2011 Mar 9	15	
STS-134	Endeavour	2011 Jun 1	15	Yes 02:34:51 EDT
STS-135	Atlantis	2011 Jul 21	15	Yes 05:57:00 EDT
Edwards (Diverted) Landings				
STS-92	Discovery	2000 Oct 24	22	
STS-98	Atlantis	2001 Feb 20	22	
STS-100	Endeavour	2001 May 1	22	
STS-111	Endeavour	2002 Jun 19	22	
STS-114	Discovery	2005 Aug 9	22	Yes 05:11:22 PDT
STS-117	Atlantis	2007 Jun 22	22	
STS-126	Endeavour	2008 Nov 30	Temporary	
STS-128	Discovery	2009 Sep 11	22L	

The Diverted Landings

The success of landing twenty-nine missions back at the Cape considerably helped the overall ground processing cycle. It also eliminated the costs of returning the Orbiter to the Cape and then flying the empty Shuttle Carrier Aircraft (SCA) back to California. Each of the eight diverted landings had a very different journey home.

The first diversion of the ISS program was STS-92 in October 2000. It spent two extra days on-orbit because of unfavorable weather, both at KSC and initially also at Edwards, before it finally landed at Edwards. There were two diversions in 2001. In February, STS-98 returned to Edwards after two opportunities at Florida were waved off on the same day owing to broken clouds and precipitation over the SLF. This was followed in May by STS-100, which also had two weather wave-offs at KSC. There was a single diversion in 2002. In June, STS-111 landed on the first of two available opportunities at Edwards, again following two days of weather concerns at the Cape. After four landing attempts were waved off in Florida, the STS-114 Return-To-Flight mission in July 2005 went to Edwards, in the process becoming the only ISS-related mission to land at night. STS-117 in June 2007 had two wave-offs at KSC because of stormy weather and landed at Edwards the next day. The problem for STS-126 at the end of November 2008 was that strong winds ruled out two options for landing at the Cape. When it was predicted that the situation would remain unsuitable into the first day of December, the decision was made to divert to Edwards and this was done on November 30. The final diverted station-related mission was STS-128 in September 2009, which had to return to Edwards when dynamic weather conditions resulted in four wave-offs at KSC.

- *STS-92 [6 days' delay]:* Finally landing at Edwards on October 24, 2000, after rain and high winds initially prevented the mission returning to either the Cape or Edwards, Discovery made the first landing in California since STS-76 over four years previously. After a week of work the Orbiter was ready to return to Florida but, on October 31, high winds and an awkward bolt (one of eight that were used to secure the aerodynamic tail cone assembly across the three main engines for the journey atop the 747) delayed the first leg of the journey. The mating of Discovery to the SCA (serial #905) was delayed until November 1. The combination left Edwards the next day, making a refueling stop at Altus AFB in Oklahoma before a forecast of bad weather necessitated an overnight stop at Whiteman AFB in Missouri. The flight was completed on November 3 and the Orbiter was towed to OPF Bay 1 the following day.
- *STS-98 [12 days' delay]:* A thick coverage of cloud over Florida on March 2, 2001, delayed the return of Atlantis after its diverted landing on February 20. The Orbiter traveled (on SCA #911) from Edwards to Oklahoma on March 1 but could not complete the journey until March 4.
- *STS-100 [8 days' delay]:* Endeavour landed in California on May 1, 2001. It was ready (on SCA # 905) for its return to the Cape on May 8, with a stop at Altus AFB, Oklahoma, and an overnighter in Little Rock, Arkansas, prior to arriving at KSC on May 9.
- *STS-111 [10 days' delay]:* STS-111 landed Edwards on June 19, 2002, but owing to the post-9/11 security policy there were no announcements about its transcontinental ferry flight. Carried atop SCA #911, Endeavour returned to the Cape on June 29 with little fanfare or details released concerning its trip, but presumably it had left the West Coast the day before.

Atlantis swoops to a landing on Runway 33 at the Shuttle Landing Facility, KSC, ending another successful mission to the ISS.

With the Orbiter secure, the STS-120 flight crew disembark the crew transfer van to begin the series of greetings, post-flight debriefs, and public affair appearances prior to entering the training flow once again.

With the crew disembarked, the smaller elements of the returned payload were removed from the crew compartment.

On the runway at the Cape, an Orbiter undergoes safing prior to being towed to the OPF to be de-processed, serviced and maintained prior to starting the processing flow all over again with its next payload.

All but nine of the forty-seven missions to Mir and the ISS landed at KSC. The others landed at the Dryden Flight Research Center at Edwards Air Force Base in California. In this image the Orbiter is being prepared for mating with the Shuttle Carrier Aircraft for the ferrying to Florida.

- *STS-114 [12 days' delay]:* With the scheduled landing of Discovery at the Cape ruled out by the weather, the August 9, 2005, diversion to California meant that the Return-To-Flight mission would also be the first to require a ferry flight post-Columbia. The SCA (#905) left Edwards ten days later. It made a refueling stop at Altus AFB, Oklahoma and then an overnight stop at Barksdale AFB, Louisiana. Bad weather precluded flying to KSC on August 20 but the trip was finished the next day. The Orbiter was promptly demated and then towed to the OPF to start its processing for the STS-121 mission.
- *STS-117 [11 days' delay]:* After its diversion to Edwards on June 22, 2007, Atlantis departed (on SCA #905) on July 1. It had stops at Amarillo AFB in Texas and Offutt AFB in Nebraska, spent the night of July 2 at Fort Campbell in Kentucky, and reached KSC on July 3, where it was transferred to the OPF.
- *STS-126 [12 days' delay]:* Weather issues diverted Endeavour to Edwards on November 30, 2008. The planned departure on December 7 was ruled out by heavy rain and sustained winds of 20 knots and gusts of up to 27 knots. It was actually December 10 before the journey began (on SCA #911) with a stop at Biggs Army Air Field of Fort Bliss at El Paso, Texas, and then at Fort Worth Naval Air Station in Texas. The following day further weather infringements delayed the next leg of the journey by 24 hr but the ferry departed Fort Worth on December 12. It flew over NASA JSC in Houston, Texas, then Shreveport in Louisiana, and landed at KSC later the same day.

- *STS-128 [10 days' delay]:* The final Shuttle to divert to Edwards landed on September 11, 2009. Nine days later, on September 20, Discovery began its journey back to KSC atop SCA #911, stopping for refueling first at the Rick Husband International Airport, Amarillo, Texas (named in honor of the late CDR of Columbia, lost on the STS-107 mission) and then at Carswell Field-Fort Worth Naval Air Station, Texas, that afternoon. The third leg of the day was to Barksdale AFB, Shreveport, Louisiana, flying about 20 min behind a NASA C-9 weather pathfinder plane. The SCA flew its last leg to Florida on September 21, thereby completing the final return of a diverted Orbiter.

Back at the Cape, the SCA is taken to the Mate-Demate Device to have the Orbiter offloaded and prepared for towing to the OPF.

Across the nine diversions of space station missions (one Mir, eight ISS) an average of 10.3 days was lost per processing cycle as a result of landing in California and then ferry flights back to the Cape. In total, this amounted to approximately $9 million and an estimated 93 lost work days at the Cape. But the team at KSC were able to make up this deficit and insert the returned Orbiters into the processing flow as soon as possible. This activity *after* a Shuttle mission is usually overlooked, but it was a key element of preparing an Orbiter for its next mission and was crucial to maintaining the reusability of the system. The original rationale for developing the Shuttle system was to create a reusable launching and landing system which could support the assembly and resupply of a space station. The delays from diversions to Edwards pushed that premise to – but fortunately not beyond – achievable limits.

Table 10.2   Shuttle Carrier Aircraft (SCA) Ferry Flights After ISS Missions 2000–2009

STS-92	
Oct 31, 2000	Bad weather and a faulty bolt delayed ferry flight. High winds delay installation of with Tail Cone and one of eight bolts to secure tail sized up
Nov 1	Mating began of Discovery to SCA
Nov 2	Departed Edwards – refueling stop at Altus AFB Oklahoma; bad weather results in an overnight stop Whiteman AFB, Missouri
Nov 3	Landed at KSC
Nov 4	Moved to OPF Bay 1
STS-98	
Feb 20, 2001	STS-98 lands; suffers rain damage on runway unable to be moved to a hanger, 500 tiles were rain soaked delaying STS-104
Mar 1	departed Edwards for overnight stop at Altus AFB Oklahoma
Mar 2	bad weather grounds Shuttle Atlantis at Altus for 4 days
Mar 5	Atlantis finally returns SLF KSC
STS-100	
May 1	STS-100 lands at Edwards, hopes to return to KSC by May 5 but delayed to May 10
May 8	Endeavour leaves Edwards AFB and completes overnight stop at Dyess AFB Abilene, Texas
May 9	Completes a refueling stop at Altus AFB, Oklahoma, and an overnight stop at Little Rock AFB, Arkansas;
May 10	Endeavour arrive back at SLF, KSC
STS-111	
Jun 19, 2002	STS-111 lands at Edwards. Under post 9/11 policy there would be no announcement of Shuttle's return from California.
Jun 29	Endeavour 'sneaked' into KSC, little media coverage
STS-114	
Aug 9, 2005	STS-114 lands at Edwards
Aug 19, 2005	Departs Edwards for an refueling stop at Altus AFB, Oklahoma then on to Barksdale AFB Louisiana for overnight stop
Aug 20	Bad weather cancels planned return to KSC for 24 hours
Aug 21	Arrives SLF and taken to Mate-Demate Device for subsequent transfer to OPF for STS-121 processing
STS-117	
Jun 22, 2007	STS-117 lands at Edwards

(continued)

Table 10.2 (continued)

STS-92	
Jul 1	SCA carrying Atlantis departs Edwards AFB and completed a refueling stop in Amarillo then on to Offutt AFB, Nebraska for an overnight stay
Jul 2	SCA carries Atlantis to Fort Campbell, Kentucky, but bad weather forces a second overnight stop
Jul 3	SCA departs Fort Campbell, Kentucky and arrives at SLF KSC, and later in the day moved to OPF
STS-126	
Nov 30, 2008	STS-126 lands Edwards
Dec 5	Problems mating tail cone delays departure by at least a day
Dec 7	Rain delays ferry flight at least 24 hrs, with sustained wind 20 knots gusts up to 27 knots
Dec 10	Weather delays threaten but ferry began from EAFB to Biggs Army Air Field, Fort Bliss, El Paso, Texas; then on to Fort Worth Naval Air Station, Texas for overnight stay
Dec 11	Ferry from Ft Worth includes a fly over NASA JSC then on to Barksdale AFB, Shreveport Louisiana
Dec 12	SCA departs Barksdale for final leg to KSC
STS-128	
Sep 11, 2009	STS-128 lands at Edwards
Sep 20	Departs EAFB, completes a refueling stop in Amarillo, Texas, followed by a short stop at Fort Worth, then on to Barksdale for an overnight stop
Sep 21	Avoiding weather fronts SCA lands back at KSC

THE 'WHAT IF' FACTOR

Each year, the six days between January 27 to February 1 can be difficult for those who have an association with, or an interest in, the American space program. This week of remembrance rekindles memories of the lost crews of Apollo 1 on January 27, 1967, in a fire in their capsule during what was expected to be a routine ground test; Challenger on January 28, 1986, during launch; and Columbia on February 1, 2003, in returning to Earth after a highly successful research mission. These three accidents left scars in the fabric of NASA, and have prompted thoughts about what could have been done to save the astronauts, or even to have prevented the tragedies. After lessons were learned and changes implemented, there was a renewed effort to face the risks of launching human into space once.

A Phoenix Rising From The Ashes

With the loss of Columbia, the future of the Shuttle program was called into question, casting serious doubts about its ability to complete assembling the ISS. However, the recommendations for corrective actions implied that with safeguards the Shuttle could resume flights. Nevertheless, after it had completed the ISS and made a final servicing call to the Hubble Space Telescope the fleet was to be retired.

Undocking An Unmanned Orbiter

Between July 1995 and July 2011, nine Shuttle missions docked at Mir and a further thirty-seven docked at the ISS; a remarkable forty-six dockings and undockings. But what would have happened if an Orbiter was damaged during launch, made it to the ISS, and was judged incapable of attempting to return safely to Earth? Could it have been undocked from the station without its crew and disposed of in a manner that did not pose undue risk to life and property on Earth? What would happen to the Shuttle crew who had intended to make a brief visit to the station and now found themselves stranded there with no immediate means of returning to Earth?

  This scenario never occurred, but like so many contingencies in space flight it was studied and modeled just in case.

ISS As A Safe Haven, From The Right Orbit

During the investigation into the root cause of the Columbia accident, the Shuttle was grounded and the ISS was placed in 'caretaker' status. A succession of crews with one Russian cosmonaut and one American astronaut performed basic housekeeping duties and limited science activities, essentially keeping the station going for the next three to four years until the Shuttle could resume the assembly process.

  Part of the post-Columbia Return-To-Flight process considered the means of adding more safety to each remaining Shuttle flight. If something went wrong on ascent there were options available to the crew with contingency landing sites or abort scenarios to attempt to safely return to Earth. But once in space aboard a damaged spacecraft, with little possibility of returning safely, the choices were limited. Bearing in mind the need to finish the assembly of the ISS, and with a resident crew already on board supported by Soyuz ferries, the option to use the station as a "safe haven" was logical. If a future Orbiter were to be sufficiently damaged during its ascent to make it unable to return to Earth but still able to reach the ISS, the station could host that crew until they could be rescued either by launching another Shuttle with lots of spare seats or by sending up a series of Soyuz ships to retrieve them.

This was assessed for baseline missions orbiting at the 51.6° inclination of the ISS, but it could not have saved the Columbia crew orbiting at an angle of 39° because the spacecraft had insufficient propellant to make such a large plane change. Afterwards, studies were made of whether it would have been possible to mount a rescue flight to recover the crew of Columbia, had the scale of the damage caused at launch been fully appreciated at the time, but the evidence suggested that there was very little that could have been done in this case. Even given adequate time to prepare a rescue flight, there were too many variables, making the proposed hypothetical rescue highly unlikely.[3]

It was therefore decided that all future Shuttle flights would be targeted for the ISS and that in each case the spacecraft for the next planned mission would be readied as a potential rescue craft for the current one. Fortunately it was never necessary to put this scheme into practice. It was also decided, after some debate, that one final solo Shuttle mission should be flown in 2008 (which later slipped into 2009) to service the Hubble Space Telescope because if the operational life of that facility could be extended to at least 2014, and perhaps to 2020, that mission was believed worth the risk. On each of the previous five Shuttle missions involving the HST (the first to deploy it and four to service it) there certainly *was* a risk. Orbiting at 28.5° to the equator and at a height of 380 miles (600 km), Hubble was operating at the very limit of an Orbiter's propellant reserves. In any case, even with full tanks, it would not have been possible to make the plane change to reach the ISS in the event of there being a problem that prevented the spacecraft from returning to Earth. As a safeguard, the ISS assembly mission that was scheduled to follow after the HST mission was first to be prepared to execute a Launch-on-Need rescue flight designated STS-400. Fortunately, STS-125, the last solo flight of a Shuttle, successfully executed the final HST service mission (SM-4) in May 2009.[4]

Contingency Shuttle Crew Support

With all the other remaining Shuttle flights between 2005 and 2011 destined to dock at the ISS, the station could be used as a safe haven should the need arise. NASA set up a working group to investigate this Contingency Shuttle Crew Support (CSCS) concept. Consisting of representatives of both the Shuttle and ISS teams, the investigation soon worked out the basics of the idea.

The first consideration was that any damage to an Orbiter incurred during launch or enroute to the ISS could not be serious enough to prevent a safe rendezvous, docking, and crew transfer over to the station. Then all Orbiter consumables would be powered down in order to sustain the resident crew and up to seven unexpected guests, certainly for several weeks and possibly several months. The challenge was to prepare a rescue mission, designated STS-300 for planning purposes, which would be launched with a minimum four-person flight crew to rendezvous with the ISS and retrieve the stranded astronauts for return to Earth.

The option of launching a series of Soyuz craft to return the stranded astronauts was also reviewed, but because the Russians built these on an as-required basis there might not be a supply available at short notice. In addition, it would have been necessary for each Shuttle to carry a Sokol suit and a form-fitting couch for each of its crewmembers to enable them to return to Earth in Soyuz capsules. This was done only in the case of the four-person crew of STS-135 because, it being the final Shuttle mission, there was no other option.

During the post-Columbia Return-To-Flight effort, a detailed study was made of the ability of the ISS to support a stranded Shuttle crew. Its baseline capability was set to sustain a resident crew of three, then six, and on occasion up to nine for short periods. These residents relied on there being three-person Soyuz ships available for transport and escape in an emergency. The situation would change dramatically if an additional seven astronauts had suddenly to remain on board the station awaiting rescue.

Safe Haven

The CSCS idea represented a 'last resort' for the unlikely event that all safety measures had failed or were unavailable, with the result that there was a Shuttle docked at the ISS which was no longer capable of returning safely to Earth. If the station was going to be used as a safe haven, then preparations on the ground would have to be sufficiently far advanced for a rescue Shuttle to be launched promptly. This contingency would have to be factored into all subsequent processing schedules.

In the CSCS scenario, the crew of a disabled Orbiter would have lived on board the station until the rescue Shuttle could be launched. During this period all of the stricken spacecraft's consumables would be transferred to the station and then the vessel would be remotely commanded to undock from the station. Its ultimate fate would be to burn up in the atmosphere.

In addition to instigating a program to reduce the possibility of debris shed by an ET causing impact damage to an ascending Orbiter, NASA developed a suite of inspection and repair techniques designed to detect critical impacts and provide a range of options to the astronauts to attempt a repair. The CSCS was one of the 'raise the bar' initiatives added to the fifteen required recommendations of the Columbia Accident Investigation Board prior to a Return-To-Flight mission. Starting with STS-114, provision was made to have another vehicle available as a rescue option. Data from STS-114 and STS-121, both of which were classified as Return-To-Flight missions, was analyzed to determine whether this practice should continue.

The next Shuttle undergoing the normal launch processing flow at KSC became the potential rescue vehicle, with care being taken in the processing to ensure that it could be launched and rendezvous with the ISS before the consumables expired. The number of days that the station could support a stranded crew depended upon the consumables already on board the station and those transferred from the stricken Orbiter, as well as upon the number of residents and their reluctant visitors. These consumables included food, water, oxygen, and spares. The lists of supplies were constantly updated between missions up to the day of the planned Shuttle launch. This defined the CSCS capability for that particular mission.

All-Out Effort

Whilst the ISS was certified to support a crew of up to six permanently or nine during Soyuz handovers (three arriving, three departing and three only part way through their period of residency) the CSCS was a non-certified option which would be used only to maximize the chances of successfully rescuing the stranded crew.

Under 'safe haven' conditions the Orbiter would be powered down to the minimum level necessary to run the life support and hygiene systems for the maximum possible docked duration, while protecting the critical equipment that would be required for the unmanned docking and disposal of the vehicle. This would minimize the strain placed on the limited station resources, and maintain comfortable conditions for the residents. Even though all of the Orbiter's food, water, lithium hydroxide and other consumables would be transferred to the ISS, the duration that the station could sustain the enlarged population would still depend on critical ISS equipment. There might not be sufficient redundancy, waste management, supplies and food available to sustain the unexpected visitors until a rescue was possible.

The CSCS had to assume it would be possible to launch the rescue mission without necessarily resolving beforehand the problem that disabled its predecessor. That is, the investigation into what triggered the rescue situation would have to wait until after the rescue flight had attempted to rescue the stranded crew.

Declaring A CSCS Situation

The CSCS process was to be initiated only after the decision was taken that the Orbiter was incapable of returning to Earth. That decision would only be made once all efforts to repair the Orbiter for a safe de-orbit and landing had failed. The initial actions would have been to power down the spacecraft and transfer its consumables across to the ISS. During the docked period the management of consumables would have to be optimized in order to maximize the CSCS capabilities. Food rationing for the visitors would have been insti- gated as necessary and, if possible, the opportunity would have been taken to remove trash and unwanted items from the ISS to the Orbiter prior to its departure. The exact stowage locations of such items would not have been quite as critical. As long as the Orbiter could be undocked safely, the location of its center of mass would not be as crucial as on a nor- mal mission, since the vehicle would be destined to break up in the atmosphere. The unmanned Orbiter would have to be undocked before it could violate cryogenic redline parameters, and once it was clear it would execute the disposal burn.

Declaring And Un-Flyable Orbiter

The CSCS and Launch-On-Need (LON) declaration could have been made in response to a number of scenarios:

- A systems failure which would prevent a safe and successful re-entry and landing.
- A systems failure indicating in a low probability of a successful re-entry, landing, and crew survival.

- A critical failure which did not provide time to assess and evaluate Orbiter system failures safely and efficiently, nor the risks of re-entry and landing from those failures.
- Suspected damage to the Orbiter thermal protection systems, until such time that the TPS damage (and repair if required) had been cleared to support re-entry. Meanwhile, on the ground, the STS-300 rescue mission would be put into motion and processing initiated.

On-orbit, the initiation of the CSCS and LON mission depended upon a number of factors. The first was that the damaged Orbiter still had the capability to achieve a safe docking that would not put the ISS at risk or prevent the resident crew from undocking their Soyuz vehicles. Similarly, once the Shuttle was docked and awaiting LON rescue it was essential that this arrangement must not jeopardize the ability of the residents to return independently. Finally, if the failure were to occur after undocking from the ISS on a flight which up to that point had gone to plan, then the CSCS-LON option wasn't protected for a re-docking because to have reserved (or in NASA parlance 'redlined') the propellant that would be required for a re-rendezvous would have interfered with nominal mission planning. However, if there *was* sufficient propellant on board and a re-rendezvous *could* be attempted, then the CSCS would become available if the pre-docking criteria already specified could be met. Redlining re-docking attempts would impact nominal mission planning, but since the CSCS-LON situation would not have allowed a safe landing, propellant meant for re-entry could have been spent trying to return to the ISS because it would not be required to attempt to return to Earth. Using the propellant in this way would have had an effect on the options for a disposal burn, but it would offer a vital second chance for the stricken crew to rendezvous and dock with the ISS.[5]

In response to the situation, the Mission Management Team [MMT] retained the final decision as to whether the CSCS or LON conditions had been invoked. If any failure had required an immediate response the flight controllers on duty would have had to evaluate those failures, the capabilities of the remaining systems, the current mission activities and those still to be performed, plus the available landing options. They would then have had to weigh up the relevant risks and select the most suitable course of action, bearing in mind that the safety of the crew was paramount.

Although the CSCS and LON options were theoretical opportunities to rescue the crew, both would have required analysis so that any actions taken would improve the situation because launching a second Shuttle crew or risking the safety of the station residents had the potential to make an already critical situation worse. Consideration would have been directed to precisely when the Orbiter failure occurred in a nominal mission and what the failure was. If there was a possibility that the issue might strike another Orbiter, the rescue mission would probably not be launched at all. All of this data would affect the availability of consumables and resources needed to sustain the safe haven on the ISS, particularly if the failure occurred after departing from the ISS and would necessitate a second rendezvous and re-docking, if indeed that was feasible given the propellant usage. In order to be deemed safe and prudent to continue to full-term CSCS duration it was also imperative to determine the minimum level of Orbiter consumables that would have to be maintained. Any rescue mission would have had a maximum of ten days and a minimum of just one day of launch preparations for LON, but the available CSCS duration at this point might not have been sufficient to support all of the preparations required for the rescue flight. As the NASA document observed, "Again a tough call, but the number one goal is to use all available options to provide the opportunity to get the Shuttle crew safely home, versus doing nothing."[6]

In these scenarios the requirement to proceed with an unmanned undocking was not mandatory. A variety of options existed for the unmanned docking of an Orbiter from the ISS, but if it was unable to be undocked, even unmanned, then once the cryogenic consumables were depleted on the stricken vessel the hatches of the Orbiter Docking System (ODS), the Pressurized Mating Adapter (PMA), and inner end of the Destiny laboratory would have been closed. As NASA said, "The bottom line is, all that can be done will be done to improve the possibilities and risks to provide the opportunity to get the stranded Shuttle crew safely home."

How To Undock An Unmanned Orbiter

The next challenge for the study team planning the options for a stranded crew was how to safely undock and dispose of an unmanned Orbiter.

Because the track of the discarded Orbiter would be inclined at 51° to the equator, it would pass over a vast expanse of inhabited terrain as well as a lot of ocean. Hence the plan required minimizing the risk of injury or damage back on Earth from any pieces of debris that might pass through the atmosphere. The plan to achieve this was both daring and challenging, as reported in a 2006 account of the study.[7]

Mounting The Insurmountable Problems

The idea to undock an unmanned Orbiter from the ISS was fine on paper but turning it into a practical proposition was a major challenge. The study team faced hurdles which initially appeared unsurmountable. It was concluded there was a theoretical possibility of attaining success, but like the rescue missions planned in support of the final Shuttle flights, it would have needed a series of circumstances and perhaps an element of good fortune to ensure it played out as envisaged.

When the Shuttle was designed, it was assumed that all of the key tasks required to fly the vehicle would be carried out under the control of an on-board crew. In the case of discarding an unmanned Orbiter from the ISS the first hurdle was to figure out how to operate the vehicle safely and successfully. The reasoning capability of a crew, and their skills and dexterity would not be available, therefore a significant fraction of the Orbiter's capabilities would simply not be available. There was never a plan to install specific automated software, or to modify the hardware to permit the spacecraft to be controlled from the ground as was the case for the 1988 maiden mission of the Soviet space shuttle Buran. For the American Orbiter such an upgrade would have been very expensive and time consuming, and by being intended for a contingency option, there was every likelihood that it would never be needed operationally. As far as this study was concerned, if the flight controllers could not fully use a system they would either have to figure out how to work around the system or do without it.

The mission-defining situation facing the team was finding a means to overcome the very parameters that defined a safe haven mission. Having purchased time and possibly averted another tragedy in the American space program, there would be a limited time available for a team of several hundred at KSC to prepare the rescue flight. On the ISS the residents and their reluctant visitors would have to work together to ensure that the Orbiter's power and life support, which could only last about sixteen nominal mission days, were safely placed into a state of hibernation in order to save the power required for the unmanned de-orbit, while still being able to bring critical systems back on line. An Orbiter couldn't be fully powered down and later restored to full operational status on-orbit, so it was necessary to determine how to keep the stricken vessel at minimum power whilst docked at the station. The decision was made to insert triple redundancy into the guidance and navigation, control, propulsion, thermal and communications to support the de-orbit maneuver. A study sub-group was established to look into which systems must keep running and which could be turned off, at what point, and for how long.

The next big challenge was to work out how to command the Orbiter from Mission Control instead of from the flight deck. Fortunately, capability had been built into the systems to enable controllers on Earth to send a variety of computer commands up to the Orbiter. This was included as a contingency for the rare event of the loss of voice communications with the Orbiter, for time-critical changes that had to be made while the crew slept, or if they had been incapacitated by some unforeseen event. In reality such commands were few and far between and this option was rarely used. The other challenge was to command the Orbiter remotely through a series of very critical tasks which normally required a significant amount of coordination prior to being executed on a precise timeline. These tasks included the unmanned undocking from the station, initiating any orbital maneuvers and translation burns, capturing essential sensor data, and ensuring a stable re-entry attitude.

In a vessel of such complexity, it seems strange to think that there was no built-in back-up to permit the Orbiter to undock from a space station without a crewmember being on board to press the Undock button on the aft flight deck. Usually, a nominal crew transfer between the Shuttle and the ISS took about an hour, so in a safe haven situation it would not be possible for someone to press the button and then race back into the station, sealing hatches along the way.

Next the Orbiter required to separate from the station at an increasing, yet safe and controlled rate. A nominal undocking was sequenced procedurally, so if necessary the PLT would initiate a manual +Z burn of the reaction control jets to achieve the desired separation rate. But this was usually not necessary, because the push from the springs and the thruster firings produced a separation rate greater than the specified minimum. With a crew on board, there was also the option to fire the jets manually to establish a safe, controlled, and smooth separation. The task for the planners was to ensure that the departure of the unmanned Orbiter shortly after separation remained in the corridor for safely clearing the station.

All automated controls on the Orbiter were disabled during a nominal undocking, to prevent accidental attitude control jet firing and potential contact with the station during separation. This required pressing the Free button on the panel for the Digital Auto Pilot (DAP). It put the vessel in free-drift, and in this case facilitated control by means of the separation springs and the already executed jet firings in conjunction with the influence of the gravity gradient and other forces. An unmanned Orbiter would also require to use this free-drift mode in order to prevent unwanted jet firings and plume impingement on the station's surfaces. The complication was that an unmanned separation would require the automated controls to remain active. Unfortunately, the control modes could not be commanded from the ground because a button on the DAP had to be physically pressed to activate or disengage. An unmanned Orbiter could therefore be preset to full attitude control or to none at all; but once set it couldn't be altered remotely. As the procedures required the Orbiter to adopt a range of attitudes, the automatic control mode would be essential. The team had to establish how few RCS firings would be required in order to suggest free-drift, whilst still having the automatic attitude control system engaged and not being able to toggle between the two modes.

The final conundrum was the mission rule that said the rate of separation must be sufficient to ensure that within one orbital period the Orbiter could not pose a risk of collision. To achieve this safely, the separation rate could only be commanded by an astronaut on the flight deck using the Translational Hand Controller. Mission Control retained the command option for a burn of the Orbital Maneuvering System from the ground, but a +X burn by the RCS would be needed prior to initiating an OMS firing when at a safe distance from the station. This was a head-scratcher for the study team because the hypothetical unmanned mode for an Orbiter lacked any suitable means of adding a +X translation to the operation.

The Process Defined

In their study report the authors explained the process that they had devised to dispose of an unmanned Orbiter during a hypothetical safe haven situation. In all likelihood, in preparation for the disposal sequence the Orbiter would have been stripped of supplies and any equipment that could be cannibalized for use by the ISS, and everything in the payload bay that could be transferred across to the station would have been moved. In the days prior to the undocking, the Shuttle crew would have carried out a number of in-flight maintenance procedures to configure the Orbiter for unmanned operations.

A team of engineers and flight controllers in the Mechanical Systems Division at JSC had figured out a way to work around the APAS Undock button issue by using the three main electrical power circuits. Disrupting power from the fuel cells to the second of three power buses, Main Bus B, would have the beneficial effect of isolating power to the APAS, rendezvous radar and forward RCS jets. To achieve this, the crew would have to enact a lengthy procedure. "To fool the APAS into thinking it's Undock button had been pressed when, in fact, it had not," the astronauts had to isolate power to Main Bus B and then physically open the specific panel on the aft flight deck in order to add jumpers that would short circuit the button. When the power was restored to the system it would interpret this as meaning that the button had been pushed. This would allow the APAS hooks to open. As the controllers could reroute power from another bus to Main B even when it was switched off, the crew could remain safely sealed inside the ISS as this was initiated.

The downside of this deception was that disrupting power to Bus B also affected the Trajectory Control Sensor (TCS), the laser that provided ranging data during docking.

When undocking the TCS also provided valuable information on the vehicle's position and motion in close proximity to the station. In the unmanned mode this data would be particularly useful. The planners of the contingency action were eager to have this data in order to monitor the separation velocity and resulting trajectory. The solution was to ensure it was switched on by the crew prior to egress by rewiring the TCS hardware to another bus, rather than Main Bus B.

Undocking Day

The preparation sequence for the Orbiter on the day of its undocking would begin by restoring the vessel from its low power, extended duration 'hibernation.' All the RCS jets would be re-enabled, particularly the upward-firing jets which are inhibited after docking. Then the rendezvous navigation function would be reactivated. This would allow the rendezvous radar to track the ISS relative to itself and downlink the data to Earth so that the flight controllers could calculate the de-orbit burn accurately. At this point the mass of the Orbiter would be far less than originally planned at undocking, there being no payload, crew, or other hardware on board.[*] Its center of mass would also differ from the norm. These differences would probably require a revision to the normal logic for using the attitude control jets and any other jet firing.

The unmanned undocking would clearly be a complicated process, but the plan was to keep things as simple as possible and thereby minimize the opportunity for things to go awry. It was therefore decided to use the nominal initial separation technique, as if a crew were aboard. Normally, in the final set ups, the crew would configure the DAP to use only the 'tail' jets. When the APAS springs eased the unmanned Orbiter away from the station, a negative pitch movement would be imparted due to the vessel's center of mass being some 33 ft (10 m) away from the push off point at the docking system. The up-firing RCS tail jets would be fired to increase the separation rate, without having to initiate any +Z translation. As the Orbiter drifted away, it was important to ensure that plumes from the jets didn't impinge the solar arrays structures of the station, therefore the autopilot would have to execute the firings in measured increments. The rate limits would have to be set to their maximum values to ensure that the autopilot remained in Auto mode for the duration of the maneuvers. This allowed the maximum duration of free drift and required fewer RCS firings, resulting in the 'simulated free-drift' mode. As a safeguard, if the Orbiter encroached on a specific attitude or rate limit during the separation, Mission Control could uplink changes to the dead bands (inhibited thruster firings) around the vehicle's current attitude to prevent unnecessary RCS firing and to allow the drift away from the station to continue.

One of the early decisions in planning the unmanned undocking was to choose the attitude of the Shuttle at the moment of undocking. Earlier experience of undockings allowed the team to determine the best configuration for the unmanned docking. The original idea was to place the Shuttle-ISS combination in the +R Bar attitude with the Orbiter

---

[*]To lessen the mass, unlike Progress and similar returning supply ships which are disposed of by a fiery re-entry. Ideally no trash or unwanted equipment would be placed in the crew compartment of the Orbiter, depending on the real-time situation.

between the station and Earth because this would enable the departing vehicle to make the best use of differential gravitational acceleration to move away from the station without requiring additional jet firing. The unoccupied Orbiter would literally 'drop' away and, by decreasing its altitude, would be drawn in front of the station by orbital dynamics. An alternative option was to use the –R Bar attitude with the station between the Orbiter and Earth, to make the craft ascend and therefore trail the station on-orbit.

Finally, the team looked at the consequences of both modes later in the timeline. It was found that a large de-orbit burn would be needed during the separation trajectory. In the +R Bar separation mode this would have seen the Orbiter rise back towards the trajectory of the ISS and into a potential collision. In the –R Bar separation mode, the burn would move the Orbiter higher and therefore away from the station. The –R Bar mode would also allow the S-band antenna of the Orbiter to make direct contact with the TDRSS relay satellites without the line of sight being temporarily blocked by the station, which would occur in the +R Bar mode.

Crew Egress

With everything ready for separation, one of the last actions by the Shuttle crew on the flight deck would be to use the Orbiter's attitude control jets to maneuver the Shuttle-ISS combination to the correct separation attitude. After a final look around the flight deck, the crew would then float down to the mid-deck and out through the connecting hatches and tunnels into the station, closing the hatches behind them. At that time they would help to execute a long series of leak checks to ensure the ISS was airtight. Then the Orbiter would be prepared for separation from the station, this historic event being documented by cameras in every available window of the station.

One point that wasn't raised in the study paper, was the risk of a systems failure on the Orbiter resulting in collision with the station or possibly punching a hole in one of the pressurized compartments – as happened during the 1997 Progress M-34 collision with Mir. All of the internal hatches would likely have been sealed except for routes to the available Soyuz escape craft, but of course those were meant for the resident crew; in an emergency evacuation there would have been no escape for the stranded Shuttle astronauts.

While awaiting the arrival of the rescue Shuttle, it is possible that another Soyuz could have been launched quickly in order to return the first members of the Shuttle crew. However, to recover a Shuttle of seven crewmembers would require four such vehicles because each would be flown to the station by a single cosmonaut with two seats vacant.

Flying a Soyuz solo was not a problem. It was done by Vladimir M. Komarov on Soyuz 1 in 1967, Georgi T. Beregovoi on Soyuz 3 in 1968, Vladimir A. Shatalov on Soyuz 4 and Boris V. Volynov on Soyuz 5 in 1969 (the latter two having docked and exchanged two crewmembers during a spacewalk). In-space rescue was an option the Soviets had instigated in September 1985 by selecting a group of pilots (Vladimir A. Lyakhov, Anatoly Berezovoi, and Yuri V. Malyshev) to train to fly the Soyuz TM in that role to Mir. And in modern times a number of ESA astronauts qualified as Soyuz return Commanders for ISS missions, to take over if the Russian Commander became incapacitated by illness.[*] If enough Soyuz vehicles had been available, the recovery of the stranded Shuttle crew could have been achieved over a period of about one year. In addition to carrying seat liners and Sokol suits as cargo on board Atlantis, the STS-135 crew had the Penguin load-bearing suits which they would need in order to stay on the ISS for a considerable time awaiting rescue.

It would have been interesting to learn how, and indeed whether, the entire Shuttle crew could have been evacuated in the event of another emergency arising on the ISS before the arrival of either the rescue Shuttle or additional Soyuz craft.[**]

Undocking And Separation

The undocking of the unmanned Orbiter would have been initiated by Mission Control commanding Main Bus B to provide power to the APAS, and then the releasing of the hooks. At the same time, the latest data on the state vectors of the Orbiter and the ISS would be uplinked. At push off, the Orbiter's aft upward-firing jets would balance the negative pitch rate imparted by the springs of the docking system. This would prevent the tail of the Orbiter from contacting the station. As the vehicle separated, commands would be uplinked to constrain its attitude and trajectory and thereby minimize engine firings.

Further data would be uplinked from the ground as the separation increased and the Orbiter initially climbed and then, by being higher, trailed behind the ISS. Meanwhile, the monitoring flight controllers would have a good indication of the rate of separation and the direction in which the unoccupied Orbiter was traveling.

Eventually the TCS would lose its laser lock on the station but this would not be so important as the Orbiter pitched up to clear the station. As the range increased, the on-board K_u-band antenna would be used to determine the precise location of the station, as well as the range, range-rate, and relative angles, all of which would be downlinked to the ground via the rendezvous navigation filter. If a small burn was necessary upon reaching the safe minimum range from the ISS, the RCS would push the Orbiter even higher and consequently further in trail of the station. Normally, any RCS burn would be initiated through the hand controller, but in 2001 extra software was added into the crew kit for the Orbiter to permit a re-boost burn to be initiated by a timer delay. This was used to avoid

---

[*] In November 1985 cosmonaut Vladimir Vasyutin became so ill aboard Salyut 7 that the residency he was commanding was curtailed. FE Viktor Savinykh took command of the Soyuz T-14 ferry for their re-entry and landing.

[**] Incredible as it may seem, prior to the loss of Challenger in 1986 it was intended that astronauts would serve aboard Space Station Freedom between Shuttle visits, without any means of escape in an emergency. Studies had been completed on effective rescue systems but no commitment had been made. Following a reduction of Shuttle flights to transport crews to Freedom and the budget to support them, together with the loss of Challenger, it was realized that had Freedom been in space with a crew on board they would have been stranded while the Shuttle fleet was grounded. In Challenger's case the STS-26 Return-To-Flight mission was launched thirty-two months after the accident. These concerns resulted in a Crew Rescue Vehicle proposal for the ISS, but this was later canceled in favor of using Russian Soyuz craft instead.

adding to the crew's workload during busy periods in the lengthy re-boost maneuvers. Initially planned for manned separations, it would also have worked in the unmanned mode although the crew would have had to specify the start time and end time prior to evacuation. The controllers would have carried out all steps to set up and maneuver the vehicle for the burn, which would be enabled by command from the ground.

De-Orbit Maneuver

On nominal missions a minimum safe range was required before making a dual OMS burn in the vicinity of the ISS to prevent plume impingement from contaminating the station and docked Progress and Soyuz vehicles. In the case of the unmanned Orbiter, ground controllers would have monitored its position during the separation and coast phase and, using ground tracking and the relative navigation supplied by the Orbiter, they would have estimated its velocity. This would have allowed the timing, attitude and velocity for the de-orbit burn to be refined using the most up to date information and then uplinked to the Orbiter.

The re-entry data for nominal missions was based upon where the landing would occur. In this case there would be no landing; the mission would end by the vehicle burning up during re-entry, akin to the destruction of Mir when it re-entered in 2001. Options for the 'debris footprint' were studied by the Flight Design and Dynamics Branch at JSC and, as with Mir, the route selected was over the largest uninhabited portion of the South Pacific. When the Orbiter was a safe distance from the ISS, the flight controllers would have commanded it to adopt the attitude required for the de-orbit burn, then the command would have been sent to fire the two OMS engines to achieve that objective. The burn would have begun over western Asia and led to the destruction of the vehicle high above the target about an hour later. It was calculated that the debris footprint would stretch 756×35 nautical miles ($1,400 \times 65$ km) in the direction of the trajectory.

Final Considerations

The safe-haven team also considered the potential for an unmanned Orbiter making an automated landing at a variety of sites. The island of Diego Garcia in the Indian Ocean was deemed to be a safe location for the Orbiter to break up and ditch if re-entry was deemed feasible but a landing was not possible. However, the cost and complications involved in adding an automatic landing capability to the Orbiter fleet specifically for this contingency were deemed prohibitive.

A nominal de-orbit burn was also considered by the team, since it would reduce the number of commands for moving the Orbiter and thus decrease the risk of error, with fewer commands required to enact re-entry. This option would hopefully lead to a fast and efficient break up, but it was discovered that if the de-orbit burn was long enough, resulting in a very steep re-entry angle, the attitude of the Orbiter would be irrelevant since the vehicle would tumble and break up irrespective of its attitude upon initiating re-entry. It was therefore decided that the optimal re-entry attitude to achieve efficient break up would be to have the payload bay doors open and pointed in the direction of travel, the tail facing towards Earth and the nose towards space. This would incur the maximum atmospheric drag and rate of destruction at an altitude of 23 nautical miles (41.4 km) and would also maintain the TDRSS relay for tracking purposes through to the onset of break up.

Simulating The Simulation

Several dedicated and highly skilled teams worked on this issue at JSC. The design phase was defined by the Flight Design and Dynamic Proximity Operations group, which specified the relatively safe separation mode from the station. The Proximity Operations Group specified the models for undocking and separation by utilizing a desktop computer that had six degrees of freedom, called the Spacecraft Proximity Operations Real-Time Simulator (SPORTS). This could simulate a wide variety of cases using original data simply by varying the initial conditions and factors.

The whole event was practiced by a team of flight controllers. Three simulations were completed, with a team of nineteen flight controllers manning the consoles in Mission Control with the assistance of their back room support teams. A number of astronauts who were working in the high-fidelity Shuttle Mission Simulator (SMS) at that time also took part, as did dozens of training and engineering support personnel across the JSC complex. Interestingly, although these were contingency simulations, important lessons were learned that proved useful for the operational program. Most notable was the lack of TCS power when the Main Bus B was unavailable. This came as a surprise to the controllers, who realized they had outdated engineering drawings. They received more up to date digital specifications. There were several instances of flight hardware or software not matching what was planned or required on-orbit. And running the procedures through a real flight control team unearthed some operational difficulties that permitted the flight controllers and engineers to refine the procedures for conducting certain operations.

SUMMARY

The study of how to undock and dispose of an unmanned Orbiter was undertaken in response to the loss of Columbia but it was not one of the official recommendations made by the investigation into that loss. However, as part of plans to use the ISS as a safe haven, the hypothetical situation of a damaged Orbiter managing to dock at the station without any prospect of being able to return home was studied by the Mission Operations Directorate at JSC. Innovative problem-solving in the tradition of NASA teamwork overcame serious technical issues to yield an ingenious solution. The study established that in theory and in simulation such an operation could be attempted, but thankfully it was never necessary to put it into practice.

With the Shuttle retired, Discovery, Endeavour and Atlantis, which visited Mir and built the International Space Station, are now on display in museums across America.[*]

---

[*]On April 23, 2012 OV-103 Discovery was flown on top of SCA #905 from KSC to Washington DC for permanent display at the Steven F. Udvar-Hazy Center, Chantilly, Virginia. On October 14, OV-105 Endeavour arrived on SCA 905 in Los Angeles, California for permanent display at the California Science Center there. Finally on November 2, OV-104 Atlantis was transferred to the KSC Visitor Complex.

Notes

1. Tanner 2015
2. AIS interview with Tom Jones, August 3, 2006
3. STS-107: Rescue or repair, in *Space Rescue, Ensuring the Safety of Manned Spaceflight*, David J. Shayler, Springer-Praxis, 2009, pp. 1–29
4. "Service Mission 4" in *Enhancing Hubble's Vision, Service Missions That Expanded Our View of the Universe*, David J. Shayler with David M. Harland, Springer-Praxis, 2016, pp149–209
5. Orbiter Systems Failures and Contingency Shuttle Crew Support (CSCS) – Launch on Need (LON) ULF1.1 (STS-121) (undated)
6. Contingency Shuttle Crew Support (CSCS)/Rescue Flight Resource Book, Mission Operations Directorate, DA8/Flight Director Office, Final July 12, 2005, NASA JSC-62900; also ULF1.1_C2 Contingency Shuttle Crew Support (CSCS) Declaration Actions (RI) Reference Rule, and NSTS 21519 LON Crew Rescue Mission MIP (undated)
7. Unmanned Orbiter Undocking: Method for disposal of a damaged Space Shuttle Orbiter, Ray A. Bigonesse, Rendezvous, Guidance and Procedures Office, USA LLC, Houston and William R. Summa, Flight Design and Dynamics, USA, Houston, in Spaceflight Mechanics 2006, Volume 124, Part 1, Advances in the Astronautical Sciences, American Aeronautical Society, 2006. Proceedings of the AAS/AIAA Space Flight Mechanics Meeting, Tampa, Florida, January 22-26, 2006, and AAS 06-171 pp. 1131-1147, courtesy BIS Library, London

Closing Comments

The Shuttle had its faults and its critics, but once on-orbit it was nearly always able to complete its tasks. Its capabilities enabled the fleet of just three Orbiters not merely to visit a space station, albeit a Russian one, but to expand it, resupply it, and deliver and retrieve resident crewmembers.

After countless designs and studies stretching back almost five decades and with the Shuttle having been flying for fifteen years, the fleet was given the chance to do what it had been designed to accomplish, namely to assemble and maintain a much larger space station as an international venture.

The legacies of the Shuttle are many and varied across the 135 missions of the thirty-year program but paramount in any assessment must be the creation of the International Space Station. Nevertheless the physical assembly of the station is not the full story. As related in these pages, that legacy took an international team many years to achieve and included devising how to prepare the hardware and deliver hundreds of tons of logistics and cargo into space, where most of it was linked together. For just a few days on each mission, crews blended their skills in robotics, spacewalking, and cargo handling into a smooth and almost seamless operation.

A great deal was learned, not only by the Russians and Americans, but by the other partners from Shuttle-Mir and various other joint ventures. This cooperative teamwork delivered significant results for each expedition flown to the ISS, even after the Shuttle was retired. Experience from past cooperation, together with new lessons learned from the fiasco of Space Station Freedom and the success of Shuttle-Mir, helped to establish firm foundations for the future. Equally, the experiences of the ISS, both on Earth and in space, are paving the way for further peaceful joint ventures that will take us back to the Moon and eventually on to new endeavors deeper in space.

The Shuttle missions to Mir and the ISS had a remarkable record of success, without a single docking failure across either program. The nine Shuttle missions that docked at Mir logged 38 days 14 hr 52 min attached to the station. The thirty-seven ISS assembly missions added another 276 days 11 hr 23 min to this total. Atlantis, Discovery and Endeavour clocked up an impressive total of 315 days 2 hr 15 min (some 45 weeks) docked at the two

© Springer International Publishing Switzerland 2017
D.J. Shayler, *Assembling and Supplying the ISS*, Springer Praxis Books,
DOI 10.1007/978-3-319-40443-1

The First and the Last. The crew of STS-1 who proved the Shuttle concept, and the crew of STS-135 who rounded out the Shuttle's involvement with the ISS and ended its thirty-year program: [l-r] Doug Hurley (STS-135 PLT), Bob Crippen (STS-1 PLT), John Young (STS-1 CDR), Chris Ferguson (STS-135 CDR), Sandra Magnus (STS-135 MS1), and Rex Walheim (STS-135 MS2).

stations, completing a vast program of joint operations with eight resident Mir resident crews and seventeen ISS resident crews from 1995 through to 2011.

Behind the scenes, there were successes in management and budget, in mission planning and control, in crew selection and training, in hardware development and processing, and in launch and recovery operations. From the missions themselves, valuable experience was gained in rendezvous and docking large vehicles together, extensive robotic operations, planning and executing multiple EVA operations and understanding how to transfer tons of hardware inside and outside a large structure.

Now the ISS is complete, the work continues to develop and expand the science program as an international laboratory that just happens to be situated in Earth orbit. Scores research papers and conference presentations have been issued since the ISS commenced scientific work, essentially in 2001. Details concerning the hundreds of investigations can be found on the websites of the space agencies and their academic and industrial partners. A number of publications focus on the daily activities on the ISS and the research being conducted there, but it is interesting to speculate how the configuration of a space station would have been different if the Shuttle had not been available to assemble it.

Ideally, had the Saturn V still been in service, that would have been used to lift far larger elements for a station. Indeed, that seems to be the preferred option even today. The station that we have was designed to be launched using the Shuttle. The irony is that although a larger vehicle is currently being developed in support of future NASA goals in space, there are no plans to expand the ISS further, nor are there any thoughts of replacing it with a more sophisticated ISS Mk-II. But the Chinese have announced their intention to construct their own Mir-class station in the 2020s.

The Shuttle was ideally suited to launch, assemble, and supply the multi-modular design of the station originally conceived as Freedom and later morphed into the ISS with something of the Russian Mir-2 proposal thrown in. It will also be fascinating to witness developments, in due course, of whatever comes after the ISS and the ultimate decommissioning of the station.

Had the Shuttle not been retired in 2011, would it have been economically viable to sustain it as a general duty delivery truck and garbage hauler? Perhaps not. We've seen the failure of the effort to develop a commercial Shuttle and with no plans to expand or replace the ISS it would have been difficult to retain the fleet simply for logistics when unmanned vehicles were available for that function. The one advantage that the Shuttle had over its current and planned rivals was its to-orbit and to-Earth mass capability but with no new large payloads to deliver it was indeed time to retire the fleet. The sad part of that decision was that the replacement for the Shuttle was nowhere near ready to fly, and so, like during the six year period between ASTP and STS-1, we will have a pause in American human space flight capability.*

The Apollo-Saturn production line by which manned space station operations could have followed on without a break after the initial lunar landings was scrapped to create the Shuttle and a new type of space station. Over forty years after the final Apollo flew the ASTP mission, we again see a complete change of direction with utter confusion of long-term plans and the timetables to execute them.

Nearly fifty years after the first Apollo astronaut stepped on to the Moon, the ISS is today providing a wealth of research which, it is to be hoped, will assist those on Earth and those lucky few who will eventually set off into deep space.

Hopefully the ISS, with its current and future fleets of resupply vehicles, will remain operational for at least another decade. This is testament to the legacy of the Shuttle and the design of the International Space Station. The Space Transportation System was the key to assembling the ISS, but was not required for its further operation. In assembling the ISS, and perhaps in servicing the Hubble Space Telescope, the Shuttle achieved its zenith with tasks that it was always meant to perform. It also offered the opportunity to greatly expand the experience and knowledge of people within the space program, not only flying missions but also on the ground, and more importantly internationally, and at a wide variety of levels of involvement. It is from that wealth of information that, in due course, future generations will benefit.

During his career as a NASA astronaut, Leroy Chiao was involved in an expedition to the ISS in 2004. As he recalled two years later, "I had flown three Shuttle missions. I had flown a science Shuttle mission (STS-65, Spacelab IML-2) and I'd gotten to fly on a mission (STS-72) during which I had my first spacewalk, testing tools and assembly techniques we ended up using on the space station. On my third Shuttle flight (STS-92) I led an EVA team to help assemble the station. And finally I got to be the Commander of the

---

*As this book goes to press in 2017 we have already seen six years pass since the retirement of the Shuttle fleet, and the first manned flight of the new Orion spacecraft is still some years in the future.

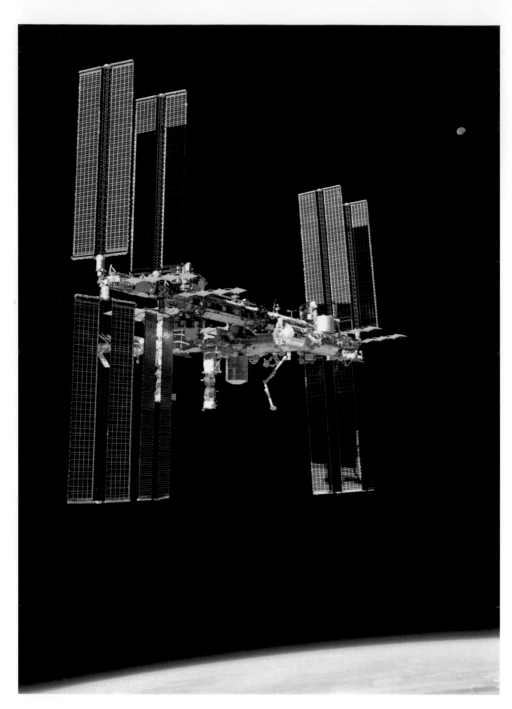

The End: July 2011, STS-135 Atlantis captures the final views of the completed ISS from a Space Shuttle.

station on a long-duration mission. I flew up and down on a Soyuz, and got to do Russian EVAs as well operate the robotic arm."[†]

Like many of his colleagues, Chiao would have really loved to fly to the Moon but he recognized that the experiences gained on Shuttle missions would contribute to the database of experience and knowledge accumulated over the long series of flights into space. That knowledge will allow future explorers to embark on voyages to the Moon and beyond. When those pioneers explore the solar system, they might reflect that the years of Shuttle operations with space stations were an essential step for international cooperation in the exploration of space.

The result of all those decades of effort was suitably summarized by the words of Chris Ferguson, STS-135 CDR, as Atlantis left the ISS on July 19, 2011, to conclude the Shuttle's involvement in the space station program:

> When a generation accomplishes a great thing it has the right to stand back for a moment, admire, and take pride in its work. From our unique vantage point right here perched above the Earth, we can see the International Space Station as a wonderful accomplishment. It was born at the end of the Cold War; it has enabled many nations to speak [as] one in space.
>
> As the ISS now enters its era of utilization, we'll never forget the role the Space Shuttle played in its construction. Like a proud parent, we anticipate great things to follow from the men and women who build, operate and live there. From this unique vantage point we can see that a great thing has been accomplished.
>
> Farewell ISS, make us proud.

Over half a century before these words were spoken from orbit author Eric Burgess wrote in *An Introduction to Rockets and Spaceflight* (1956): "If we sent up twenty or thirty [large] rockets…we would…be able to place quite a large amount of material in space. If it consisted of the pre-fabricated components for a large manned station, men equipped with spacesuits could assembly the structure in space. We would be building a terminal in space which…may hold the key to interplanetary flight."

This is exactly what happened on the three dozen Shuttle missions flown to assemble and supply the International Space Station. Hopefully we will soon be able to state with certainty that this station did indeed "hold the key to interplanetary flight."

[†] AIS interview with Leroy Chiao, June 6, 2006.

Afterword

My second visit to the International Space Station, STS-110, came three and a half years after STS-88. The ISS in 2002 was much larger than it was when I last saw the mated Russian Control Module Zarya and the US connecting module Unity in 1998. I was amazed to see the transformation. The first sights of the ISS from close in as we approached to dock reminded me of the thrills I felt years before as we approached the Mir station on STS-74. I thought, "We are going to visit a place where human beings are living and conducting science and research in space." While docked, I once again ventured outside on two spacewalks to undertake tasks which supported the station's future expansion.

After the tragic loss of Columbia in 2003, I was fully involved in the arduous effort on the ground, supported by teams of searchers, investigators, and engineers, to ensure that the Space Shuttle returned to safe flight and completed its objectives. Although I remained at NASA until 2012, through the end of the Shuttle program and its series of missions to finish the assembly of the ISS, I unfortunately never had the opportunity to see the International Space Station close up again.

Today, nearly twenty years after the crew of STS-88 started the assembly of the ISS, we regularly see a rotation of international crews conducting a wide range of scientific research, building on what we pioneered onboard Spacelab-D2 all those years ago. The station is now so large that it can be viewed from Earth as it passes over in the evening and early morning skies, a beacon for those who wish to follow a career in science and technology. The International Space Station is man's home and research laboratory in space and a stepping stone to ever more explorative programs which will take humans farther into space than we could go with the Shuttle.

In the future, it will be recalled that it was the Space Shuttle that made possible the creation of the International Space Station and enabled humans to develop the skills to embark on those new adventures.

This is a story which David J. Shayler has reconstructed from the inception of the space program, from almost-forgotten plans to visit Salyut and Skylab space stations, through the difficult years of Space Station Freedom to the creation of the ISS.

© Springer International Publishing Switzerland 2017
D.J. Shayler, *Assembling and Supplying the ISS*, Springer Praxis Books,
DOI 10.1007/978-3-319-40443-1

On the ISS during STS-110, Jerry Ross displays a memento of his record breaking seven Shuttle missions between November 1985 and April 2002.

Wherever I am in the world, I can look up at the night sky and recall my adventures there. If I see the International Space Station fly across the twilight as a bright pinprick of light, I smile and remember that I was fortunate to be a member of the large, global team that completed the greatest international construction project in history, offering the next generations the opportunity to push the boundaries of space exploration even further.

Colonel Jerry L. Ross USAF (retired)
NASA Astronaut (1980–2012)
Mission Specialist, STS-61B, -27, -37, -55, -74, -88 & -110

Author of *Spacewalker: My Journey in Space and Faith as NASA's Record-Setting Frequent Flyer* (2013) and *Becoming a Spacewalker: My Journey to the Stars* (2014).

Abbreviations

AAP	Apollo Applications Program
ACCESS	Assembly Concept for Construction of Erectable Space Structures
AFB	Air Force Base
AHMS	Advanced Health Management System
ALS	Advanced Logistics System
ALT	Approach and Landing Tests
AM	Alpha Magnetic Spectrometer
APAS	Androgynous Peripheral Attachment System
APCU	Auxiliary Power Control Unit
APU	Auxiliary Power Unit
ASCAN	Astronaut Candidate
ASSEM	Assembly of Space Station by EVA Methods
ASTP	Apollo Soyuz Test Project
ASVS	Advanced Space Vision System
BUp	Back-Up (astronaut)
Capcom	Capsule Communicator
CB	Astronaut Office, NASA JSC (CB = Directorate Mail Code)
CBM	Common Berthing Mechanism
CDR	Commander
CDR	Critical Design Review
CEIT	Crew Equipment Interface Test
CETA	Crew Equipment Transfer Assembly
CG	Center of Gravity
CM	Command Module (Apollo)
CMG	Control Moment Gyro
CNES	Centre National d'Etudes Spatiales (French National Space Agency)
COAS	Crew Optical Alignment Sight
COLBERT	Combined Operational Load Bearing External Resistance Treadmill
CSA	Canadian Space Agency

© Springer International Publishing Switzerland 2017
D.J. Shayler, *Assembling and Supplying the ISS*, Springer Praxis Books,
DOI 10.1007/978-3-319-40443-1

CSCS	Contingency Shuttle Crew Support
CSM	Command and Service Module (Apollo)
CTB	Crew Transfer Bag
DAP	Digital Auto Pilot
DM	Docking Module
DOR	Director of Operations - Russia
DSO	Detailed Supplementary Objective
DTO	Detailed Test Objective
EAFB	Edwards Air Force Base
EAP	Education Astronaut Project
EASE	Experimental Assembly of Structures for EVA
EDFT	EVA Demonstration Flight Test
EDO	Extended Duration Orbiter
EDVT	EVA Development and Verification Test
ELC	Express Logistics Carrier
ELM-P	Experiment Logistics Module – Pressurized (Kibo)
EMTT	EVA Maintenance Task Team (Freedom)
EMU	Extravehicular Mobility Unit
EO-xx	Space station main expedition crew (*Ekspeditsiya Osnovnaya)*
ESA	European Space Agency
ESRO	European Space Research Organization
ET	External Tank
EVA	Extra Vehicular Activity (spacewalking)
FCR	Flight Control Room
FD	Flight Director ('Flight')
FE	Flight Engineer
FGB	Functional Cargo Blok (*Funktsionalno-gruzovoy blok)*
FY	Financial Year (Fiscal)
GAS	Get Away Special
GPS	Global Positioning Satellite
GRO	Gamma Ray Observatory (Compton)
GSFC	(Robert F.) Goddard Space Flight Center, Greenbelt, Maryland
HB	High Bay
H-Bar	Horizontal Bar (rendezvous mode)
HHL	Hand Held Laser
HST	Hubble Space Telescope
HTV	H-II Transfer Vehicle (Japanese)
ICC	Integrated Cargo Carrier
ICC-G	ICC Generic
ICC-GD	ICC Generic Deployable
ICC-L	ICC Light (Lite)
ICC-VLD	ICC Vertical Light (Lite)
IFA	In-Flight Anomaly
IFM	In-Flight Maintenance
IMOC	Integrated Mission Operations Control

IMU	Inertial Measurement Unit
ISS	International Space Station
IST	Integrated Systems Test
ITS	Integrated Truss Structure
IVA	Intra Vehicular Activity
JAXA	Japan Aerospace eXploration Agency
JEM	Japanese Experiment module (Kibo)
JSC	(Lyndon B.) Johnston Space Center (NASA, Houston, Texas)
JWG	Joint Working Group
KSC	(John F.) Kennedy Space Center (Florida)
LCC	Launch Control Center
LDEF	Long Duration Exposure Facility
LES	Launch and Entry Suit
LF	Logistics Flight (Shuttle, ISS)
LON	Launch-on-Need
Low	Rendezvous approach mode
LSEAT	Launch Systems Evaluation Advisory Team
LVLH	Local Vertical Local Horizontal
MBS	Mobile Base System
MCC-M	Mission Control Center – Moscow
MCC-H	Mission Control Center - Houston
MEIT	Multi-Element Integration Testing
MER	Mission Evaluation Room
MIT	Massachusetts Institute of Technology
MLP	Mobile Launch Platform
MMT	Mission Management Team
MOD	Mission Operations Director
MPLM	Multi-Purpose Logistic Module
MPRESS	Mission Peculiar Equipment Support Structure
MPS	Main Propulsion System (STS)
MRM	Mini Research Module
MS	Mission Specialist
MSFC	(George C.) Marshall Space Flight Center, Huntsville, Alabama
MSIS	Man-Systems Integration Standards
MST	Mission Sequence Test
MT	Mobile Transporter (ISS)
NASA	National Aeronautics and Space Administration
NASDA	NAtional Space Development Agency of Japan
NBL	(Sonny L. Carter) Neutral Buoyancy Laboratory, Houston
NBS	Neutral Buoyancy Simulator (MSFC, Huntsville)
NC-bur	Nominal Corrective burn
NIH	Nickel-Hydrogen
NSC	National Security Council
OBSS	Orbiter Boom Sensor System
O&C	Operations and Checkout

ODS	Orbiter Docking System
OETF	Operations Engineering Training Facility (Canada)
OFT	Orbital Flight Test
O&M	Operations and Maintenance
OMB	Office of Management and Budget
OMDP	Orbiter Maintenance Down Period
OMP	Operational Maintenance and Inspection Program
OMRF	Orbiter Maintenance and Refurbishment Facility
OMRSD	Operations and Maintenance Requirements and Specifications Document
OMS	Orbital Maneuvering System
OPCU	Orbiter Power Conversion Unit (Freedom which became the SSPTS for ISS)
OPF	Orbiter Processing Facility (KSC, Florida)
ORBT	Optimized R-Bar Target Rendezvous
ORU	Orbital Replacement Unit
OSK	Official Souvenir Kit
OV	Orbital Vehicle (Space Shuttle)
PCT	Post Contact Thrusting
PDR	Payload Deployment and Retrieval
PFR	Portable Foot Restraint
PGHM	Payload Ground Handling Mechanism
PLT	Pilot
PMA	Pressurized Mating Adapter (ISS)
PMM	Permanent Multipurpose Module
POHS	Position Orientation Hold Selection (RMS software)
PPK	Personal Preference Kit
RAM	Research and Application Module (forerunner of Spacelab)
R-Bar	Rendezvous approach mode
RCS	Reaction Control System
RME	Risk Mitigation Experiment
RMS	Remote Manipulator System (Canadarm, Space Shuttle)
ROEU	Remotely Operated Electrical Umbilical (Freedom)
RPM	R-Bar Pitch Manoeuver
RPOP	Rendezvous and Proximity Operations Program
RSA	Russian Space Agency
RSSA	Recumbent Seat System Assembly
RTAS	Rocketdyne Truss Attachment System
SAFER	Simplified Aid For EVA Rescue
SAIL	Shuttle Avionics Integration Laboratory
SARJ	Solar Alpha Rotary Joint
SAVE	Structural Assembly Verification Experiment
SCA	Shuttle Carrier aircraft (Boeing 747)
SDTO	Station Detailed Test Objective (ISS)
SEM	Space Experiment Module (educational)
SFOC	Space Flight Operations Contract
SHOSS	Spacehab and Oceaneering Space Systems

S-IVB	Saturn IB second, Saturn V third stage (Apollo)
SLF	Shuttle Landing Facility (KSC, Florida)
SM	Service Module (Apollo)
SMS	Shuttle Mission Simulator
SNIP	Shuttle Nose In Plane
SNOOPy	Shuttle Nose Out Of Plane
SOC	Space Operations Contract
SPAS	Shuttle Pallet Applications Satellite (free-flyer)
SPC	Shuttle Processing Contract
SPDM	Special Purpose Dexterous Manipulator
SPORTS	Spacecraft Proximity Operations Real-Time Simulator
SRB	Solid Rocket Booster (STS)
SSF	Space Station Freedom
SSME	Space Shuttle Main Engine
SSPP	Shuttle Small Payloads Project
SSPSG	Shuttle Salyut Payload Study Group
SSPTS	Station-to-Shuttle Power Transfer System
SSRMS	Space Station Remote Manipulator System (Canadarm2, ISS)
STG	Space Task Group
SSTG	Space Shuttle Task Group
STA	Shuttle Training Aircraft (Gulfstream)
STEAM	Science Technology Engineering Art Math (education)
STEM	Science Technology Engineering Math (education)
STS	(National) Space Transportation System (Space Shuttle)
TACAN	Tactical Air Control And Navigation
TAL	Trans-Atlantic Landing (Shuttle abort mode)
TCS	Trajectory Control Sensor
TDRSS	Tracking and Data Relay Satellite System
TI	Terminal Phase Initial Burn
TM	*Transportni Modifitsirovannyi* (Transport, Modified Soyuz variant)
TMA	*Transportni Modifitsirovannyi Antrpometricheskii* (Transport, Modified, Anthropometric a Soyuz variant)
TPS	Thermal Protection System (Shuttle)
TsPK	Cosmonaut Training Center named for Yuri A. Gagarin
UARS	Upper Atmosphere Research Satellite
UBA	Unpressurized Berthing Adapter (Space Station Freedom)
UF	Utilization Flight (Shuttle, ISS)
UHF	Ultra High Frequency
UIPT	Utilization Integrated Product Team
ULF	Utilization Logistics Flight (Shuttle, ISS)
US	United States
USA	United Space Alliance
USAF	United States Air Force
USMC	United States Marine Corps
USN	United States Navy

USSR	Union of Soviet Socialistic Republics (1917-1991)
V axis	Rendezvous approach mode
VAB	Vehicle Assembly Building
V-Bar	Rendezvous approach mode
VHF	Very High Frequency
VR	Virtual Reality
WAD	Work Authorization Document
WETF	Weightless Environment Training Facility
X axis	Rendezvous approach mode
Y axis	Rendezvous approach mode
Z axis	Rendezvous approach mode
Z-Bar	Rendezvous approach mode

Appendix 1

SHUTTLE-ISS DOCKING AND ASSEMBLY MISSION CREWMEMBERS 1998–2011

Space Agency Key: CSA (Canadian); CNES (French); ESA (European); RSA (Russian); JAXA (Japanese); NASA (American)

Note: This list does not include those crewmembers that were part of a Mir or ISS resident crew. Resident crewmembers flown to or from the ISS onboard Shuttle are listed separately in Table 5.

Acaba, Joseph M., NASA; (March 15–28, 2009) MS1 STS-119

Altman, Scott D., NASA; (September 8–20, 2000) PLT STS-106

Anderson, Clayton C., NASA; (April 5–20, 2010) MS5 STS-131

Antonelli, Dominic A., NASA; (March 15–28, 2009) PLT STS-119; (May 14–26, 2010) PLT STS-132

Archambault, Lee J., NASA; (June 8–22, 2007) PLT STS-117; (March 15–28, 2009) CDR STS-119

Ashby, Jeffrey S., NASA; (April 19–May 1, 2001) PLT STS-100; (October 7–18, 2002) CDR STS-112

Barratt, Michael R., NASA; (February 24–March 9, 2011) MS3 STS-133

Barry, Donald T., NASA (27 May–June 6, 1999) MS3 STS-96; (August 10–22, 2001) MS2 STS-105

Behnken, Robert L., NASA; (March 11–26, 2008) MS1 STS-123; (February 8–21, 2010) MS4 STS-130

Bloomfield, Michael J., NASA; (November 30–December 11, 2000) PLT STS-97; (April 8–20, 2002) CDR STS-110

Boe, Eric A., NASA; (November 14–30, 2008) PLT STS-126; (February 24–March 9, 2011) PLT STS-133

Bowen, Stephen G., NASA; (November 14–30, 2008) MS2 STS-126; (May 14–26, 2010) MS3 STS-132; (February 24–March 9, 2011) MS2 STS-133

© Springer International Publishing Switzerland 2017
D.J. Shayler, *Assembling and Supplying the ISS*, Springer Praxis Books,
DOI 10.1007/978-3-319-40443-1

Bresnik, Randolph J., NASA; (November 16–27, 2009) MS2 STS-129

Burbank, Daniel C., NASA; (September 8–20, 2000) MS3 STS-106; (September 9–21, 2006) MS2 STS-115

Cabana, Robert D., NASA; (December 4–15, 1998) CDR STS-88

Cadwell, Tracy E., NASA; (August 8–21, 2007) MS1 STS-118

Camarda, Charles J., NASA; (July 26–August 9, 2005) MS5, STS-114

Cassidy, Christopher J., NASA; (July 15–31, 2009) MS2 STS-127

Chamitoff, Gregory E. NASA; (May 16–June 1, 2011) MS4 STS-134

Chang-Diaz, Franklin R., NASA; (June 2–12, 1998) MS1 STS-91; (June 5–19, 2002) MS1 STS-111

Chiao, Leroy, NASA; (October 11–24, 2000) MS1 STS-92

Cockrell, Kenneth D., NASA; (February 7–20, 2001) CDR STS-98; (June 5–19, 2002) CDR STS-111

Collins, Eileen M., NASA; (26 July–9 August, 2005) CDR STS-114

Curbeam Jr., Robert L., NASA; (February 7–20, 2001) MS1 STS-98; (December 9–22, 2006) MS2 STS-116

Currie, Nancy J., NASA; (December 4–15, 1998) MS2 STS-88

Doi, Takao., JAXA; (March 11–26, 2008) MS3 STS-123

Drew, Benjamin A., (August 8–21, 2007) MS5 STS-118; (February 24–March 9, 2011) MS1 STS-133

Duffy, Brian., NASA; (October 11–24, 2000) CDR STS-92

Dunbar, Bonnie J., NASA; (June 27–July 7, 1995) MS3 STS-71; (January 22–31, 1998) MS3 STS-89

Dutton Jr., James P. NASA; (April 5–20, 2010) PLT STS-131

Ferguson, Christopher J., NASA; (September 9–21, 2006) PLT STS-115; (November 14–30, 2008) CDR STS-126; (July 8–21, 2011) CDR STS-135

Feustal, Andrew J., NASA; (May 16–June 1, 2011) MS3 STS-134

Fincke, Edward M., NASA; (May 16–June 1, 2011) MS1 STS-134

Ford, Kevin A., NASA; (August 28–September 12, 2009) PLT STS-128

Foreman, Michael J., NASA; (March 11–26, 2008) MS2 STS-123; (November 16–27, 2009) MS3 STS-129

Forrester, Patrick G., NASA; (August 10–22, 2001) MS1 STS-105; (June 8–22, 2007) MS1 STS-117; (August 28–September 12, 2009) MS1 STS-128

Fossum, Michael E., NASA; (July 4–21, 2006) MS1 STS-121; (May 31–June 14, 2008) MS3 STS-124

Frick, Stephen N., NASA; (April 8–20, 2002) PLT STS-110; (February 7–20, 2008) CDR STS-122

Fuglesang, A. Christer., ESA; (December 9–22, 2006) MS3 STS-116; (August 28– September 12, 2009) MS4 STS-128

Garan Jr., Ronald J., NASA; (May 31–June 14, 2008) MS2 STS-124

Garneau, Marc J. J-P., CSA; (November 30–December 11, 2000) MS2 STS-97

Gernhardt, Michael L., NASA; (July 12–24, 2001) MS1 STS-104

Godwin, Linda M., NASA; (December 5–17, 2001) MS1 STS-108

Good, Michael T., NASA; (May 14–26, 2010) MS2 STS-132

Gorie, Dominic L. P., NASA; ((December 5-17, 2001) CDR STS-108

Guidoni, Umberto, ESA; (April 19–May 1, 2001) MS4 STS-100

Hadfield, Chris A., CSA; (April 19–May 1, 2001) MS1 STS-100

Halsell Jr., James D., NASA; (19–29 May, 2000) CDR STS-101

Ham, Kenneth T., NASA; (May 31–June 14, 2008) PLT STS-124; (May 14–26, 2010) CDR STS-132

Helms, Susan J., NASA; (May 19–29, 2000) MS4 STS-101

Hernandez, José D., NASA; (August 28–September 12, 2009) MS2 STS-128

Herrington, John B., NASA; (November 23–December 7, 2002) MS2 STS-113

Higginbotham, Joan E., NASA; (December 9–22, 2006) MS4 STS-116

Hire, Kathryn P., NASA; (February 8–21, 2010) MS1 STS-130

Hobaugh, Charles O., NASA; (July 12–24, 2001) PLT STS-104; (August 8-21, 2007) PLT STS-118; (November 16–27, 2009) CDR STS-129

Horowitz, Scott J., NASA; (May 19–29, 2000) PLT STS-101; (August 10–22, 2001) CDR STS-105

Hoshide, Akihiko JAXA; (May 31–June 14, 2008) MS4 STS-124

Hurley, Douglas G.; NASA; (July 15–31, 2009) PLT STS-127; (July 8–21, 2011) PLT STS-135

Husband, Rick D., NASA (May 27–June 6, 1999) PLT STS-96

Ivins, Marsha S., NASA, (February 7–20, 2001) MS2 STS-98

Jernigan, Tamara E., NASA; (May 27–June 6, 1999) MS1 STS-96

Jett Jr., Brent W., NASA; (November 30–December 11, 2000) CDR STS-97; (September 9–21, 2006) CDR STS-115

Johnson, Gregory H., NASA; (March 11–26, 2008) PLT STS-123; (May 16–June 1, 2011) PLT STS-134

Jones, Thomas D., NASA; (February 7–20, 2001) MS3 STS-98

Kavandi, Janet L., NASA; (July 12–24, 2001) MS3 STS-104

Kelly, James M., NASA; (March 8–21, 2001) PLT STS-102; (26 July–9 August, 2005) PLT, STS-114

Kelly, Mark E., NASA; (December 5–17, 2001) PLT STS-108; (4–21 July, 2006) PLT, STS-121; (May 31–June 14, 2008) CDR; (May 16–June 1, 2011) CDR STS-134

Kelly, Scott J., NASA; (August 8–21, 2007) CDR STS-118

Kimbrough, Robert S., NASA; (November 14–30, 2008) MS4 STS-126

Krikalev, Sergei K., RSA; (February 3–11, 1994) MS4 STS-60; (December 4–15, 1998) MS5 STS-88

Lawrence, Wendy B., NASA; (July 26–August 9, 2005) MS4, STS-114

Lindsay, Steven W., NASA; (July 12–24, 2001) CDR STS-104; (July 4–21, 2006) CDR STS-121; (February 24–March 9, 2011) CDR STS-133

Linnehan, Richard M., NASA; (March 11–26, 2008) MS1 STS-123

Lockhart, Paul S., NASA; (June 5–19, 2002) PLT STS-111; (November 23–December 7, 2002) PLT STS-113

Lonchakov, Yuri V., RSA; (April 19–May 1, 2001) MS5 STS-100

Lopez-Alegria, Michael E., NASA; (October 11–24, 2000) MS4 STS-92; (November 23–December 7, 2002) MS1 STS-113

Love, Stanley G., NASA; (February 7–20, 2008) MS4 STS-122

Lu, Edward T., NASA, NASA; (September 8–20, 2000) MS1 STS-106

Magnus, Sandra H., NASA; (October 7–18, 2002) MS2 STS-112 (July 8–21, 2011) MS1 STS-135

Malenchenko, Yuri I., RSA; (September 8–20, 2000) MS4 STS-106

Marshburn, Thomas H., NASA; (July 15–31, 2009) MS4 STS-127

Mastracchio, Richard A., NASA; (September 8–20, 2000) MS2 STS-106; (August 8–21, 2007) MS2 STS-118; (April 5–20, 2010) MS1 STS-131

MacLean, Steven Glenwood, CSA; (September 9–21, 2006) MS4 STS-115

McArthur Jr., William S., NASA (October 11–24, 2000) MS2 STS-92

Melroy, Pamela A., NASA; (October 11–24, 2000) PLT STS-92; (October 7–18, 2002) PLT STS-112; (October 23–November 7, 2007) CDR STS-120

Melvin, Leland D., (February 7–20, 2008) MS1 STS-122; (November 16–27, 2009) MS1 STS-129

Metcalf-Lindenburger, Dorothy M. NASA; (April 5–20, 2010) MS2 STS-131

Morgan, Barbara R., NASA; (August 8–21, 2007) MS4 STS-118

Morin, Lee M. E., NASA; (April 8–20, 2002) MS3 STS-110

Morukov, Boris V., RSA; (September 8–20, 2000) MS5 STS-106

Nespoli, Paolo., ESA; (October 23–November 7, 2007) MS4 STS-120

Newman, James H., NASA; (December 4–15, 1998) MS3 STS-88

Noguchi, Soichi, JAXA; (26 July–9 August, 2005) MS1, STS-114

Noriega, Carlos I., NASA; (November 30–December 11, 2000) MS3 STS-97

Nowak, Lisa M., (NASA); (4–21 July, 2006) MS2 STS-121

Nyberg, Karen L., NASA; (May 31–June 14, 2008) MS1 STS-124

Ochoa, Ellen L., NASA (May 27–June 6, 1999) MS2 STS-96

Oefelein, William A., NASA; (December 9–22, 2006) PLT STS-116

Olivas, John D., NASA; (June 8–22, 2007) MS3 STS-117; (August 28–September 12, 2009) MS3 STS-128

Parazynski, Scott E., NASA; (April 19–May 1, 2001) MS3 STS-100; (October 23–November 7, 2007) MS1 STS-120

Patrick, Nicolas J. M., NASA; (December 9–22, 2006) MS1 STS-116; (February 8–21, 2010) MS3 STS-130

Payette, Julie, CSA; (May 27–June 6, 1999) MS4 STS-96; (July 15–31, 2009) MS3 STS-127

Perrin, Phillipe, CNES; (June 5–19, 2002) MS2 STS-111

Pettit, Donald R., NASA; (November 14–30, 2008) MS1 STS-126

Phillips, John L., NASA; (April 19–May 1, 2001) MS2 STS-100; (March 15–28, 2009) MS4 STS-119

Polansky, Mark L., NASA; (February 7–20, 2001) PLT STS-98; (December 9–22, 2006); CDR STS-116; (July 15–31, 2009) CDR STS-127

Poindexter, Alan G., NASA; (February 7–20, 1998) PLT STS-122

Reilly II., James F., NASA; (July 12–24, 2001) MS2 STS-104; (June 8–22, 2007) MS4 STS-117

Reisman, Garrett E., NASA; (May 14–26, 2010) MS1 STS-132

Richards, Paul W., NASA; (March 8–21, 2001) MS2 STS-102

Robinson, Stephen K., NASA; (26 July–9 August, 2005) MS2, STS-114; (February 8–21, 2010) MS2 STS-130

Rominger, Kent V., NASA; (May 27–June 6, 1999) PLT STS-96; (April 19–May 1, 2001) CDR STS-100

Ross, Jerry L., NASA; (December 4–15, 1998) MS1 STS-88; (April 8–20, 2002) MS4 STS-110

Satcher Jr., Robert L., NASA; (November 16–27, 2009) MS4 STS-129

Schlegel, Hans W., ESA; (February 7–20, 2008) MS3 STS-122

Sellers, Piers J., NASA; (October 7–18, 2002) MS3 STS-112; (July 4–21, 2006) MS3 STS-121; (May 14–26, 2010) MS4 STS-132

Smith, Steven L., NASA; (April 8–20, 2002) MS5 STS-110

Stefanyshyn-Piper, Heidemarie M., NASA; (September 9–21, 2006) MS3 STS-115; (November 14–30, 2008) MS3 STS-126

Stott, Nicole M. P., NASA; (February 24–March 9, 2011) MS4 STS-133

Sturckow, Frederick W., NASA; (December 4–15, 1998) PLT STS-88; (August 10–22, 2001) PLT STS-105; (June 8–22, 2007) CDR STS-117; (August 28–September 12, 2009) CDR STS-128

Swanson, Steven R., NASA; (June 8–22, 2007) MS2 STS-117; (March 15–28, 2009) MS2 STS-119

Tani, Daniel M., NASA; (December 5–17, 2001) MS2 STS-108

Tanner, Joseph R., NASA; (November 30–December 11, 2000) MS1 STS-97; (September 9–21, 2006) MS1 STS-115

Thomas, Andrew S. W., NASA; (March 8–21, 2001) MS1 STS-102; (26 July–9 August, 2005) MS3, STS-114

Tokarev, Valeri I., RSA; (May 27–June 6, 1999) MS5 STS-96

Usachev, Yuri V., RSA; (May 19–29, 2000) MS5 STS-101

Virts Jr., Terry W. NASA; (February 8–21, 2010) PLT STS-130

Vittori, Roberto, ESA; (May 16–June 1, 2011) MS2 STS-134

Voss, James S., NASA; (May 19–29, 2000) MS3 STS-101

Wakata, Koichi, JAXA; (October 11–24, 2000) MS5 STS-92

Walheim, Rex J., NASA; (April 8–20, 2002) MS1 STS-110; (February 7–20, 2008) MS2 STS-122; (July 8–21, 2011) MS2 STS-135

Weber, Mary E., NASA; (May 19–29, 2000) MS1 STS-101

Wetherbee, James D., NASA; (March 8–21, 2001) CDR STS-102; (November 23–December 7, 2002) CDR STS-113.

Wheelock, Douglas H., NASA; (October 23–November 7, 2007) MS3 STS-120

Wilcutt, Terrance W., NASA, (September 8–20, 2000) CDR STS-106

Williams, Daffyd (David) R., CSA; (August 8–21, 2007) MS3 STS-118

Williams, Jeffrey N., NASA; (May 19–29, 2000) MS2 STS-101

Wilmore, Barry E., NASA; (November 16–27, 2009) PLT STS-129

Wilson, Stephanie D., NASA; (4–21 July, 2006) MS4 STS-121; (October 23–November 7, 2007) MS2 STS-120; (April 5–20, 2010) MS3 STS-131

Wisoff, P. Jeff K., NASA; (October 11–24, 2000) MS3 STS-92

Wolf, David A., NASA; (October 7–18, 2002) MS1 STS-112; (July 15–31, 2009) MS1 STS-127

Yamazaki, Naoko, JAXA; (April 5–20, 2010) MS4 STS-131

Yurchikin, Fyodor N., RSA, (October 7–18, 2002) MS4 STS-112

Zamka, George D., NASA (October 23–November 7, 2007) PLT STS-120; (February 8–21, 2010) CDR STS-130

Appendix 2

Shuttle-ISS Assembly Mission Crew Members Summary

This table summarizes the each of the visits by all Shuttle Orbiter crewmembers (US and International) to the ISS during the assembly period of 1998-2011. Details of earlier, non ISS missions are included in addition to any re-visits to the ISS *after* serving on an assembly mission crew or subsequent visits to the Hubble Space Telescope.

© Springer International Publishing Switzerland 2017
D.J. Shayler, *Assembling and Supplying the ISS*, Springer Praxis Books,
DOI 10.1007/978-3-319-40443-1

Shuttle-ISS Assembly Mission Crew Members

Name	Prev Flts	Agency	Nationality	1st Visit	Des.	Year	2nd Visit	Des.	Year	3rd Visit	Des.	Year	4th Visit	Des.	Year
Acaba	0	NASA	American	STS-119	MS1	2009	ISS-31/32 Flight Engineer								
Altman	1	NASA	American	STS-106	PLT	2000	*CDR STS-109 & STS-125 Hubble SM 3B & 4*								
Anderson C.	1	NASA	American	STS-131	MS5	2010									
Antonelli	0	NASA	American	STS-119	PLT	2009	STS-132	PLT	2010						
Archambault	0	NASA	American	STS-117	PLT	2007	STS-119	CDR	2009						
Arnold	0	NASA	American	STS-119	MS3	2009									
Ashby	1	NASA	American	STS-100	PLT	2001	STS-112	CDR	2002						
Barratt	1	NASA	American	STS-133	MS3	2011									
Barry	1	NASA	American	STS-96	MS3	1999	STS-105	MS2	2001						
Behnken	0	NASA	American	STS-123	MS1	2008	STS-130	MS4	2010						
Bloomfield*	1	NASA	American	STS-97	PLT	2000	STS-110	CDR	2002						
Boe	0	NASA	American	STS-126	PLT	2008	STS-133	PLT	2011						
Bowen	0	NASA	American	STS-126	MS2	2008	STS-132	MS3	2010	STS-133	MS2	2011			
Bresnik	0	NASA	American	STS-129	MS2	2009									
Burbank	0	NASA	American	STS-106	MS3	2000	STS-115	MS2	2006	ISS-29 Flight Engineer, ISS-30 Commander					
Cabana	3	NASA	American	STS-88	CDR	1998									
Caldwell (Dyson)	0	NASA	American	STS-118	MS1	2007	ISS-23/24 Flight Engineer								
Camarda	0	NASA	American	STS-114	MS5	2005									
Cassidy	0	NASA	American	STS-127	MS2	2009	ISS-35/36 Flight Engineer								
Chamitoff	1	NASA	American	STS-134	MS4	2011									
Chang-Diaz*	6	NASA	American	STS-111	MS1	2002									
Chiao	2	NASA	American	STS-92	MS1	2000	ISS-10 Commander								
Cockrell	3	NASA	American	STS-98	CDR	2001	STS-111	CDR	2002						
Collins, E.*	3	NASA	American	STS-114	CDR	2005									
Curbeam	1	NASA	American	STS-98	MS1	2001	STS-116	MS2	2006						
Currie	2	NASA	American	STS-88	MS2	1998	*MS STS-109 Hubble Service Mission 3B*								
Doi	1	JAXA	Japanese	STS-123	MS3	2008									
Drew	0	NASA	American	STS-118	MS5	2007	STS-133	MS1	2011						
Duffy	3	NASA	American	STS-92	CDR	2000									

Name	Prev Flts	Agency	Nationality	1st Visit	Des.	Year	2nd Visit	Des.	Year	3rd Visit	Des.	Year	4th Visit	Des.	Year
Dutton	0	NASA	American	STS-131	PLT	2010									
Ferguson	0	NASA	American	STS-115	PLT	2006	STS-126	CDR	2008	STS-135	CDR	2011			
Feustel	1	NASA	American	STS-134	MS3	2011									
Fincke	2	NASA	American	STS-134	MS1	2011									
Ford	0	NASA	American	STS-128	PLT	2009	ISS-33 Flight Engineer/ISS-34 Commander								
Foreman	0	NASA	American	STS-123	MS2	2008	STS-129	MS3	2009						
Forrester	0	NASA	American	STS-105	MS1	2001	STS-117	MS1	2007	STS-128	MS1	2009			
Fossum	0	NASA	American	STS-121	MS1	2006	STS-124	MS3	2008	ISS-28 Flight Engineer/ ISS-29 Commander					
Frick	0	NASA	American	STS-110	PLT	2002	STS-122	CDR	2008						
Fuglesang	0	ESA	Swedish	STS-116	MS3	2006	STS-128	MS4	2009						
Garan	0	NASA	American	STS-124	MS2	2008	ISS-27/28 Flight Engineer								
Garneau	2	CSA	Canadian	STS-97	MS2	2000									
Gernhardt	3	NASA	American	STS-104	MS1	2001									
Godwin*	3	NASA	American	STS-108	MS1	2001									
Good	1	NASA	American	STS-132	MS2	2010									
Gorie*	2	NASA	American	STS-108	CDR	2001	STS-123	CDR	2008						
Guidoni	1	ESA	Italian	STS-100	MS4	2001									
Hadfield*	1	CSA	Canadian	STS-100	MS1	2001	ISS-34 Flight Engineer/ISS-35 Commander								
Halsell*	4	NASA	American	STS-101	CDR	2000									
Ham	0	NASA	American	STS-124	PLT	2008	STS-132	CDR	2010						
Helms	3	NASA	American	STS-101	MS4	2000	ISS-2 Flight Engineer								
Hernandez	0	NASA	American	STS-128	MS2	2009									
Herrington	0	NASA	American	STS-113	MS2	2002									
Higginbotham	0	NASA	American	STS-116	MS4	2006									
Hire	1	NASA	American	STS-130	MS1	2010									
Hobaugh	0	NASA	American	STS-104	PLT	2001	STS-118	PLT	2007	STS-129	CDR	2009			
Horowitz	2	NASA	American	STS-101	PLT	2000	STS-105	CDR	2001						
Hoshide	0	NASA	Japanese	STS-124	MS4	2008	ISS-32/33 Flight Engineer								
Hurley	0	NASA	American	STS-127	PLT	2009	STS-135	PLT	2011						
Husband	0	NASA	American	STS-96	PLT	1999	Died Feb 1 2003 in loss of Columbia (STS-107)								
Ivins*	4	NASA	American	STS-98	MS2	2001									

(continued)

Shuttle-ISS Assembly Mission Crew Members (continued)

Name	Prev Flts	Agency	Nationality	1st Visit	Des.	Year	2nd Visit	Des.	Year	3rd Visit	Des.	Year	4th Visit	Des.	Year
Jernigan	4	NASA	American	STS-96	MS1	1999									
Jett*	2	NASA	American	STS-97	CDR	2000	STS-115	CDR	2006						
Johnson, G.H.	0	NASA	American	STS-123	PLT	2008	STS-134	PLT	2011						
Jones, T.	3	NASA	American	STS-98	MS3	2001									
Kavandi*	2	NASA	American	STS-104	MS3	2001									
Kelly S.	1	NASA	American	STS-118	CDR	2007	ISS-25 Flight Engineer/ISS-26 Commander						ISS-43 FE/ISS-44/45 Commander		
Kelly, J.	0	NASA	American	STS-102	PLT	2001	STS-114	PLT	2005						
Kelly, M...	0	NASA	American	STS-108	PLT	2001	STS-121	PLT	2006	STS-124	CDR	2008	STS-134	CDR	2011
Kimbrough	0	NASA	American	STS-126	MS4	2008									
Krikalev	3	RSA	Russian	STS-88	MS4	1998	ISS-1 Flight Engineer; ISS-11 Commander								
Lawrence*	3	NASA	American	STS-114	MS4	2005									
Lindsey	2	NASA	American	STS-104	CDR	2001	STS-121	CDR	2006	STS-133	CDR	2011			
Linnehan	3	NASA	American	STS-123	MS4	2008									
Lockhart	0	NASA	American	STS-111	PLT	2002	STS-113	PLT	2002						
Lonchakov	0	RSA	Russian	STS-100	MS5	2001	VC-4 Flight Engineer; ISS-18 Commander								
Lopez-Alegria	1	NASA	American	STS-92	MS4	2000	STS-113	MS1	2002						
Love	0	NASA	American	STS-122	MS4	2008									
Lu*	1	NASA	American	STS-106	MS1	2000	ISS-7 Flight Engineer								
MacLean	1	CSA	Canadian	STS-115	MS4	2006									
Magnus	0	NASA	American	STS-112	MS2	2002	ISS-18 Flight Engineer			STS-135	MS1	2011			
Malenchenko	1	RSA	Russian	STS-106	MS4	2000	ISS-7 Commander; ISS-16 Flight Engineer; ISS-32/33 Flight Engineer; ISS-45/46 FE								
Marshburn	0	NASA	American	STS-127	MS4	2009	ISS-34/35 Flight Engineer								
Mastracchio	0	NASA	American	STS-106	MS2	2000	STS-118	MS2	2007	STS-131	MS1	2010	ISS-38/39 Flight Engineer		
McArthur*	2	NASA	American	STS-92	MS2	2000	STS-112	PLT	2002	STS-120	CDR	2007			
Melroy	0	NASA	American	STS-92	PLT	2000									
Melvin	0	NASA	American	STS-122	MS1	2008	STS-129	MS1	2009						
Metcalf-Lindenburger	0	NASA	American	STS-131	MS2	2010									

Name		Agency	Nationality	Mission 1	Role 1	Year 1	Mission 2	Role 2	Year 2	Mission 3	Role 3	Year 3
Morgan	0	NASA	American	STS-118	MS4	2007						
Morin	0	NASA	American	STS-110	MS3	2002						
Morukov	0	RSA	Russian	STS-106	MS5	2000						
Nespoli	0	ESA	Italian	STS-120	MS4	2007	ISS-26/27 Flight Engineer					
Newman	2	NASA	American	STS-88	MS3	1998	MS STS-109 Hubble Service Mission 3B					
Noguchi	0	JAXA	Japanese	STS-114	MS1	2005	ISS-22/23 Flight Engineer					
Noriega*	1	NASA	American	STS-97	MS3	2000						
Nowak	0	NASA	American	STS-121	MS2	2006						
Nyberg	0	NASA	American	STS-124	MS1	2008	ISS-36/37 Flight Engineer					
Ochoa	2	NASA	American	STS-96	MS2	1999	STS-110	MS2	2002			
Oefelein	0	NASA	American	STS-116	PLT	2006						
Olivas	0	NASA	American	STS-117	MS3	2007	STS-128	MS3	2009			
Parazynski*	3	NASA	American	STS-100	MS3	2001	STS-120	MS1	2007			
Patrick	0	NASA	American	STS-116	MS1	2006	STS-130	MS3	2010			
Payette	0	CSA	Canadian	STS-96	MS4	1999	STS-127	MS3	2009			
Perrin	0	CNES	French	STS-111	MS2	2002						
Pettit	1	NASA	American	STS-126	MS1	2008	ISS-30/31 Flight Engineer					
Phillips	0	NASA	American	STS-100	MS2	2001	ISS-11 Flight Engineer			STS-119	MS4	2009
Poindexter	0	NASA	American	STS-122	PLT	2008	STS-131	CDR	2010			
Polansky	0	NASA	American	STS-98	PLT	2001	STS-116	CDR	2006	STS-127	CDR	2009
Reilly*	1	NASA	American	STS-104	MS2	2001	STS-117	MS4	2007			
Reisman	1	NASA	American	STS-132	MS1	2010						
Richards, P.	0	NASA	American	STS-102	MS2	2001						
Robinson	2	NASA	American	STS-114	MS2	2005	STS-130	MS2	2010			
Rominger	3	NASA	American	STS-96	CDR	1999	STS-100	CDR	2001			
Ross*	5	NASA	American	STS-88	MS1	1998	STS-110	MS4	2002			
Satcher	0	NASA	American	STS-129	MS4	2009						
Schlegel	1	ESA	German	STS-122	MS3	2008						
Sellers	0	NASA	American	STS-112	MS3	2002	STS-121	MS4	2006	STS-132	MS4	2010
Smith, S.	3	NASA	American	STS-110	MS5	2002						

(continued)

Shuttle-ISS Assembly Mission Crew Members (continued)

Name	Prev Flts	Agency	Nationality	1st Visit	Des.	Year	2nd Visit	Des.	Year	3rd Visit	Des.	Year	4th Visit	Des.	Year
Stefanyshyn-Piper	0	NASA	American	STS-115	MS3	2006	STS-126	MS3	2008						
Stott	1	NASA	American	STS-133	MS4	2011									
Sturckow	0	NASA	American	STS-88	PLT	1998	STS-105	PLT	2001	STS-117	CDR	2007	STS-128	CDR	2009
Swanson	0	NASA	American	STS-117	MS2	2007	STS-119	MS2	2009	ISS-39 Flight Engineer/ISS-40 Commander					
Tani	0	NASA	American	STS-108	MS2	2001									
Tanner	2	NASA	American	STS-97	MS1	2000	STS-115	MS1	2006						
Thomas, A.	2	NASA	American	STS-102	MS1	2001	STS-114	MS3	2005						
Tokarev	0	RSA	Russian	STS-96	MS5	1999	ISS-12 Flight Engineer								
Usachev	2	RSA	Russian	STS-101	MS5	2000	ISS-2 Commander								
Virts	0	NASA	American	STS-130	PLT	2010	ISS-42/43 Flight Engineer								
Vittori	1	ESA	Italian	STS-134	MS2	2011	ISS-2 Flight Engineer								
Voss J. S.	3	NASA	American	STS-101	MS3	2000	ISS-18/19/20 Flight Engineer; ISS-38 Flight Engineer/ISS-39 Commander								
Wakata	1	JAXA	Japanese	STS-92	MS5	2000									
Walheim	0	NASA	American	STS-110	MS1	2002	STS-122	MS2	2008	STS-135	MS2	2011			
Weber	1	NASA	American	STS-101	MS1	2000									
Wetherbee*	4	NASA	American	STS-102	CDR	2001	STS-113	CDR	2002						
Wheelock	0	NASA	American	STS-120	MS3	2007	ISS-24/25 Flight Engineer								
Wilcutt*	3	NASA	American	STS-106	CDR	2000									
Williams J.	0	NASA	American	STS-101	MS2	2000	ISS-13 Flight Engineer; ISS-21 Flight Engineer/ISS-22 Commander								
Williams, D.	2	CSA	Canadian	STS-118	MS3	2007									
Wilmore	0	NASA	American	STS-129	PLT	2009									
Wilson	0	NASA	American	STS-121	MS3	2006	STS-120	MS2	2007	STS-131	MS3	2010			
Wisoff*	3	NASA	American	STS-92	MS3	2000	ISS-12 Commander								
Wolf	2	NASA	American	STS-112	MS1	2002	STS-127	MS1	2009						
Yamazaki	0	JAXA	Japanese	STS-131	MS4	2010									
Yurchikin	0	RSA	Russian	STS-112	MS4	2002	ISS-15 CDR; ISS-24/25 FE; ISS-36 FE/ISS-37 CDR								
Zamka	0	NASA	American	STS-120	PLT	2007	STS-130	CDR	2010						

Appendix 3

Shuttle-ISS Space Station Manifest 1993–2011

Continuing the list in the companion volume *Linking the Space Shuttle and Space Stations: Early Docking Technologies from Concept to Implementation*, here is the evolving manifest for Shuttle flights associated with Mir and the International Space Station.

The majority of this data derives from issues of the "NASA Mixed Fleet Payload Flight Assignments" compiled mainly to assist the aerospace community. They were published by the Customer Service Division of NASA Headquarters and reflected the assignments at the date of issue. In addition to pending flights and payloads they listed missions that had already been flown. The manifests are very useful for following the development and fates of payloads that were intended to be launched by Shuttle. From the late 1990s they appeared more on-line than in print, replacing the earlier manifests which, unless downloaded or printed out, became harder to find.

Manifest Dated April 29, 1993

(Courtesy Space Shuttle Almanac)

In this manifest the first flight to Mir (Phase-I) of the ISS program was listed and those associated to Space Station Freedom removed. The manifest also listed the first Space Station launch opportunities including a return to Mir by a Shuttle carrying a Spacelab Long Module payload configuration.

March 16, 1995, STS-71 Spacelab Mir Long Module #2
May 23, 1996, STS-79 Space station flight opportunity 1
September 19, 1996, STS-82 Space Station flight opportunity 2
January 20, 1997, STS-85 Space Station flight opportunity 3
March 6, 1997, STS-86 Space Station flight opportunity 4
August 28, 1997, STS-90 Space Station flight opportunity 5
November 20, 1997, STS-92 Space Station flight opportunity 6
March 5, 1998, STS-94 Space Station flight opportunity 7

© Springer International Publishing Switzerland 2017
D.J. Shayler, *Assembling and Supplying the ISS*, Springer Praxis Books,
DOI 10.1007/978-3-319-40443-1

April 30, 1998, STS-95 Space Station flight opportunity 8
June 11, 1998, STS-96 Space Station flight opportunity 9
July 16, 1998, STS-97 Spacelab Mir 2 (Revisit) Long Module
October 1, 1998, STS-99 Space Station flight opportunity 10
January 21, 1999, STS-101 Space Station flight opportunity 11
February 25, 1999, STS-102 Space Station flight opportunity 12
April 15, 1999, STS-103 Space Station flight opportunity 13
May 13, 1999, STS-104 Space Station flight opportunity 14
June 24, 1999, STS-105 Space Station flight opportunity 15

Manifest Dated March 1, 1994

(Courtesy Space News March 7–13, 1994)
A year later the full Shuttle-Mir program of ten dockings is listed, together with a program of thirty Shuttle station assembly and resupply missions.

STS-71, Atlantis, March 30, 1995, Mir Docking #1 (Spacelab long module)
STS-74, Atlantis, October 26, 1995, Mir docking #2
STS-77, Atlantis, March 21, 1996, Mir docking #3
STS-79, Discovery, June 27, 1996, Mir docking #4
STS-81, Atlantis, September 6, 1996, Mir docking #5 & CHRISTA-SPAS
STS-82, Discovery, November 7, 1996, Mir docking #6
STS-84, Atlantis, January 30, 1997, Mir docking #7
STS-86, Discovery, April 17, 1997, Mir docking #8
STS-88, Atlantis, June 26, 1997, Mir docking #9
STS-90, Discovery, October 2, 1997, Mir docking #10
STS-91, Endeavour, December 4, 1997, Space Station 1 (Docking node)
STS-93, Discovery, February 26, 1998, Space Station 2 (Airlock Module)
STS-95, Endeavour, May 25, 1998, Space Station 3 (U.S. Laboratory)
STS-96, Atlantis, July 30, 1998, Space Station 4 (assembly flight)
STS-97, Discovery, September 24, 1998, Space Station 5 (ACRV Soyuz)
STS-98, Endeavour, October 29, 1998, Space Station 6 (utilization flight #1)
STS-99, Atlantis, December 3, 1998, Space Station 7 (assembly flight)
STS-100, Discovery, February 18, 1999, Space Station 8 (assembly flight)
STS-101, Endeavour, March 18, 1999, Space Station 9 (utilization flight #2)
STS-102, Atlantis, April 29, 1999, Space Station 10 (assembly flight)
STS-104, Endeavour, July 29, 1999, Space Station 11 (assembly flight)
STS-105, Atlantis, September 9, 1999, Space Station 12 (assembly flight)
STS-107, Endeavour, December 12, 1999, Space Station 13 (assembly flight)
STS-108, Discovery, January 29, 2000, Space Station 14 (Japan flight #1)
STS-109, Atlantis, March 2, 2000, Space Station 15 (Japan flight #2/JEM)
STS-111, Discovery, December 3, 1998, Space Station 16 (utilization flight #3)
STS-112, Atlantis, July 20, 2000, Space Station 17 (Japan flight #3)
STS-114, Discovery, November 2, 2000, Space Station 18 (utilization flight #4)
STS-115, Atlantis, November 30, 2000, Space Station 19 (assembly flight)

STS-118, Discovery, April 6, 2001, Space Station 20 (utilization flight #5)

STS-120, Endeavour, June 14, 2001, Space Station 21 (ESA flight #1 Columbus laboratory)

STS-122, Discovery, October 4, 2001, Space Station 22 (ESA flight #2)

STS-123, Endeavour, November 15, 2001, Space Station 23 (utilization flight #6)

STS-125, Discovery, February 28, 2002, Space Station 24 (utilization flight #7)

STS-127, Atlantis, May 9, 2002, Space Station 25 (assembly flight)

STS-128, Discovery, July 18, 2002, Space Station 26 (assembly flight)

STS-129, Endeavour, August 29, 2002, Space Station 27 (assembly flight)

STS-132, Endeavour, February 20, 2003, Space Station 28 (utilization flight #8)

STS-133, Atlantis, March 13, 2003, Space Station 29 (utilization flight #9)

STS-135, Discovery, July 3, 2003, Space Station 30 (utilization flight #10)

Payload Flight Assignments: NASA Mixed Fleet, April 1994

This was an interesting release connected to the March 1, 1994 manifest above. Only three ISF flights are included and without any indication of launch date. The Shuttle-Mir missions had more detail, but the Space Station Assembly Flights had less detail than before. Note that the CY98-02 Shuttle manifest planning was omitted, pending final resolution of the required Space Station assembly sequence launch dates.

Industrial Space Facility (ISF)

ISF-01 July 1997
ISF-02 January 1998
ISF-03 January 1999

Shuttle/Mission (S/MM)

S/MM-01 STS-71, May 1995, Atlantis 9 + 1 days, 7 crewmembers Spacelab Mir Long Module

S/MM-02 STS-74, October 1995, Atlantis 6 + 1 days, 5 crewmembers, Docking Module

S/MM-03 STS-77, April 1996, Atlantis 10 days, 7 crewmembers, Spacelab Long Module

S/MM-04 STS-80, August 1996, Atlantis 10 days, 7 crewmembers

S/MM-05 STS-81, September 1996, Discovery 10 days, 7 crewmembers, Spacelab Long Module

S/MM-06 STS-83, December 1996, Atlantis 10 days, 7 crewmembers

S/MM-07 STS-84, March 1997, Discovery 10 days, 7 crewmembers, Spacelab Long Module

S/MM-08 STS-87, July 1997, Discovery 10 days, 7 crewmembers, Solar Dynamic; JFD

S/MM-09 STS-89, September 1997, Atlantis 10 days, 7 crewmembers, Spacelab Long Module

S/MM-10 STS- 91, November 1997, Discovery 10 days, 7 crewmembers

Space Station Assembly Flight

A = *American element*
E = *European element*
J = *Japanese element*
J/A = *Japanese and American element*

SSAF01-1A STS-92, December 1997, Endeavour 9 + 1 days, 5 crewmembers (Node 1,
    PMA1, PMA2, 2 lab sys, 2 ISPR racks)
SSAF02-2A, January 1998
SSAF03-3A, May 1998k
SSAF04-4A, July 1998
SSAF05-5A, October 1998
SSAF06-6A, February 1999
SSAF07-7A, June 1999
SSAF08-8A, July 1999
SSAF09-9A, September 1999
SSAF10-10A, November 1999
SSAF11-1J/A, February 2000
SSAF12-1J, March 2000
SSAF13-11A, August 2000
SSAF14-2J/A, November 2000
SSAF15-1E, March 2001
SSAF16-2E, May 2001
SSAF17-12A, October 2001
SSAF18-13A, February 2002
SSAF19-14A, May 2002
SSAF20-15A, June 2002

Payload Flight Assignments: NASA Mixed Fleet, February 1995

The three ISF flights retained their dates and the total number of Shuttle-Mir missions was
officially listed as seven dockings with three others under review. While the first ISS
assembly flight was manifested before the tenth and final Mir docking mission the planned
launch of ESA's Columbus module under SSAF 1E was not in this manifest. The CY99-02
Shuttle manifest planning was omitted pending final resolution of the required Space
Station assembly sequence launch dates.

Industrial Space Facility (ISF)

ISF-01, July 1997
ISF-02, January 1998
ISF-03, January 1999

Shuttle/Mir Mission (S/MM)

S/MM-01 STS-71, June 1995, Atlantis 9 + 1 days, 7 crewmembers Up 8 down, Spacelab
    Mir Long Module

S/MM-02 STS-74, November 1995, Atlantis 6 + 1 days, 5 crewmembers, Docking Module

S/MM-03 STS-76, April 1996, Atlantis 9 days, 6 crewmembers, Spacehab Single Module; EDFT-04 (under review)

S/MM-04 STS-79, August 1996, Atlantis 10 days, 6 crewmembers, Spacehab 5 (under review) Double Module

S/MM-05 STS-81, December 1996, Atlantis 10 days, 6 crewmembers, Spacehab Double Module

S/MM-08 STS-82, February 1997, Discovery (under review OR HST SM-02) 10 days, 7 crewmembers,

S/MM-06 STS-84, May 1997, Atlantis 10 days, 7 crewmembers Spacehab Double Module

S/MM-09 STS-85, July 1997, Discovery (under review or Spacehab 6) 11 days, 5 crewmembers, Spacehab 6; JFD

S/MM-07 STS-86, September 1997, Atlantis 10 days, 6 crewmembers, Energy Module, EDFT-06

(SSAF-01-2A STS-88, December 1997, Endeavour (Node 1 PMA1, PMA2; see below)

S/MM-10 STS-90, April 1998, Discovery (Under Review OR Spacehab 8/WSF-03) 10 days, 6 crewmembers

Space Station

SSAF-01-2A STS-88, December 1997, Endeavour (Node 1 PMA1, PMA2)

(S/MM-10 STS-90, April 1998, Discovery (Under Review OR Spacehab 8/WSF-03) 10 days, 6 crewmembers; see above if authorized)

SSAF-02-3A STS-91, June 1998, Endeavour (Z1 Truss w/ CMGs, PMA3) 9 + 1 days, 7 crewmembers

SSAF-03-4A STS-93, September 1998, Discovery (P6 PV Module) 8 + 1 days, 5 crewmembers

SSUF-01, October 1998

SSAF-04-5A, STS-94 November 1998, Endeavour (US Laboratory) 9 + 1 days, 5 crewmembers

SSAF-05-6A, STS-95 December 1998, Atlantis (MPLM (Lab O/F) SLP (SSRMS) 12 + 1 days 5 crewmembers

SSAF-06-7A, February 1999

SSUF-02, April 1999

SSAF-07-8A, June 1999

SSAF-08-9A, July 1999

SSAF-09-10A, September 1999

SSUF-03, October 1999

SSAF-10-11A, November 1999

SSAF-12-1J/A, February 2000

SSAF-13-1J, March 2000

SSUF-04, June 2000

SSAF-11-12A, August 2000

SSAF-15-2J/A, November 2000
SSUF-05, January 2001
[SSAF-??-1E]
SSAF-18-15A, March 2001
SSAF-16-2E, May 2001
SSAF-14-13A, October 2001
SSUF-06, October 2001
SSAF-17-14A, February 2002
SSAF-19-16A, May 2002
SSAF-20-17A, June 2002

Payload Flight Assignments: NASA Mixed Fleet, June 1996, and NASA Shuttle Schedule Sheet as of July 1996

By the time this manifest was released all mention of the ISF had been removed. By now three Shuttle missions to Mir had been completed and the manifest finished with the ninth docking and no mention of the originally planned tenth docking. Full details of the now designated ISS program were presented but details of missions 21 through 34 were not finalised.

Shuttle/Mir Mission (S/MM)

S/MM-04 STS-79, July 31, 1996, Spacehab DM Atlantis, 6 crew, NASA resident crew exchange, 9 + 1 days
S/MM-05 STS-81, December 5, 1996, Spacehab DM Atlantis, 6 crew, NASA resident exchange 9 + 1 days
S/MM-06 STS-84, May 1, 1997, Spacehab DM Atlantis, 6 crew, NASA resident crew exchange, 9 + 1 days
S/MM-07 STS-86, September 11, 1997, Spacehab DM Atlantis, 6 crew, NASA resident crew exchange, 9 + 1 days
ISS-01-2A STS-88, December 4, 1997, (Node 1 PMA-1 & PMA-2) 5 crew 7 + 2 days
S/MM-08 STS-89, January 15, 1998, Spacehab DM Discovery, 6 crew, NASA resident crew exchange 9 + 1 days
S/MM-09 STS-91, May 29, 1998, Spacehab DM Discovery, 6 crew, 5 up 6 down, NASA resident crew return 9 + 1 days
S/MM-10 [no details listed]

Space Station Assembly Flight

A = American element
E = European element
J = Japanese element
J/A = Japanese and American element
UF = Utilization Flight

ISS-01-2A, December 4, 1997 planned launch December 1998, STS-88 Endeavour, 5 crew 7 + 2 days (Node 1, PMA-1 & PMA-2)

ISS-02-3A, June 1998 planned launch July 2, 1998, STS-92 Spacelab pallet, Endeavour 5 crew 9 + 1 days (Z1 Truss, CMGs, PMA-3)

ISS-03-4A, September 1998, planned November 5, 1998, STS-94 Discovery, 5 crew 8 + 1 days (P6 PV Module)

ISS-04-5A, November 1998, planned launch December 3, 1998, STS-95 Endeavour, 7 crew 9 + 1 day, (US Laboratory)

ISS-05-6A, December 1998, planned launch January 14, 1999, STS-96 MPLM, Spacelab pallet, Atlantis, 5 crew 11 + 1 days, (Lab out fitting, SSRMS)

ISS-06-UF1, February 1999, planned launch March 25,1999, STS-97 MLPM, Spacelab pallet, Discovery 5 crew 10 + 1 days (ISPRS, PV battery)

ISS-07-7A, March 1999, planned launch date April 22, 1999, STS-98 Spacelab pallet, Endeavour, 5 crew 11 + 1 days, (Airlock, HP Gas)

ISS-08-8A, June 1999, planned launch date June 10, 1999, STS-99 Atlantis, 5 crew 8 + 1 days, (S0 Truss)

ISS-09-UF2, July 1999, planned launch date August 5, 1999, STS-100 MPLM, Discovery, 5 crew 9 days, (MBS, ISPRS)

ISS-10-9A, July 1999, planned launch date September 10, 1999, STS-101 Endeavour 5 crew 7 + 1 days (S1 Truss)

Note: Starting in FY2000 (from September 1999) there would be fourteen additional Space Shuttle/ISS Assembly flights which concluded with the completion of the ISS assembly planned for June 2002. These include the launch assembly of the Japanese Experiment Module, the European Module, and the Russian Solar Arrays. In addition there are four additional utilization flights to the ISS prior to completion of assembly.

ISS-11-9A.1, September 1999, STS-102, Atlantis, November 4, 1999 (SPP)

ISS-12-11A, November 1999, STS-104, Discovery, January 13, 2000 (P1 Truss)

ISS-13-12A, August 2000, STS-105, Endeavour, February 10, 2000 (P3/4, PV Module)

ISS-14-10A, October 2001, STS-106, Atlantis, March 23, 2000 (Node 2)

ISS-15-1J/A, February 2000, STS-107, Endeavour, June 22, 2000 (JEM ELM PS, SPDM, ULC)

ISS-16-13A, February 2002, STS-108, Atlantis, August 3, 2000 (S3/4, PV Module)

ISS-17-1J, March 2000, STS-109, Endeavour, November 9, 2000 (JEM PM, JEM RMS)

ISS-18-UF3, October 1999, STS-110, Discovery, December 7, 2000 (MPLM)

ISS-19-UF4, June 2000, STS-111, Atlantis, January 11, 2001 (ULC x 2)

ISS-20-2J/A, November 2000, STS-112, May 3, 2001, (JEM EF, ELM-ES, ULC)

ISS-21-UF5, January 2001

ISS-22-14A, March 2001

ISS-23-2E, May 2001

ISS-24-15A, May 2002

ISS-25-16A, June 2002

ISS-26-UF6, October 2001

ISS-27-17A, May 2002

ISS-28-19A, June 2002, planned launch date June 2002

ISS-29 TBD, No payload assigned

ISS-30 TBD, No payload assigned
ISS-31 TBD, No payload assigned
ISS-32 TBD, No payload assigned
ISS-33 TBD, No payload assigned
ISS-34 TBD, No payload assigned

Payload Flight Assignments: NASA Mixed Fleet, June 1997

This manifest summarized Shuttle flights through the calendar year 2003. Those for calendar years 2002–2003 were under review pending the resolution of details in the assembly sequence of the ISS, therefore the European Columbus Module, Habitation Module, Centrifuge Module were not listed.

Shuttle-Mir

STS-86 (S/MM-07), September 1997, Atlantis, 6 crew, 10 + 1 days, NASA resident crew exchange, Spacelab DM

STS-89 (S/MM-08), January 1998, Endeavour, 6 crew, 9 + 1 days, NASA resident crew exchange, Spacelab DM

STS-91 (S/MM-09), June 1998, Discovery, 6 crew, 10 days, 1 NASA resident crew down, Spacelab SM, AMS-01

International Space Station (ISS)

ISS-01-2A, July 1998, STS-88 Endeavour, Node 1 PMA1/PMA2, 7 crew 11 days

ISS-02-2A.1, December 1998, STS-96 Endeavour, (ICM or Spacehab SM logistics) 5 crew 11 + 1 days

ISS-03-3A, January 1999, STS-92 Atlantis, (SL Pallet Z1 Truss, CMGs) 5 crew 9 + 2 days

ISS-04-4A, March 1999, STS-97 Discovery, (PV Module P6), 5 crew 8 + 2 days

ISS-05-5A, May 1999, STS-98 Endeavour, (US Lab) 5 crew 9 + 2 days

ISS-06-6A, June 1999, STS-99 Atlantis, MPLM 1 (P) -01 (lab outfitting) Spacelab pallet; SSRMS crew rotation, 5 crew 11 + 2 days

ISS-07-7A, August 1999, STS-100 Discovery, Airlock Spacelab pallet, HP Gas.5 crew 11 + 2 days

ISS-08-7A.1, October 1999, STS-102 Atlantis, 5 days

ISS-09-UF1, January 2000, STS-104 Discovery, 5 crew, MPLM 2 (P) 01 Spacelab pallet

ISS-10-8A, February 2000, STS-105 Endeavour, 5 crew, ITS S0

ISS-11-UF2, March 2000, STS-106, 5 crew, MPLM 1–02 (MBS)

ISS-12-9A, June 2000, STS-108 Endeavour, 5 crew, (ITS S1, CETA)

ISS-13-9A.1, July 2000, STS-109 Atlantis, 5 crew, (Science Power Platform - Russian)

ISS-14-11A, October 2000, STS-111 Endeavour, 5 crew, (ITS P1, CETA)

ISS-15-12A, November 2000, STS-112 Atlantis, 5 crew, (ITS P3, PV Module P4)

ISS-16-13A, March 2001, STS-114 Endeavour, 5 crew, (ITS S3/4, PV Array)

ISS-17-10A, April 2001, STS-115 Atlantis, 5 crew, (Node 2)

ISS-18-1J/A, May 2001, STS-116 Discovery, 5 crew, (JEM ELM PS)

ISS-19-1J, August 2001, STS-118 Atlantis, 5 crew (JEM PM, JEM RMS)

ISS-20-UF3, September 2001, STS-119 Discovery, 5 crew, MPLM 3–01

ISS-21-UF4, January 2002, STS-121 Atlantis, 5 crew, (Spacelab pallet (SPDM) Express Pallet 1, AMS-02 (TBD))

ISS-22-2J/A, February 2002, STS-122 Discovery, 5 crew, (JEM EF, JEM ELM-ES Spacelab pallet)

ISS-23-14A, May 2002, STS-124 Atlantis, 5 crew, (EDO Pallet, Spacelab Pallet, Cupola)

ISS-24-UF5, June 2002, STS-125 Discovery, 5 crew, (MPLM-2-02, Express Pallet 2)

ISS-25-TBD, July 2002, STS-126 Endeavour, 5 crew

ISS-26-TBD, October 2002, STS-128 Discovery, 5 crew

ISS-27-TBD, November 2002, STS-129 Endeavour, 5 crew

ISS-28-TBD, March 2003, STS-131 Discovery, 5 crew

ISS-29-TBD, April 2003, STS-132 Endeavour, 5 crew

ISS-30-TBD, July 2003, STS-134 Discovery, 5 crew

ISS-31-TBD, August 2003, STS-135 Atlantis, 5 crew

ISS-32-TBD, October 2003, STS-136 Endeavour, 5 crew

ISS-33-TBD, November 2003, STS-137 Discovery, 5 crew

International Space Station Assembly Sequence

October 1998: Planning Reference, NASA JSC

Published just prior to the beginning of ISS assembly, this release pointed out: "Flights beyond the launch of the Service Module in July 1999 are under review and will not be finalized until a meeting of all international partners in December. The timetable shown for flights through July 2000 are planning dates only for the Space Shuttle manifest."

2A, STS-88, December 1998, Node 1 (Unity) PMA- & PMA-2

2A.1, STS-96, May 1999, Spacehab DM (logistics)

2A.2, STS-101, August 1999, Spacehab DM (logistics)

3A, STS-92, October 1999, ITS Z1, PMA-3, CMGs

4A, TBD, December 1999, P6

5A, TBD, February 2000, U.S. Laboratory Module

5A.1, TBD, March 2000, MPLM

6A, TBD, April 2000, MPLM; SSRMS (Canadarm2)

7A, TBD, July 2000, Joint Airlock

Completion of Early Assembly Phase

7A.1, MPLM

UF1, MPLM

8A, ITS S0, MT

UF2, MPLM, MBS

9A, ITS S1, CETA Cart A

9A.1, Science Power Platform (SPP) – Russian; ESA ERA

11A, ITS P1, CETA Cart B

12A, ITS P3/P4
12A.1, ITS P5
13A, ITS S3/S4
10A, Node 2
1J/A, JEM ELM PS
1J, JEM, JEM-RMS
UF3, MPLM
UF4, SPDM, AMS-2
2J/A, JEM EF
14A, Cupola
UF5, MPLM
20A, Node 3
17A, MPLM
1E, ESA Columbus Laboratory
18A, U.S. Crew Return Vehicle
19A, MPLM; ITS S5
15A, S6
UF6, MPLM
UF7, Centrifuge Accommodations Module
16A, U.S. Habitation Module

ISS Assembly Sequence

Revision E (March 2000 Planning Reference); updated October 30, 2000.

With the launch of the Service Module imminent in July 2000 the Shuttle manifest was revised to incorporate the delay but without details until the Russian module was safely on-orbit and attached to Zarya. Hence missions beyond STS-109 were not yet assigned an STS number.

STS-88, December 4, 1988, 2A, Unity Node 1 (PMA-1 & PMA-2)
STS-96, May 27, 1999, 2A.1, Spacehab logistics flight
STS-101, April 24, 2000, 2A.2a, Spacehab maintenance flight
STS-106, August 19, 2000, 2A.2b, Spacehab logistics flight
STS-92, September 21, 2000, 3A, ITS Z1, PMA-3, CMGs
STS-97, November 30, 2000, 4A, ITS P6
STS-98, January 18, 2001, 5A, U.S. Destiny Laboratory Module
STS-102, February 15, 2001, 5A.1, MLPM *Leonardo* logistics flight
STS-100, April 19, 2001, 6A, MPLM *Rafaello* logistics flight; SSRMS (*Canadarm2*)
STS-104, May 17, 2001, 7A, Joint Airlock (*Quest*)
STS-105, June 21, 2001, 7A.1, MPLM *Donatello* logistics flight
STS-109, August 23, 2001, UF1, MPLM logistics flight
STS-xxx, October 2001, 8A, ITS S0; MT
STS-xxx, January 2002, UF2, MPLM logistics flight; MBS
STS-xxx, February 2002, 9A, ITS S1; CETA Cart A
STS-xxx, May 2002, 11A, ITS P1; CETA Cart B

STS-xxx, June 2002, 9A.1, Science Power Platform (Russian hardware)
STS-xxx, September 2002, 12A, ITS P3/P4
STS-xxx, October 2002, 12A.1, ITS P5; MPLM logistics flight
STS-xxx, January 2003, 13A, ITS S3/S4
STS-xxx, February 2003, 10A, U.S. Node 2
STS-xxx, May 2003, 10A.1, Propulsion Module
STS-xxx, June 2003, 1J/A, JEM ELM PS
STS-xxx, September 2003, 1J, Kibo Japanese Experiment Module (JEM); Japanese RMS
STS-xxx, October 2003, UF3, MPLM logistics flight; Express Pallet
STS-xxx, January 2004, UF4, Express Pallet; SPDM (Dextre)
STS-xxx, February 2004, 2J/A, JEM-EF
STS-xxx, May 2004, 14A, Cupola; SPP arrays with truss
STS-xxx, June 2004, UF5, MLPM logistics flight; Express Pallet
STS-xxx, September 2004, 20A, U.S. Node 3
STS-xxx, October 2004, 1E, European ESA Laboratory Columbus
STS-xxx, January 2005, 17A, MPLM logistics flight
STS-xxx, February 2005, 18A, Crew Return Vehicle (CRV)
STS-xxx, March 2005, 19A, MPLM logistics flight
STS-xxx, May 2005, 15A, S6
STS-xxx, June 2005, UF7, Centrifuge Accommodation Module (CAM)
STS-xxx, July 2005, UF6, MPLM logistics flight
STS-xxx, September 2005, 16A, U.S. Habitation Module

ISS Assembly Sequence, October 30, 2000

With the docking of Zvezda with Zarya, resident crewing could begin, and it was now possible to make detailed planning to expand the station. Missions which had not been assigned a specific STS number in the earlier manifest were now given that (albeit the details were still subject to change).

STS-110, 2002, 8A, ITS S0; MT
STS-111, 2002, UF2, MPLM logistics flight; MBS
STS-112, 2002, 9A, ITS S1; CETA Cart A
STS-114, 2002, ULF1, MPLM logistics flight
STS-115, 2002, 11A, ITS P1; CETA Cart B
STS-116, 2002, 9A.1, Science Power Platform (Russian hardware)
STS-118, 2003, 12A, ITS P3/P4
STS-119, 2003, 12A.1, ITS P5; Spacehab Single Module logistics flight
STS-121, 2003, 13A, ITS S3/S4
STS-122, 2003, 13A.1, ITS S5; Spacehab Single Module logistics flight
STS-124, 2003, 10A, U.S. Node 2
STS-125, 2004, 1J/A, JEM ELM PS
STS-127, 2004, 10A.1, Propulsion Module
STS-126, 2004, 1J, Kibo Japanese Experiment Module (JEM); Japanese RMS
STS-128, 2004, UF3, MPLM logistics flight; Express Pallet.

STS-TBD, 2004, UF4, Express Pallet; SPDM (Dextre)
STS-132, 2005, 2J/A, JEM-EF; Cupola
STS-133, 2005, UF5, MPLM logistics flight
STS-130, 2005, 1E, European ESA Laboratory Columbus.
STS-136, 2005, UF6, MPLM logistics flight
STS-TBD, 2005, 14A, SPP arrays with truss
STS-TBD, 2005, 20A, U.S. Node 3
STS-138, 2005, 16A, U.S. Habitation Module
STS-139, 2005, 17A, MPLM logistics flight
STS-140, 2006, 18A, Crew Return Vehicle (CRV) #1
STS-141, 2006, 19A, MPLM logistics flight
STS-142, 2006, 15A, S6 arrays
STS-143, 2006, UF7, Centrifuge Accommodation Module (CAM)

STS-ISS Manifest, June 1, 2005

Following the loss of Columbia in February 2003 the Shuttle assembly missions were grounded pending the investigation and subsequent Return-To-Flight program. In the interim, some planned ISS hardware elements (including the U.S. Habitation Module, Centrifuge Module, and Node 4) were removed from the manifest. This new manifest was released just before the STS-114 Return-To-Flight mission with all missions after STS-120 "under review."

STS-114, LF1, Return to Flight test mission #1; MPLM logistics flight; CMGs
STS-121, ULF1.1, Return to Flight test mission #2; MPLM logistics flight
STS-115, 12A, ITS P3/P4 truss; solar arrays
STS-116, 12A.1, P5 truss; Spacehab single module; logistics and supplies
STS-117, 13A, S3/S4 truss, solar arrays
STS-118, 13A.1, S5 truss, Spacehab Single Module
STS-119, 15A, S6 truss, solar arrays
STS-120, 10A, U.S. Node 2 (ISS Core Complete Mission)
STS-xxx, ULF2, MPLM, logistics
STS-xxx, 1E, Columbus European laboratory
STS-xxx, UF3, MPLM, logistics
STS-xxx, UF4, SPDM (Dextre); EDO pallet
STS-xxx, UF5, MPLM, logistics
STS-xxx, UF4.1, Express pallet; S3 attached payload
STS-xxx, UF6, MPLM, logistics
STS-xxx, 1J/A, JEM Logistics Module; Express Pallet
STS-xxx, 1J, Kibo Japanese Laboratory Module (JEM); Japanese RMS
STS-xxx, ULF3, MPLM, logistics
STS-xxx, 9A.1, Science Power Platform solar arrays with truss, MPLM logistics
STS-xxx, UF7, Centrifuge Accommodation Module
STS-xxx, 2J/A, JEM Exposed Facility; SPP solar arrays
STS-xxx, ULF5, MPLM, logistics
STS-xxx, 14A, Cupola, Express Pallet, EDO Pallet

STS-ISS Manifest, September 2006

With the successful Return-To-Flight missions (STS-114 in July 2005 and STS-121 a year later) NASA revised the Shuttle manifest to include missions that were meant to complete ISS assembly prior to the retirement of the fleet in 2010 (later delayed until 2011). A final Hubble servicing mission (SM-4/STS-125) was also reinserted into the sequence. The names in brackets are the final vehicles/missions/payloads as assigned after this manifest was released. In fact this was the final major manifest projection in the Shuttle-ISS assembly program and ending a process that began over twenty years earlier with the targeting of Shuttle missions to support the assembly of Space Station Freedom.[1]

Dec 14, 2006, STS-116, Discovery, 12A.1, P5 truss; Spacehab SM; ICC

Feb 22, 2007, STS-117, Atlantis, 13A, S3/S4 truss

Jun 11, 2007, STS-118, Endeavour, 13A.1, S5 truss; Spacehab SM; ESP-3

Aug 9, 2007, STS-120, Atlantis (Discovery), 10A, Node 2 *(Harmony)*

Oct 2007, STS-122, (Atlantis), 1E, Columbus European laboratory

Dec 2007, STS-(123), (Endeavour), 1J/A, Kibo Japanese Experiment Logistics Module Pressurized Section (ELM-PS); Spacelab pallet; Canadian Special Purpose Dextrous Manipulator (Dextre)

Feb 2008, STS-(124), (Discovery), 1J, Kibo Japanese Experiment Module – Pressurized Module (JEM-PM); Japanese RMS

Jun 2008, STS-119, (Discovery), 15A, S6 truss

Aug 2008, STS-(126), (Endeavour), ULF2, MPLM logistics

Oct 2008, STS-(127), (Endeavour), 2J/A, Kibo Japanese Experiment Module Exposed Facility (JEM-EF); Japanese Experiment Logistics Module – Exposed Section (ELM-ES); Spacelab Pallet

Jan 2009, STS-(128), (Discovery), 17A, MPLM, logistics

Six person resident crew capabilities established

Apr 2009, STS-(129), (Atlantis), ULF3, Express Logistics Carrier 1 (ELC-1); Express Logistics Carrier 2 (ELC-2)

Jul 2009, STS-(131), (Discovery), 19A, MPLM, logistics

Oct 2009, STS-(132), (Atlantis), ULF4, Russian Mini Research Module (MRM1) *Rassvet*

Jan 2010, STS-(130), (Endeavour), 20A, Node 3 *(Tranquility)*, Cupola

Jul 2010, STS-(133), (Discovery), ULF5, Permanent Multi-purpose Module (PMM); Express Logistics Carrier 4 (ELC-4)

ISS Assembly Complete – Shuttle fleet retires (delayed to 2011 & extra missions added)

2011, STS-134, Endeavour, ULF6, Express Logistics Carrier 3 (ELC-3); AMS-02

2011, STS-135, Atlantis, ULF7, MPLM, logistics

JULY 2011, ISS Assembly Complete – Shuttle fleet retires its final objective achieved

---

[1] Praxis Manned Spaceflight Log 1961–2006, p775–777.

Bibliography

As with previous projects, I referred to a wide range of material over several years in support of both books which make up this project and the primary sources have been listed in footnotes. In compiling the tables, some data was found to conflict or simply could not be found. I would welcome any additional information to fill in the gaps in the tables for future reference and completeness.

Interviews

A number of personal interviews were conducted in connection with the titles of this project and in support of related research and writings. In several cases, I undertook supplementary correspondence via E-mails. The interviews and correspondence that directly relate to this project included:

Name	Date
Akers, Tom	November 11, 2013, plus E-mail, December 11, 2015
Chiao, Leroy	June 6, 2006
Clervoy, Jean-François	August 24, 2006; December 9, 2011, plus December 9, 2015, E-mail December 2015
Crippen, Robert	February 5, 2013
Hawley, Steve	March 1, 2012, E-mail March 8, 2016
Jones, Thomas	August 3, 2006
Kavandi, Janet	October 11, 2015, plus E-mail November 13, 2015
McArthur, William	May 31, 2006
Nelson, George	July 23, 2013
Newman, James	December 6, 2013
Ochoa, Ellen	March 2, 2004
Richards, Paul	November 24, 2013
Rominger, Kent	May 23, 2006
Ross, Jerry	E-mail April 28, 2015
Smith, Steve	May 23, 2006, February 15, 2013
Tanner, Joe	February 28, 2012, plus E-mail December 13, 2015

© Springer International Publishing Switzerland 2017
D.J. Shayler, *Assembling and Supplying the ISS*, Springer Praxis Books,
DOI 10.1007/978-3-319-40443-1

Periodicals

Aviation Week and Space Technology
Capcom
Countdown
Flight International
Journal of the British Interplanetary Society (JBIS)
Orbiter (AIS)
Spaceflight (BIS)

Newspapers

Florida Today
Houston Chronicle
Houston Post
The Daily Telegraph, London
The Times, London
Washington Post

NASA publications

1994–2011   Extensive use of the NASA Shuttle Press Kits and mission information in hardcopy form, on line or from the AIS Shuttle Mission Archive Collection
1984   *Space Transportation System Facilities and Operations, Kennedy Space Center, Florida,* K-STSM-01, Appendix A, April 1984 Revision A
1988   *National Space Transportation System Overview,* September 1988
1993   *Orbiter Processing Facility Payload Processing Support Capabilities,* K-STSM-14.1.13-REVD-OPF, October 1993
1993–2011   *Astronauts and Aeronautics, A Chronology,* NASA SP various editions
1993–2011   *Chronology of KSC and KSC Related Events,* NASA TM various editions
1993   *EVA Tools and Equipment Reference Book* NASA TM-109350/JSC-24066 Rev-B, November 1993
1997   *Walking to Olympus: An EVA Chronology,* David S.F. Portree and Robert C. Treviño, NASA Monographs in Aerospace History #7
2006   *Reference Guide to the International Space Station,* Editor Gary Kitmacher, NASA SP-2006-557
2007   *Survey and Evaluation of NASA-owned Historical Facilities and Properties in the Context of the U.S. Space Shuttle Program, John F. Kennedy Space Center, Brevard County, Florida,* Archaeological Consultants Inc., Sarasota, Florida, Volume 1, *October 2007*
2010   *Historical Survey and Evaluation of the Space Station Processing Facility, John F. Kennedy Space Center, Brevard County, Florida.* Archaeological Consultants, Inc., Sarasota, Florida, September 2010
2011   *Space Shuttle Mission Summary,* Robert D. ('Bob') Legler and Floyd V. Bennett, Mission Operations, DA8, NASA JSC, Houston, Texas, NASA TM-2011-216142, September 2011
2016   *Walking to Olympus, an EVA Chronology, 1997–2011, Volume 2,* Julie B. Ta and Robert C. Treviño, Monographs in Aerospace History, #50, NASA SP-2016-4550

Media publications

1984   *Space Transportation System Press Information, Rockwell International, January 1984*

1998–2000   *International Space Station, Press Information Book, Mission Modules and Station Overview* through September 2000, Boeing

British Interplanetary Society books and articles

1991   *The Proposed USSR Salyut and US Shuttle Docking Mission c.1991*, David J. Shayler, JBIS, Volume 44

2000   *The History of Mir 1986–2000*, Editor Rex Hall

2002   *The International Space Station, From Imagination to Reality*, Editor Rex Hall

2005   *The International Space Station, From Imagination to Reality*, Vol. 2, Editor Rex Hall

Springer-Praxis Series in Space Exploration

2001   *Skylab, America's Space Station*, David J. Shayler

2002   *The Continuing Story of the International Space Station*, Peter Bond

–   *Creating the International Space Station*, David M. Harland and John E. Catchpole

2004   *Walking in Space*, David J. Shayler

–   *The Story of the Space Shuttle*, David M. Harland

2005   *Russia's Cosmonauts, Inside the Yuri Gagarin Training Center*, Rex D. Hall, David J. Shayler and Bert Vis.

–   *The Story of Space Station Mir*, David M. Harland

2007   *Praxis Manned Spaceflight Log 1961-2006*, Tim Furniss, David J. Shayler with Michael D. Shayler

2008   *The International Space Station, Building for the Future*, John E. Catchpole

2010   *Prepare for Launch, The Astronaut Training Process*, Erik Seedhouse

2012   *U.S. Spacesuits*, 2nd Edition, Kenneth S. Thomas and Harold J. McMann

2013   *Manned Spaceflight Log II 2006-2012*, David J. Shayler and Michael D. Shayler

2014   *To Orbit and Back Again, How the Space Shuttle Flew in Space*, Davide Sivolella

–   *Partnership in Space, The Mid to Late Nineties*, Ben Evans

2015   *The Twenty-First Century in Space*, Ben Evans

2016   *Hubble Space Telescope, From Concept to Success*, David J. Shayler with David M. Harland.

–   *Enhancing Hubble's Vision, Service Missions That Expanded Our View of the Universe*, David J. Shayler with David M. Harland

Other books

1985	*The Space Station, An Idea Whose Time Has Come, Editor Theodore R. Simpson,* IEEE Press
1987	*The Space Station, A Personal Journey*, Hans Mark, Duke University Press
1990	*The Space Station Decision, Increment Policy and Technological Choice,* Howard E. McCurdy, New Series in NASA History, John Hopkins University Press
2001	*Space Shuttle, The History of the National Space Transportation System, The First 100 Missions,* Dennis R. Jenkins, Midland Publishing
2002	*The Space Shuttle Decision 1965–1972,* Volume 1, T. A. Heppenheimer,
–	*Development of the Space Shuttle 1972-1981*, Vol. 2, T. A. Heppenheimer, Smithsonian Institute Press
2006	*Sky Walking, An Astronaut's Memoir*, Tom Jones, Smithsonian Books.
2007	*Space Shuttle Fact Archive,* Pocket Space Guide, Robert Godwin, Apogee Books
2012	*International Space Station 1998–2011 (All Stages), Owner's Workshop Manual,* David Baker, Haynes Publishing
2013	*An Astronaut's Guide to Life on Earth,* Chris Hadfield, Macmillan.
–	*Spacewalker, My Journey in Space and Faith as NASA's Record-Setting Frequent Flyer,* Jerry L. Ross, with John Norberg, Purdue University Press.
–	*Great Endeavour, The Missions of the Space Shuttle Endeavour,* Robert A. Adamcik, Apogee Prime
2015	*The Orbital Perspective, An Astronaut's View,* Ron Garan, John Blake Publishing Ltd.
–	*The Ordinary Spaceman, From Boyhood Dreams to Astronaut*, Clayton C. Anderson, Nebraska University Press.

PDF books

1992–2011	*Space Shuttle Almanac - A Comprehensive Overview of 40 Years of Space Shuttle Development & Operations,* Joel W. Powell and Lee Robert Brandon-Cremer, Microgravity Productions. (2011 edition)

About the Author

Space historian David J. Shayler, FBIS (Fellow of the British Interplanetary Society or – as Dave likes to call it – Future Briton In Space!) was born in England in 1955. His lifelong interest in space exploration began by drawing rockets aged five, but it was not until the launch of Apollo 8 to the Moon in December 1968 that this interest in human space exploration became a passion. He recalls staying up all night with his grandfather to watch the Apollo 11 moonwalk. Dave joined the British Interplanetary Society as a Member in January 1976, became an Associate Fellow in 1983, and a Fellow in 1984. He was elected to the Council of the BIS in 2013. His first articles were published by the Society in the late 1970s, and then in 1982 he set up Astro Info Service in order to focus his research efforts (*www.astroinfoservice.co.uk*). His first book was published in 1987, and has been followed by over twenty other titles on the American and Russian space programs, spacewalking, women in space, and the human exploration of Mars. His authorized biography of Skylab 4 astronaut Jerry Carr was published in 2008.

In 1989 Dave applied as a cosmonaut candidate for the UK's Project Juno program in cooperation with the Soviet Union (now Russia). The mission was to spend seven days in space aboard the Mir space station. Dave did not reach the final selection, but progressed further than he expected. The mission was flown in May 1991 by Helen Sharman.

In support of his research, Dave has visited NASA field centers in Houston and Florida in the USA and the Yuri Gagarin Cosmonaut Training Center in Russia. It was during these trips that he was able to conduct in-depth research, interview many space explorers and workers, tour training facilities, and handle real space hardware. He also gained a valuable insight into the activities of a space explorer, as well as the realities of not only flying and living in space but also what goes into preparing for a mission and planning future programs.

Dave is a friend of many former astronauts and cosmonauts, some of whom have accompanied him on visits to schools all across the UK. For over thirty years, he has delivered space-themed presentations and workshops to members of the public in an effort to increase popular awareness of the history and development of human space exploration.

© Springer International Publishing Switzerland 2017
D.J. Shayler, *Assembling and Supplying the ISS*, Springer Praxis Books,
DOI 10.1007/978-3-319-40443-1

Dave has a particular desire to help the younger generation to develop an interest in science and technology and the world around them.

Dave lives in the West Midlands region of the UK and enjoys spending time with his wife Bel, a youthful and enormous white German Shepherd that answers to the name of Shado, and indulging in his loves of cooking, fine wines, and classical music. His other interests are in reading, especially about military history, in particular the Napoleonic Wars, visiting historical sites and landmarks, and following Formula 1 motor racing.

Other Works by the Author

Other space exploration books by David J. Shayler

Challenger Fact File (1987), ISBN 0-86101-272-0
Apollo 11 Moon Landing (1989), ISBN 0-7110-1844-8
Exploring Space (1994), ISBN 0-600-58199-3
All About Space (1999), ISBN 0-7497-4005-X
Around the World in 84 Days: The Authorized Biography of Skylab Astronaut Jerry Carr
  (2008), ISBN 9781-894959-40-7

With Harry Siepmann

NASA Space Shuttle (1987), ISBN 0-7110-1681-X

Other books by David J. Shayler in this series

Disasters and Accidents in Manned Spaceflight (2000), ISBN 1-85233-225-5
Skylab: America's Space Station (2001), ISBN 1-85233-407-X
Gemini: Steps to the Moon (2001), ISBN 1-85233-405-3
Apollo: The Lost and Forgotten Missions (2002), ISBN 1-85233-575-0
Walking in Space (2004), ISBN 1-85233-710-9
Space Rescue (2007), ISBN 978-0-387-69905-9
Linking the Space Shuttle and Space Stations - Early Docking Technologies from Concept
  to Implementation (2017), ISBN 978-3-319-49768-6

With Rex Hall

The Rocket Men (2001), ISBN 1-85233-391-X
Soyuz: A Universal Spacecraft (2003), ISBN 1-85233-657-9

With Rex Hall and Bert Vis

Russia's Cosmonauts (2005), ISBN 0-38721-894-7

© Springer International Publishing Switzerland 2017
D.J. Shayler, *Assembling and Supplying the ISS*, Springer Praxis Books,
DOI 10.1007/978-3-319-40443-1

With Ian Moule

Women in Space: Following Valentina (2005), ISBN 1-85233-744-3

With Colin Burgess

NASA Scientist Astronauts (2006), ISBN 0-387-21897-1
The Last of NASA's Original Pilot Astronauts: Expanding the Space Frontier in the Late Sixties (2017), ISBN 978-3-319-51012-5

With David M. Harland

The Hubble Space Telescope: From Concept to Success (2016), ISBN-978-1-4939-2826-2
Enhancing Hubble's Vision: Service Missions That Expanded Our View of the Universe (2016), ISBN 978-3-319-22643-9

Other books by David J. Shayler and Michael D. Shayler in this series

Manned Spaceflight Log II – 2006-2012 (2013), ISBN 978-1-4614-4576-0

With Andy Salmon

Marswalk One: First Steps on a New Planet (2005), ISBN 1-85233-792-3

With Tim Furniss

Praxis Manned Spaceflight Log: 1961–2006 (2007), ISBN 0-387-34175-7

Index

© Springer International Publishing Switzerland 2017
D.J. Shayler, *Assembling and Supplying the ISS*, Springer Praxis Books,
DOI 10.1007/978-3-319-40443-1